A-level Study Guide

Physics

Revised and updated for 2008 by

Wendy Brown and Tony Winzor

Revision Express

Series Consultants: Geoff Black and Stuart Wall
Project Manager: Hywel Evans

Pearson Education Limited
Edinburgh Gate, Harlow
Essex CM20 2JE, England
and Associated Companies throughout the world

British Library Cataloguing in Publication Data
A catalogue entry for this title is available from the British Library.

ISBN 978-1-4082-0664-5

First edition 2000
Second edition 2004
Third edition 2008
Reprinted 2008
Set by Juice Creative Ltd
Printed and bound by Ashford Colour Press Ltd, Gosport, Hampshire

Contents

How to use this book

Specification map
Provides a quick and easy overview of the topics that you need to study for the specification you are studying (see pages 6–7)

Exam themes
At the beginning of each chapter, these give a quick overview of the key themes that will be covered in the exam

Exam themes

- *Applied physics* Extending ideas in dynamics and thermodynamics
- *Further electricity* More about electromagnetism and alternating currents

Action point
A suggested activity linked to the content

Checkpoint
Quick question to check your understanding with full answers given at the end of the chapter

Links
Cross-reference links to other relevant sections in the book

Examiner's secrets
Hints and tips for exam success

Watch out!
Flags up common mistakes and gives hints on how to avoid them

Momentum and impulse

In life, we may often 'act on impulse', but in physics impulse more often acts on us! Impulse is the product of force and time. Impulses always change the momentum of the body they act on.

Action point
Adding scalar quantities is simple, *providing* they all have the same units. For example, find the sum of 2 kg, 4.5 kg, 300 g and 750 g.

Checkpoint 1
Why does the size of the drawing matter?

Links
See *Newton's first law of motion*, pages 14–15 for more information on moving objects in equilibrium.

Examiner's secrets
Don't forget to *show* how you calculated both the magnitude and the direction of the resultant vector. There will be at least one mark for a clear method

Watch out!
These equations of motion assume constant acceleration. If acceleration varies, they won't work!.

Momentum

If a body of mass m has a velocity v, then its **momentum p** is:

$$p = mv$$

The units of momentum are kg m s^{-1} or newton seconds (N s).

Law of conservation of momentum

Momentum is conserved in *all* collisions, explosions and interactions! There are no exceptions to this law.

→ The total momentum of a system before any interaction is exactly equ to the total momentum after it, provided no external forces act (extern forces would allow momentum to be transferred to external bodies).

When two objects collide, the changes in their momenta will be equal in size, but opposite in direction. The momentum gained by one body equal the momentum lost by the other.

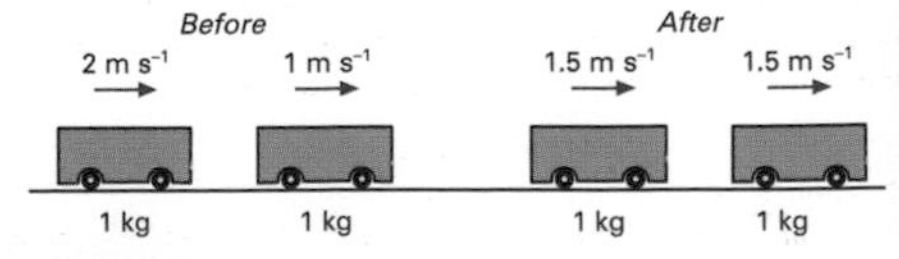

Tackling problems

1. Choose which direction is positive.
2. Draw before and after sketches of the objects involved.
3. Calculate every momentum you can.
4. Apply the law of conservation of momentum for collisions involving two bodies:

$$(m_1v_1 + m_2v_2)_{\text{before}} = (m_1v_1 + m_2v_2)_{\text{after}}$$

2D collisions

You may have to solve problems about two-dimensional collisions, for example, when a moving subatomic particle collides with a stationary one. The law of conservation of momentum still applies but here you have to apply it to the components of momenta along the initial direction of movement and also at right angles to this direction.

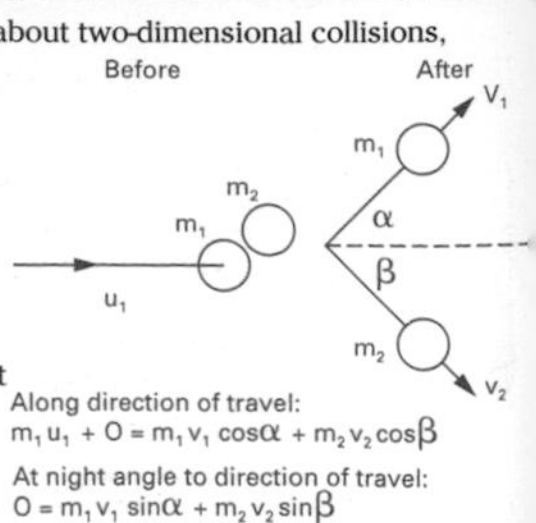

Along direction of travel: $m_1u_1 + 0 = m_1v_1\cos\alpha + m_2v_2\cos\beta$

At night angle to direction of travel: $0 = m_1v_1\sin\alpha + m_2v_2\sin\beta$

Topic checklist
A topic overview of the content covered and how it matches to the specification you are studying

Topic checklist

	Edexcel		AQA/A		OCR/A		WJEC		CCEA	
	AS	A2	AS	A2	AS	A2	AS	A2	AS	A2
Numbers and maths	○	●	○	●	○	●	○	●	○	●
Errors and uncertainties	○	●	○	●	○	●	○	●	○	●

Recoil and explosions

Guns and cannons *recoil* when fired because of the law of conservation of momentum. The positive momentum gained by the bullet or cannon ball is equal to the negative recoil momentum of the gun or cannon, and so the total momentum before and after the explosion is zero.

Elastic and inelastic collisions

→ Kinetic energy is conserved in **elastic** collisions.
→ Kinetic energy is not conserved in **inelastic** collisions.
→ Momentum is conserved in *all* collisions.
→ Total energy is conserved in *all* collisions.

In inelastic collisions some kinetic energy is converted to other forms of energy (usually mainly heat). Large scale collisions are inelastic: collisions between hard steel spheres are nearly elastic: some particle collisions are elastic.

Impulse and Newton's second law of motion

Most bodies have constant mass, so we normally (rightly) note that unbalanced forces cause acceleration. At a more fundamental level we can state that unbalanced forces cause a *change in momentum*. The change in momentum depends on the size and direction of the force and the period of time over which it is applied; i.e. it depends on its *impulse*.

→ Impulse is the product of force and time.
→ Impulse = change in momentum.
→ $Ft = m\boldsymbol{v} - m\boldsymbol{u}$ (Where $m\boldsymbol{u}$ is initial and $m\boldsymbol{v}$ is final momentum).
→ Force = rate of change of momentum = $(m\boldsymbol{v} - m\boldsymbol{u})/t$. This is another version of Newton's second law of motion.

Impulse is measured in either newton seconds (N s) or kg m s^{-1} (exactly the same units as momentum).

The jargon

Subscripts are very handy. Rather than writing out *horizontal component of velocity*, we can write $\boldsymbol{v}_x$

Test yourself

If you need to, *learn* the equations of motion given in *equations of motion*, page 10. Make a conscious effort now.
$\boldsymbol{v}$ = ?
$\boldsymbol{s}$ = ?
$\boldsymbol{v}^2$ = ?

Check the net

To test your understanding of projectiles go to www.crocodile-clips.com/absorb/AP5/sample/010105.html where you will find questions based on animations.

Grade booster

To get you to really think, how would you explain that an object can be accelerating *and* moving at constant speed?

Take note

Michael Faraday was appointed at the Royal Institution only after a fight in the main lecture theatre led to the dismissal of his predecessor!

Exam practice answers: page 41

1 A bullet of mass 40 g is fired with a horizontal velocity of 500 m s^{-1} from a rifle of mass 2.5 kg.
(a) Find: (i) the bullet's forward momentum, (ii) the bullet's kinetic energy, (iii) the speed of recoil of the rifle, (iv) the rifle's kinetic energy after the explosion.
(b) Can explosions ever be perfectly elastic? Explain your answer. (15 min)

2 What driving force is necessary to accelerate a car of mass 1 400 kg from rest to a speed of 35 m s^{-1} in 20 s? (5 min)

3 Two skaters are skating together at a steady velocity of 8 m s^{-1}. Their masses are 80 kg and 50 kg. The lighter skater is pushed forwards and accelerates to 10 m s^{-1}. Calculate the new speed of her partner. (10 min)

4 Hard snowballs bounce back off you when they hit at 90°; soft snowballs don't. Explain why these hard snowballs exert the greater force. (10 min)

The jargon
A clear outline of what subject-related and examination-related jargon means

Test yourself
Quick memory recall activities

Check the net
Suggestions for useful websites related to the topic

Grade booster
Examiner suggestions on how to get the top grade

Take note
Extension notes on the core content

Exam practice
Exam-style questions to check your understanding of the topic content with full answers given at the end of the chapter

Specification map

		Edexcel		AQA/A		AQA/B		OCR/A		OCR/B		WJEC		CCEA	
		AS	A2	AS	A2	AS	A2	AS	A2	AS	A2	AS	A2	AS	A2
Mechanics	Scalars and vectors	○		○		○		○		○		○		○	
	Forces and moments in equilibrium	○		○		○		○				○		○	
	Ways of describing motion	○		○		○		○		○		○		○	
	Equations of motion	○		○		○		○		○		○		○	
	Projectiles	○		○		○		○		○		○		○	
	Newton's laws of motion	○		○		○		○		○		○		○	
	Some important forces	○				○	●	○							
	Moving through liquids	○		○	●	○	●	○				○			
	Work, energy and power	○		○		○		○		○		○		○	
	Momentum and impulse		●		●	○	●		●	○			●		●
	Stress, strain and Hooke's law	○		○		○		○		○		○	●	○	
	Vibrations and resonance		●		●		●		●		●		●		●
	Circular motion		●		●		●		●		●		●		●
	Simple harmonic motion		●		●		●		●		●		●		●
Radioactivity, Nuclear Particle Physics	The atom and its nucleus		●	○		○			●		●	○			●
	Elements and isotopes		●	○		○			●		●	○	●		●
	Nuclear instability		●	○		○			●	○	●				●
	Properties of ionizing radiation		●		●	○			●		●	○	●		●
	Radioactive decay		●		●	○			●		●	○	●		●
	Binding energy and mass defect		●		●		●		●		●		●		●
	Nuclear fission and fusion		●		●		●		●		●				●
	Other applications of radioactivity	○	●		●		●		●		●		●		●
	Probing matter	○	●	○		○			●		●		●		●
	Particles – production and patterns		●	○		○			●		●	○			●
	More about leptons and quarks		●	○		○			●		●	○			●
	Forces/interactions and conservation laws		●	○		○			●		●	○			●
Electricity and electromagnetism	Current as a flow of charge	○		○		○		○		○		○		○	
	Current, p.d. and resistance	○		○		○		○		○		○		○	
	Resistors and resistivity	○		○		○		○		○		○		○	
	Electrical energy and power	○		○		○		○		○		○		○	
	Kirchhoff's laws	○		○		○		○				○		○	
	Potential dividers and their uses	○		○		○		○		○		○			
	EMF and internal resistance	○		○		○		○		○		○			
	Capacitors		●		●		●		●	○			●	○	
	Electromagnetism		●		●		●		●		●		●		●
	Electromagnetic induction		●		●	○	●		●		●		●	○	
	Alternating currents		●	○	●				●		●		●		●
Kinetic theory	Behaviour of gases – Experiment		●		●		●		●		●	○			●
	Behaviour of gases – Theory		●		●		●		●		●	○			●
	Behaviour of gases – in bulk		●		●		●		●		●	○			●
	Internal energy		●		●		●		●		●	○			●
	Specific and latent heat capacities		●		●		●		●		●	○			●
Waves and oscillations	Types of waves and their properties	○		○		○		○		○		○		○	
	Electromagnetic spectrum	○		○		○		○		○		○		○	
	Reflection and refraction	○		○		○		○		○		○		○	
	Applications of reflection and refraction	○		○		○		○		○		○		○	
	Diffraction	○		○		○		○		○		○		○	
	Superposition	○		○		○		○		○		○		○	
	Interference	○		○		○		○		○		○		○	
	Standing waves	○		○		○		○		○		○		○	
	Photoelectric effect	○			●		●	○		○		○		○	
	Atomic line spectra		●	○	●		●	○		○		○		○	
	De Broglie's equation and atomic models		●	○	●		●	○	●			○		○	

		Edexcel		AQA/A		AQA/B		OCR/A		OCR/B		WJEC		CCEA	
		AS	A2	AS	A2	AS	A2	AS	A2	AS	A2	AS	A2	AS	A2
Fields	Newton's law of universal gravitation		●		●		●		●		●		●		●
	Gravitational fields		●		●		●		●		●		●		●
	Electric forces and fields		●		●		●		●		●		●		●
	Electric potential* and charged particle acceleration		●		●		●		●		●		●		●
	Comparisons: gravitational and electric fields		●		●		●		●		●		●		●
	Synoptic skills		●		●		●		●		●		●		●
Options	Astrophysics 1*		●*		●	○*			●*			○	●*		
	Astrophysics 2		●*		●	○*			●*						
	Astrophysics 3				●										
	Astrophysics 4		●*		●	○*			●*	○*			●*		
	Medical and health physics 1				●	○*				○*			●	○*	
	Medical and health physics 2	○*			●	○*	●*		●*	○*			●	○*	
	Medical and health physics 3					○	●*		●*				●		
	Materials 1	○*		○*				○*					●		
	Materials 2												●		
	Turning points in physics 1	○*		○*	●	○*		○*		○*				○*	
	Turning points in physics 2				●		●*						●		
	Energy and the environment 1					○*							●		
	Energy and the environment 2					○*							●		
	Applied physics 1				●		●*								
	Applied physics 2				●		●*								
	Further electricity												●		

Mechanics

Mechanics – the study of forces and their effects – is the backbone of classical physics.

Exam themes

- *Measurements and units* All ideas in physics are tested by experiment (by measuring things). Measurements need appropriate, clearly stated units. Familiarity with SI units is required. Equations and definitions always give clues to the correct units to use.
- *Vector quantities* Direction often matters in mechanics. You need to know how forces acting in different directions add up. You need to be confident in dealing with other vector quantities too.
- *Forces in equilibrium* In static structures, the forces and moments on each and every point must be balanced. Moving bodies tend to keep moving with constant velocity.
- *Unbalanced forces* can cause acceleration, changes in momentum, circular motion, simple harmonic motion etc.
- *Use of free-body force diagrams* You will be expected to be able to draw force diagrams to isolate the forces acting on a body.
- *Interpreting graphs and gradients* You should be able to interpret graphs of motion – in particular, to understand the significance of gradient and area.

Topic checklist

	Edexcel		AQA/A		AQA/B		OCR/A		OCR/B		WJEC		CCEA	
	AS	A2	AS	A2	AS	A2	AS	A2	AS	A2	AS	A2	AS	A2
Scalars and vectors	○		○		○		○		○		○		○	
Forces and moments in equilibrium	○		○		○		○				○		○	
Ways of describing motion	○		○		○		○		○		○		○	
Equations of motion	○		○		○		○		○		○		○	
Projectiles	○		○		○		○		○		○		○	
Newton's laws of motion	○		○		○		○		○		○		○	
Some important forces	○				○	●	○							
Moving through liquids	○		○	●	○	●	○				○			
Work, energy and power	○		○		○		○		○		○		○	
Momentum and impulse		●		●	○	●		●	○			●		●
Stress, strain and Hooke's law	○		○		○		○		○		○	●	○	
Vibrations and resonance		●		●		●		●		●		●		●
Circular motion		●		●		●		●		●		●		●
Simple harmonic motion		●		●		●		●		●		●		●

Scalars and vectors

Scalar quantities like mass and energy have *size* which is a number (*how much?*) and a *unit* (*of what?*) For some quantities like force and velocity, the direction is important as well as the size – when you hit a ball with a bat it is important *where* you hit it as much as *how hard*. Quantities with size, a unit *and* direction are called vector quantities.

Checkpoint 1

Sort the following list into scalar and vector quantities: mass, weight, temperature, energy, acceleration.

Vector addition by scale drawing

Vectors can be represented by arrows pointing in the direction of the vector. The length of the arrow gives the size of the vector. The sum or **resultant** of any number of vectors can be found by joining vector arrows together in any order. The vectors must all follow on from start to finish, with the tail of each arrow starting from the tip of the last.

Action point

Convince yourself that the order of addition does not matter by adding **a**, **b** and **c** in a different order and measuring the resultant.

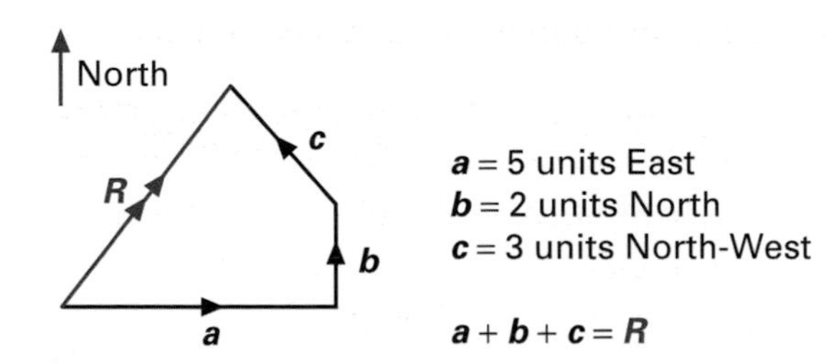

The resultant *R*

- The resultant ***R*** is the single vector that could replace all the others.
- ***R*** points from the starting point to the finishing point.
- ***R*** is normally given a double arrow to distinguish it from its components.
- The size (magnitude) and direction of ***R*** are measured from the diagram using a ruler and a protractor.
- Don't forget to use the appropriate units. If you are adding forces, your answer must be in newtons (not centimetres!).

Vector notation

Vector quantities are given symbols written in bold type (***a***), or underlined ($\underline{a}$), or with an arrow over the top ($\overrightarrow{OA}$). This notation should give you a clue that simple addition may not be appropriate.

Vector subtraction

The negative version of a vector is simply the vector reversed, so ***a*** – ***a*** = 0. To subtract any vector, reverse its direction and add it as shown.

Checkpoint 2

Why does the size of the drawing matter?

Action point

Adding scalar quantities is simple, *providing* they all have the same units. For example, find the sum of 2 kg, 4.5 kg, 300 g and 750 g.

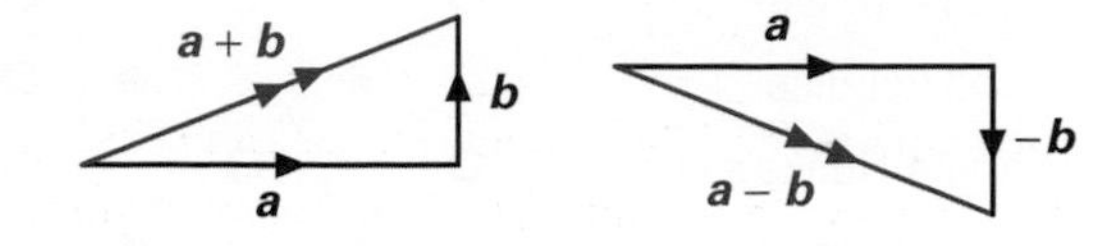

The parallelogram of forces

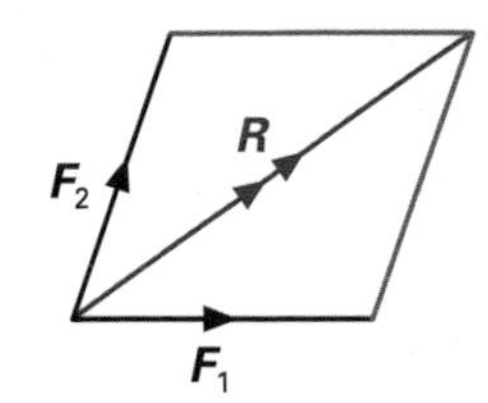

If two forces act on the same body at the same time, they can be added by making them the adjacent sides of a parallelogram. The resultant is given by the diagonal of the parallelogram. Be sure to choose the right diagonal! The resultant must start from the point of action of the forces. This method is entirely equivalent to the previous one applied to just two vectors. Its benefit is that the angle between the forces (which is normally given) is more easily marked and measured so it's less easy to make mistakes.

Grade booster

The whole point about vector quantities is that they have both *magnitude* and *direction*. Make sure you give *both* in your answers.

Examiner's secrets

Don't forget to *show* how you calculated both the magnitude and the direction of the resultant vector. There will be at least one mark for a clear method.

Watch out!

Don't forget to give the answer! A drawing is not enough. You must show that you can use your scale diagram to find the resultant. Use appropriate units and state the angle relative to one (or more) of the forces given in the question.

Vector addition by calculation

You can use Pythagoras' theorem to calculate the resultant of two perpendicular vectors. For anything more complicated, stick to scale drawing. The bigger the drawing, the better.

Resolution of vectors

A vector can be *resolved* into two perpendicular components, such that the sum of the components equals the vector itself. This is a very useful trick. It allows you to separate horizontal motion, which is not affected by gravity, from vertical motion, which is. It allows you to separate out the forces acting along a slope from the forces acting at right angles to it.

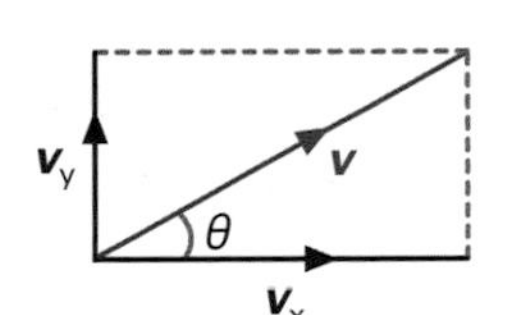

$\mathbf{v}_x = \mathbf{v} \cos \theta$
$\mathbf{v}_y = \mathbf{v} \sin \theta$
$\mathbf{v}_x + \mathbf{v}_y = \mathbf{v}$

Checkpoint 3

Learn your trigonometry. *SOHCAHTOA!* In the figure opposite, $\mathbf{v}_x$ is clearly *adjacent* to angle θ, $\mathbf{v}_y$ is (not quite so clearly) equal to the side *opposite* θ. Check the formulae given for $\mathbf{v}_x$ and $\mathbf{v}_y$ (*derive them*).

Exam practice answers: page 38

1 The diagram below shows the two forces produced by a spoiler on a racing car when it moves.

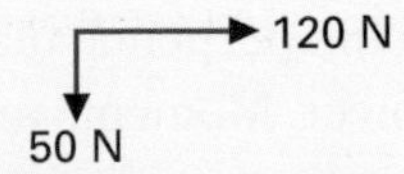

(a) Calculate the magnitude of the resultant force on the car.
(b) Calculate the direction the resultant force makes with the 120 N force.
(5 min)

2 A rope is used to pull a narrow boat along a canal. The rope is pulled by a horse. The tension in the rope is 600 N and the rope makes an angle of 30° with the canal bank. What force must be provided (by the rudder and keel) to keep the boat travelling parallel to the bank? (15 min)

Forces and moments in equilibrium

Forces can make things speed up, slow down, change direction, change shape, or spin faster or slower. Forces can make things happen! In this section, however, we are interested in situations where forces do none of the above – because they are in equilibrium.

Links

See *Newton's first law of motion*, pages 20–1 for more information on moving objects in equilibrium.

Equilibrium

An object is in **equilibrium** if the forces acting on it are balanced.

- → All the forces on the object must be balanced so the object is not accelerating in any direction.
- → All the moments about each point of the object must be balanced so the object is not changing its rotation.

Static objects are always in equilibrium. We can often use this fact to calculate the sizes and directions of unknown forces.

Links

See *vector addition*, page 10.

The jargon

A *free-body force diagram* is simply a diagram of the object of interest with all the forces acting on it.

Examiner's secrets

Follow these steps to avoid disaster:
(a) Sketch the set-up described. Find the angles at which all forces act.
(b) Draw a *free-body force diagram* with all the forces acting away from the body.
(c) Draw the force polygon to solve the problem.

Force polygons

If you add a series of force vectors by drawing them nose to tail, the resultant is found by measuring the line that joins the start to the finish. If the forces are balanced, you will end up with a closed polygon. The arrows will lead back to the start because the resultant equals zero. The most common situations dealt with in A-level Physics exams involve three forces (in which case the polygon is a triangle).

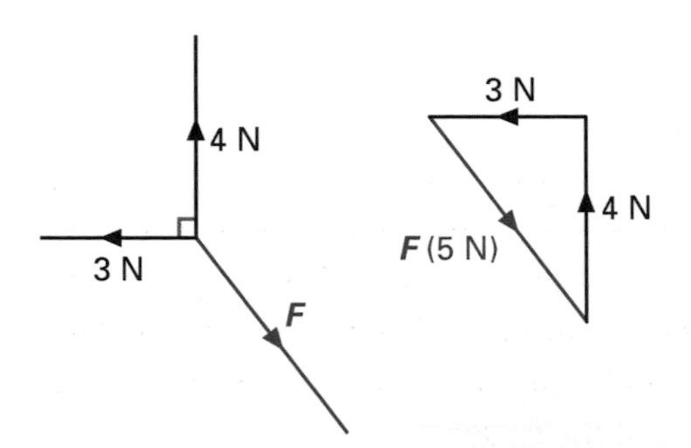

The jargon

Torque is another word for moment.

Checkpoint 1

The product of force and perpendicular distance is exactly the same as the product of perpendicular force and actual distance to the point of action. Check for yourself.

Moments

The turning effect of a force is called its **moment**. The size of a moment depends on the size and direction of the force and the distance from its point of action to the axis of rotation.

moment = Fd

Where F is the force and d is the perpendicular distance from the line of action of the force to the pivot. Moments are measured in newton metres (N m).

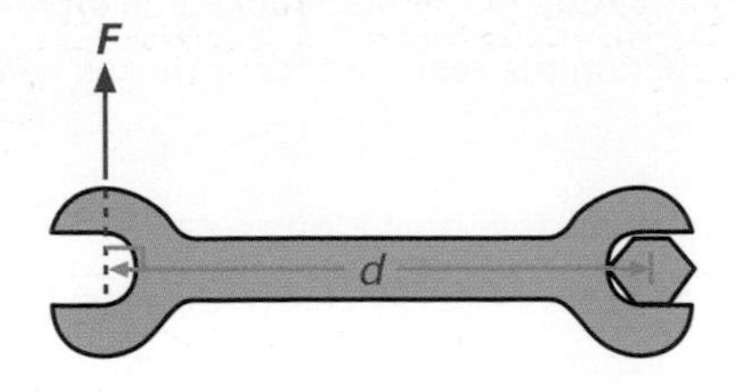

Principle of moments

If a body is in equilibrium, the sum of clockwise moments is equal (and opposite to) the sum of anticlockwise moments *about each and every point.*

Couples

A single force can cause a body to turn, but it will also cause linear acceleration. A **couple** is a pair of equal and opposite forces along different lines of action. Couples cause angular acceleration, but not linear acceleration.

→ The moment of a couple is equal to one of the forces multiplied by the perpendicular separation of the lines of action of the forces.

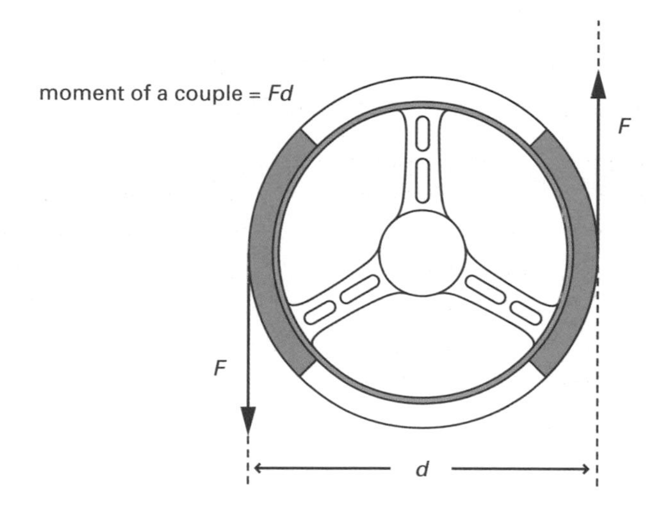

Centre of gravity

The **centre of gravity** of a rigid body is the point at which its weight appears to act. This allows us to simplify questions where we need to work out the moments caused by a body's weight. For simple symmetrical objects like ladders, beams and doors (the kinds of objects that appear most often in A-level Physics papers), the centre of gravity is also the geometrical centre of the body.

Exam practice

answers: page 38

1 A strong, light rope spans a 15 m gap. A tightrope walker weighing 500 N sets off across it. The rope sags a little due to the walker's weight. The backward and forward tension forces in the rope (T_1 and T_2) change as she makes her journey. Half way across, T_1 equals T_2 in magnitude.

Given that the centre of the rope dips 0.6 m below the horizontal, calculate the tension in the rope. (15 min)

2

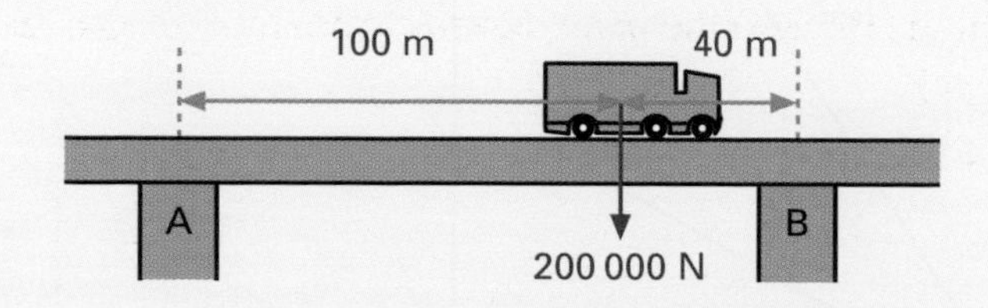

A bridge is made by resting a concrete beam on two pillars (A and B). A lorry crosses the bridge. Using the weights and dimensions given in the diagram above, calculate the extra load on each pillar due to the lorry. (30 min)

"Give me but one firm place on which to stand, and I will move the Earth."

Archimedes, who discovered the principle of moments

Grade booster

Moments are vector quantities. You must show clockwise and anticlockwise moments with opposite signs.

The jargon

Direction A moment can cause clockwise *or* anticlockwise acceleration. The net turning effect of all moments acting about a point is found by adding up all clockwise moments and subtracting all anticlockwise moments.

Don't forget

The *centre of mass* of a body is the average position of its mass (rather than its *weight*). A body's centre of mass is in exactly the same position as its centre of gravity – provided the gravitational field strength doesn't vary within the body itself.

Action point

When an object is tilted so its centre of gravity is no longer above its base, the moment causes it to fall. Wide bases and a low centre of gravity makes an object more stable – can you explain why?

Don't forget

For an object in static equilibrium, if the forces are either side of the pivot, they act in the same direction. If the forces are on the same side of the pivot, they act in opposite directions.

Ways of describing motion

The easiest way to describe motion is in the language of mathematics, using symbols, graphs and equations.

The jargon

The prefix *SI* stands for *Système International* – an internationally agreed set of standard units.

SI units have prefixes to indicate multiples and sub-multiples. For example:
pico (p) = 10^{-12} nano (n) = 10^{-9}
micro (μ) = 10^{-6} milli (m) = 10^{-3}
centi (c) = 10^{-2}
kilo (k) = 10^{3} mega (M) = 10^{6}
giga (G) = 10^{9} tera (T) = 10^{12}

Checkpoint 1

A racing car may have an average speed of 150 mph over the course of a lap, but its average velocity over one lap is always zero. Explain why.

Definitions, symbols and units

Quantity (symbol)	*Description*	*SI unit*
Displacement (***s***)	The vector version of distance. The distance an object has moved in a certain direction.	m
Velocity (***v***)	The vector version of speed. An object's speed or rate of progress in a given direction.	m s^{-1}
Acceleration (***a***)	The rate of increase in velocity; how many metres per second faster an object gets each second (in a given direction).	m s^{-2}

Speed and velocity

Speed measures how fast something travels but the direction is not taken into account. So speed is a scalar quantity and cannot be negative. Since many objects rarely travel at the same speed for long, the average speed is the total distance travelled divided by the total time taken. **Velocity** is how fast something travels in a given direction so equals the change in displacement, $\Delta \mathbf{s}$, divided by the time taken, Δt. So, $\mathbf{v} = \Delta \mathbf{s} / \Delta t$

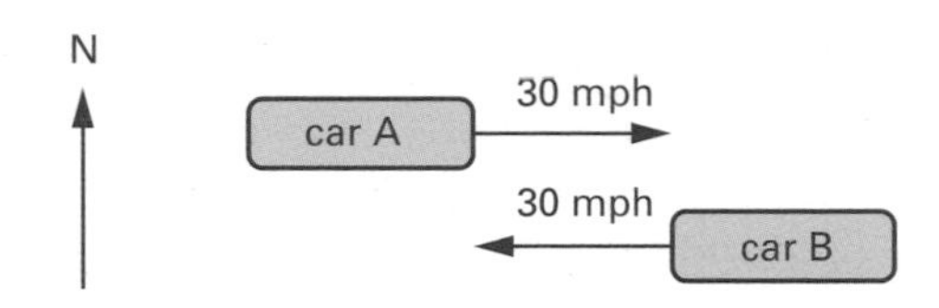

These cars both have the same speed but their velocities are different because A is travelling East and B, West. If A has velocity 30 mph then B's velocity is –30 mph.

Grade booster

If you are asked to *sketch* a graph: *don't* relax, get out your charcoal, and demonstrate your understanding of light and shade; *do* draw a graph showing all the information you are given. Show the shape of the graph, show where the line cuts the axes (if it does), label axes properly.

Displacement–time graphs

The gradient of a displacement–time graph is equal to the velocity of the body. For a body whose velocity is changing, the gradient of the tangent at any point gives the instantaneous velocity at that point.

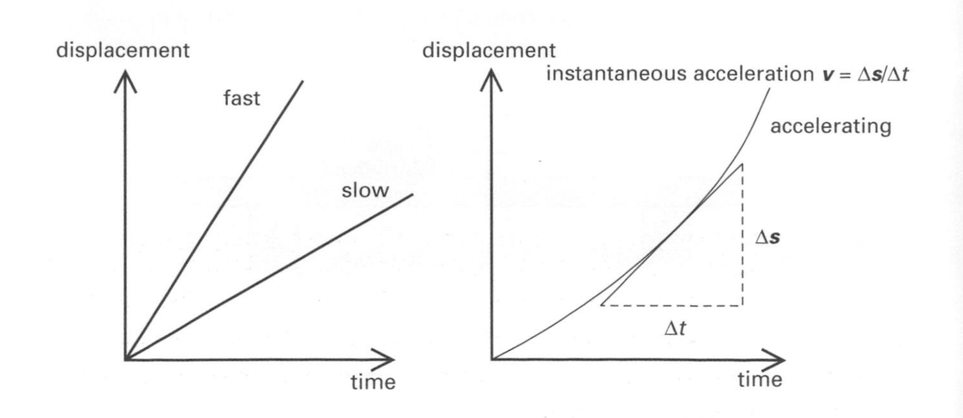

Speed–time graphs

A *tachograph* is a type of speed–time graph used in lorries to check on working hours.

➜ The area under a speed–time graph is equal to the distance travelled.

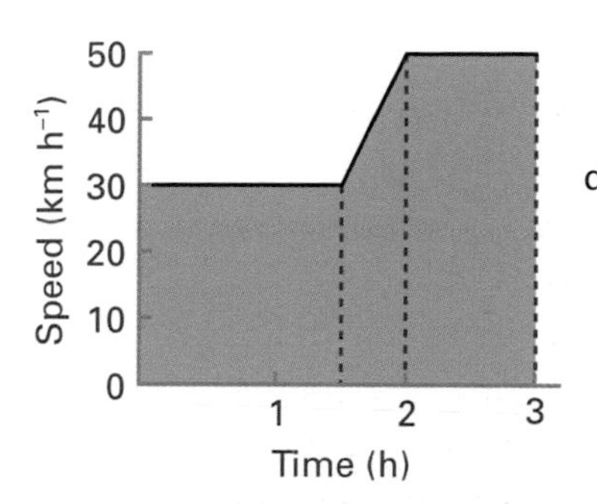

$$\text{distance} = 30 \times 1.5 + 40 \times 0.5 + 50 \times 1 = 115\ \text{km}$$

Velocity–time graphs

You can use these to find the displacement, the velocity and the acceleration of a body at any given instant.

Displacement

The displacement (from a given starting point) is equal to the area under a velocity–time graph.

Acceleration

Acceleration is change in velocity divided by the time taken, so $\boldsymbol{a} = \Delta \boldsymbol{v}/\Delta t$. The gradient of a velocity–time graph is equal to the acceleration of the body. For a body whose acceleration is changing, the gradient of the tangent at any point gives the instantaneous acceleration at that point.

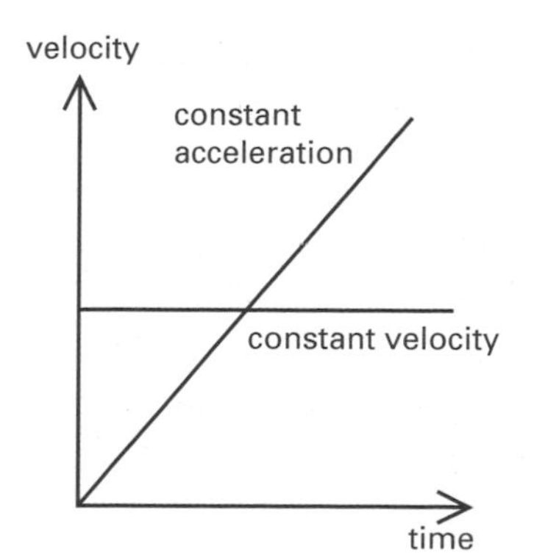

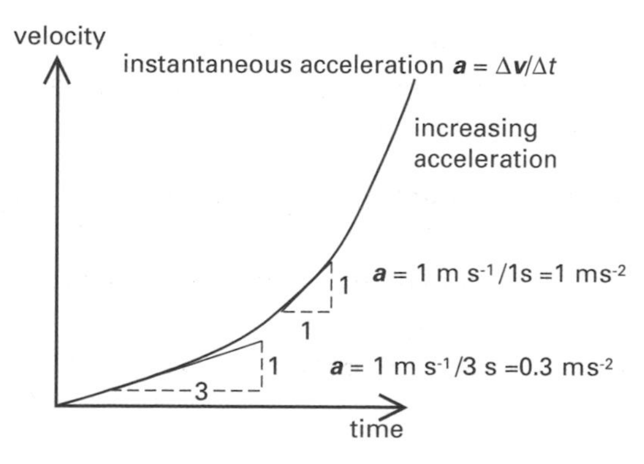

Exam practice answers: page 38

A train accelerates from rest at a constant rate of $0.2\ m\ s^{-2}$ for the first minute of its journey; it then carries on at constant speed for the next two minutes. Graphically or otherwise, find the train's top speed and the distance it travels in the first three minutes. (15 min)

Examiner's secrets

You do not need to know how to do calculus for A-level Physics, but you should be familiar with the notation. If distance travelled is s and time taken is t, the speed at any instant is written ds/dt said 'ds by dt' in calculus notation.

The jargon

The *differential* of any graph is equal to its gradient. The *integral* of any graph is equal to the area under the graph. The differential of distance with respect to time is speed. The integral of speed with respect to time is distance.

Checkpoint 2

Sketch a velocity–time graph for a stone thrown upward. How is its displacement represented by this graph?

Links

See also *simple harmonic motion*, pages 36–7, where the links between displacement, velocity and acceleration are explored in more detail.

Examiner's secrets

To find the gradient of a graph, you need to draw the tangent at a point on the graph. You may find it easier to draw a line at right angles to the graph at this point and then use a protractor to draw the tangent. Always draw a long tangent and construct a large triangle from it. Small triangles are less accurate and lose marks.

Equations of motion

The equations of motion form a useful tool kit for solving problems involving constant acceleration.

Watch out!

These equations of motion assume constant acceleration. If acceleration varies, they won't work!

Grade booster

You must get used to using **u** for initial velocity, **v** for final velocity and **s** for displacement. Some questions will expect you to manipulate the equations of motion without using numbers.

The equations

$\boldsymbol{v} = \boldsymbol{u} + \boldsymbol{a}t$ [1]

$\boldsymbol{s} = \boldsymbol{u}t + \frac{1}{2}\boldsymbol{a}t^2$ [2]

$\boldsymbol{v}^2 = \boldsymbol{u}^2 + 2\boldsymbol{as}$ [3]

Where

$\boldsymbol{s}$ = displacement
$\boldsymbol{u}$ = initial velocity
$\boldsymbol{v}$ = final velocity
$\boldsymbol{a}$ = acceleration
t = time taken

Checkpoint 1

The three equations given are all you need to learn provided you are reasonably competent at algebra. Each equation contains four of the five quantities. Two others may be learned if you want to avoid the prospect of having to deal with simultaneous equations. They are:

$\boldsymbol{s} = \boldsymbol{v}t - \frac{1}{2}\boldsymbol{a}t^2$ [4]

$\boldsymbol{s} = (\boldsymbol{u} + \boldsymbol{v})t/2$ [5]

Try to derive them from the others. If you succeed, you needn't bother to learn them!

Derivation

A body accelerates steadily from an initial velocity $\boldsymbol{u}$ to a final velocity $\boldsymbol{v}$ in time t seconds. Its acceleration is $\boldsymbol{a}$ and the distance it travels is $\boldsymbol{s}$. We want to derive equations that link these quantities.

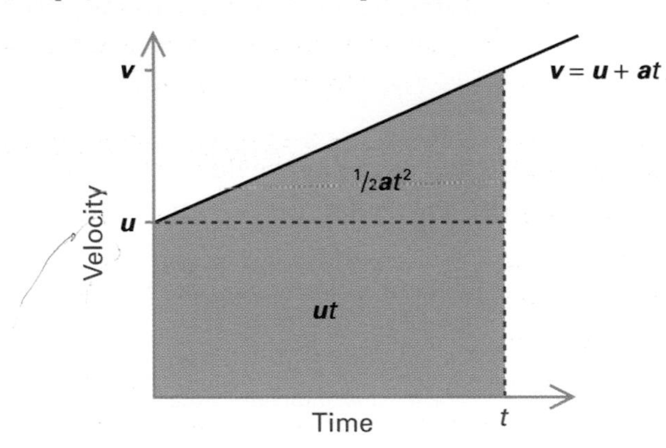

Checkpoint 2

Referring to the graph opposite, if $\boldsymbol{u} = 10\ \text{m s}^{-1}$, $\boldsymbol{v} = 18\ \text{m s}^{-1}$ and $t = 4$ s, work out the rate of acceleration and the distance travelled.

- *Equation 1* comes from the definition of acceleration. Acceleration is the rate of increase in velocity, i.e. the increase in velocity ($\boldsymbol{v} - \boldsymbol{u}$) divided by the time taken (t).
- *Equation 2* comes from the graph. Distance travelled, $\boldsymbol{s}$, equals the area under the graph (from time zero to time t). The area of the rectangle equals $\boldsymbol{u}t$. The triangle's area (half base × height) equals $\frac{1}{2}t(\boldsymbol{v} - \boldsymbol{u})$. Equation 1 tells us that $\boldsymbol{v} - \boldsymbol{u} = \boldsymbol{a}t$, so we get the $\frac{1}{2}\boldsymbol{a}t^2$ term.
- *Equation 3* is derived by substituting $(\boldsymbol{v} - \boldsymbol{u})/\boldsymbol{a}$ for t in equation 2 and shuffling terms.

Checkpoint 3

A ball is thrown vertically upwards at $10\ \text{m s}^{-1}$, how high will it get? Assume $\boldsymbol{g} = 10\ \text{m s}^{-2}$.

You need to be able to use the equations. Check whether you also have to derive them or learn them.

Watch out!

If an object decelerates (slows down), it has *negative* acceleration. You must use *negative* values for $\boldsymbol{a}$ in your calculations.

Changing the subject

Basic algebra is all you need when tackling problems involving the equations of motion. You must know how to change the subject of an equation. It pays to be slow and methodical. Work one step at a time and write everything down. There are no prizes for getting the wrong answer in record time!

The rules

You can do anything to one side of the equation, provided you do exactly the same thing to the other side.

- To move a term from one side to the other, subtract it from both sides

  e.g. $\mathbf{v} = \mathbf{u} + \mathbf{a}t \Rightarrow \mathbf{v} - \mathbf{u} = \mathbf{a}t$

- To remove an unwanted factor, divide both sides of the equation by it

  e.g. $\mathbf{v} - \mathbf{u} = \mathbf{a}t \Rightarrow \mathbf{a} = (\mathbf{v} - \mathbf{u})/t$

- To remove an unwanted denominator, multiply both sides by it

  e.g. $\mathbf{s} = \mathbf{u}t + {}^1\!/_2\mathbf{a}t^2 \Rightarrow 2\mathbf{s} = 2\mathbf{u}t + \mathbf{a}t^2$

If you find yourself making mistakes, slow down! Try enclosing each side of the equation in brackets before you do anything. This makes it clear that any multiplication or division must involve every term.

Problem solving

- Write down everything you have been given (there are five quantities: $\mathbf{s}$, $\mathbf{u}$, $\mathbf{v}$, $\mathbf{a}$ and t) and identify the unknowns.
- Look for equations that contain just one unknown quantity – preferably the one asked for in the question – and solve them.
- *An equation cannot be solved if there is more than one unknown factor.*
- If all else fails use simultaneous equations, i.e. find one unknown factor in terms of another and then substitute into a second equation to get rid of one of the two unknowns!

Exam practice answers: pages 38–9

1 Calculate the steady acceleration needed to take a cyclist from rest to $10\ \text{m s}^{-1}$ in a distance of 20 m. (5 min)

2 A sprinter accelerates at $2.5\ \text{m s}^{-2}$ for the first 4 s of a 100 m race, then keeps going at a steady pace.
(a) What is her top speed?
(b) What is her average speed over the 100 m race? (10 min)

3 A stone is thrown up into the air. It reaches a height of 8 m. Assuming $\mathbf{g} = 10\ \text{m s}^{-2}$, calculate its initial upward velocity and the time taken before it returns to its starting position. (10 min)

4 A car approaches a junction at $20\ \text{m s}^{-1}$. Traffic lights at the junction suddenly turn red when the car is 15 m away. The driver's reaction time is 0.5 s and the brakes decelerate the car steadily at $10\ \text{m s}^{-2}$. Where does the car stop? (15 min)

Action point

Skip this section if you already feel confident about manipulating equations. You can always come back to it if you hit problems.

The jargon

Algebra has its own symbolic shorthand:
$\Rightarrow$ means *implies*
$\therefore$ means *therefore*.

Checkpoint 4

Displacement equals average velocity × time, $\mathbf{s} = \left(\frac{\mathbf{u}+\mathbf{v}}{2}\right)t$

Acceleration is change in velocity divided by time, $\mathbf{a} = \frac{\mathbf{v}-\mathbf{u}}{t}$.

Combine these two equations to give $\mathbf{s} = \mathbf{u}t + {}^1\!/_2\mathbf{a}t^2$.

Watch out!

$\mathbf{s}$, $\mathbf{u}$, $\mathbf{v}$ and $\mathbf{a}$ are all vector quantities. Choose a positive direction before you start on a problem. This is particularly important when dealing with projectiles (rising and) falling under gravity.

Examiner's secrets

There is very often more than one way to solve problems using the equations of motion. The most important thing to remember is to follow through your calculation and show a clear method.

The jargon

The data given in the question is usually only what is required for the answers. Do look carefully for clues in the question, such as accelerates from rest, which means $\mathbf{u} = 0\ \text{m s}^{-1}$.

Projectiles

Projectiles move in two dimensions under gravity. As they move along, they also accelerate downwards. This added complexity makes projectiles questions a favourite among examiners.

Gravitational acceleration

Action point

If you are at all sceptical, check for yourself by dropping objects simultaneously from the same height. Provided the objects you drop are heavy enough that air resistance is insignificant, they will hit the ground at the same time.

Gravitational acceleration is directed towards the centre of the Earth. It has no effect at all on horizontal motion. So projectile motion must always be split into horizontal and vertical components which are considered separately. As before, air resistance can be ignored (very handy)! To solve problems

- ➜ horizontal velocity is constant for the duration of the projectile's flight, and the only equation that applies is
  horizontal speed = horizontal distance travelled / time
- ➜ vertical motion is subject to a constant downward acceleration of 9.81 ms^{-2}, and the equations of motion are used.

Compare the paths of the two falling balls shown below.

Vertical fall
$V_x = 0$

Fall with constant horizontal velocity
$V_x = 1\ m\ s^{-1}$

V_x V_y

ground 1m 1m 1m 1m

The jargon

Free fall is a term used to describe bodies falling and accelerating (due to gravity) at a steady rate. Air resistance is ignored and the equations of motion can be applied to free-fall problems.

Watch out!

You should be prepared for some problems to yield two possible solutions (a squared term can have positive and negative roots). In such cases, you should give both answers and mention the significance of each (for example, positive and negative velocity might signify upward and downward motion).

The first step in every projectiles question is to calculate initial vertical and horizontal velocities:

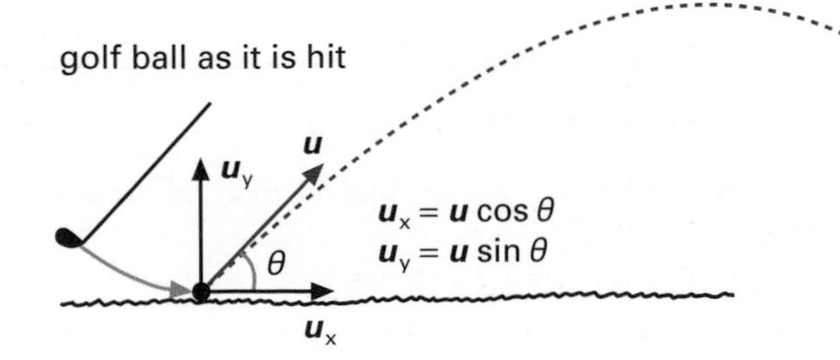

The next steps depend on the question you have been asked.

Don't forget

The curved path of a projectile is called a *parabola*.

Action point

Can you show that the time of flight of a projectile depends on the vertical velocity?
Can you also show that range of a projectile depends on the horizontal velocity and the time of flight?

Solving projectiles problems

You may be asked to:

- *find the maximum height reached by a projectile*
  The maximum height attained has nothing to do with horizontal motion, so just work out initial vertical velocity $\boldsymbol{u}_y$ and then apply the equations of motion, with $\boldsymbol{v}_y = 0\ \text{m s}^{-1}$ and $\boldsymbol{a} = -9.81\ \text{m s}^{-2}$ to find out the height reached
- *find the range of a projectile*
  The range is the product of horizontal velocity and time in the air. The tricky bit is working out the time from the vertical (gravitationally accelerated) motion.

  Split the problem into two halves:
  Going up The projectile decelerates from $\boldsymbol{u}_y$ to $0\ \text{m s}^{-1}$ at its peak ($\boldsymbol{a} = -9.81\ \text{m s}^{-2}$).
  Going down The projectile accelerates from $0\ \text{m s}^{-1}$ (at $9.81\ \text{m s}^{-2}$).

  If the starting height is the same as the stopping height, the path will be symmetrical and the total time in the air will simply be twice the time taken to reach the peak. If not, you will have to calculate the distance the projectile must fall and work out the time taken from that!
- *find the trajectory of a projectile at some point*
  The trajectory of a projectile is found using trigonometry. The tangent of the angle, relative to the horizontal, is equal to $\boldsymbol{v}_y/\boldsymbol{v}_x$. $\boldsymbol{v}_x$ is constant, but $\boldsymbol{v}_y$ varies over time, so you will have to use the equations of motion to find it!

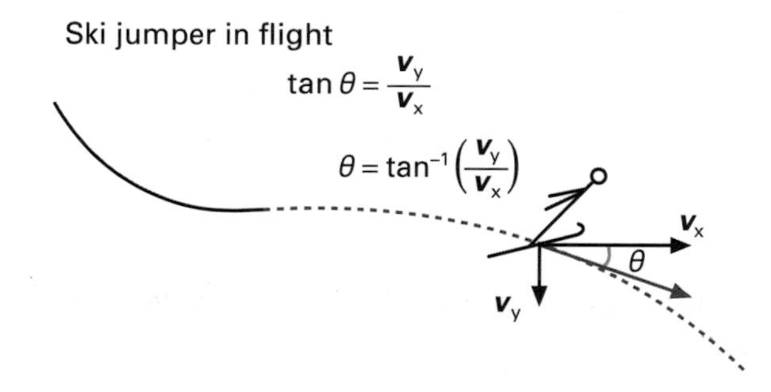

Test yourself

If you need to, *learn* the equations of motion given in *equations of motion*, page 10. Make a conscious effort now.
$\boldsymbol{v}$ = ?
$\boldsymbol{s}$ = ?
$\boldsymbol{v}^2$ = ?

Examiner's secrets

You may be expected to state any assumptions you make such as ignoring air resistance.

The jargon

Subscripts are very handy. Rather than writing out *horizontal component of velocity*, we can write $\boldsymbol{v}_x$

Check the net

To test your understanding of projectiles go to www.crocodile-clips.com/absorb/AP5/sample/010105.html where you will find questions based on animations.

Exam practice

answers: page 39

1 A car is accidentally driven off a (level) cliff at $8\ \text{m s}^{-1}$. It lands (in the sea) 3 s later.
(a) How far out to sea does it 'land'?
(b) How high is the cliff? (Assume $\boldsymbol{g} = 10\ \text{m s}^{-2}$)
(c) At what angle (relative to the horizontal) does it land? (15 min)

2 Find the range of a stone fired at an angle of 45° to the horizontal at a speed of $20\ \text{m s}^{-1}$. (Assume air resistance is negligible.) (10 min)

3 A helicopter flying horizontally at 500 m altitude and at a constant velocity of $50\ \text{m s}^{-1}$ drops a package. If air resistance is insignificant,
(a) how long does it take for the package to hit the ground?
(b) how far does it travel?
(c) at what *speed* does the package hit the ground? (Difficult) (20 min)

Newton's laws of motion

Isaac Newton was a genius who changed the world. Every physicist in the world climbs up first on Newton's shoulders before scrambling onwards and upwards!

Newton's first law of motion

Newton's first law tells us what happens when the forces acting on a body are balanced (i.e. when the resultant force is zero).

→ A body will remain at rest, or will keep travelling with constant velocity, (that is at the same speed in a straight line) unless acted on by unbalanced forces.

In exams, you may be told that a body is travelling with constant velocity; this is always a clue to use Newton's first law to find the size of any unknown forces.

Newton's second law of motion

Newton's second law describes what happens to a body acted upon by unbalanced forces.

→ unbalanced forces cause acceleration
→ the acceleration is directly proportional to the net force
→ the acceleration is in exactly the same direction as the net force
→ the acceleration is inversely proportional to the body's mass

Newton's second law can be expressed mathematically as:

$$\boldsymbol{F} = m\boldsymbol{a}$$

Where $\boldsymbol{F}$ is the net force in newtons (N), m is the body's mass in kg and $\boldsymbol{a}$ is its acceleration in m s^{-2}.

→ 1 N is defined as the force necessary to accelerate a 1 kg mass by $1\ \text{m s}^{-2}$ – with any other units, a conversion factor would be required

Newton's third law of motion

Newton realized that you cannot have an isolated force. Pushes or pulls *always* come in pairs.

Newton's third law states that when body A exerts a force on body B, body B exerts an equal and opposite force on body A.

→ The pair of forces are equal in size, opposite in direction and act on different bodies so there is no way they can cancel each other out!

Newton's third law becomes most obvious when we are dealing with two bodies of comparable mass; when one object is much more massive than the other, its acceleration may be too small to notice.

Watch out!

The Greek philosopher, Aristotle, believed that things on Earth needed a force to keep them moving and without a force they would be at rest. Two thousand years later, Galileo realised that without a net force, things could either be at rest or moving with a constant velocity. On Earth there are frictional and drag forces which oppose motion and cause moving things to slow down and stop

"Nature and nature's laws lay hid in night; God said 'Let Newton be!' and all was light."

Alexander Pope

"If I have seen farther, it is by standing on the shoulders of giants."

Isaac Newton

Links

Newton's second law can also be expressed in terms of a body's change in momentum. See *momentum and impulse*, pages 28–9.

Checkpoint 1

What is the pair to each of these forces?
(a) the force of a person on the floor
(b) the pull of the Earth's gravity on a person
(c) the force of a bat on a ball

Watch out!

Spotting a force pair is not always easy. Use this checklist:
(a) They have the same magnitude (size).
(b) They act along the same line but in opposite directions.
(c) They act on different objects.
(d) They are the same type of force.

Rockets and jets

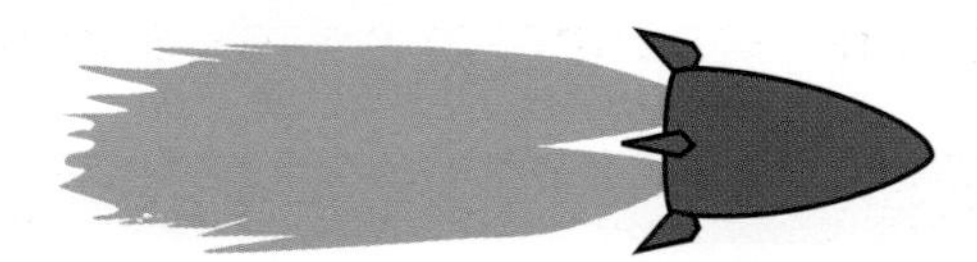

Backward force of rocket on exhaust gases = Forward force of exhaust gases on rocket

The rocket gets pushed forwards by pushing the exhaust jet backwards.

Links

See *momentum and impulse*, pages 28–9.

Free-body force diagrams

A free-body force diagram should be the starting point of every exam answer on forces. It should simplify a problem, by showing *only* the forces acting on the body you are interested in.

Parachutist

Free body force diagram for parachutist

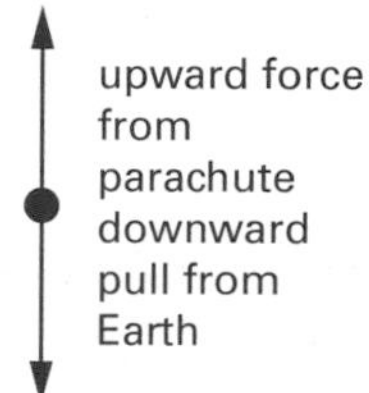

The moon in orbit

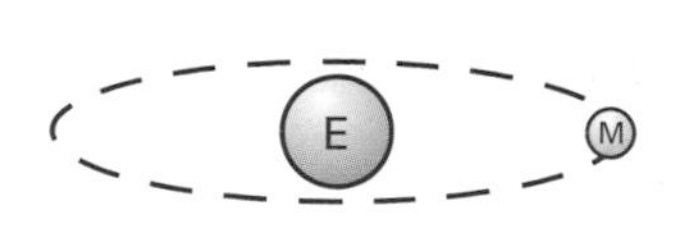

Free body force diagram for the moon pull of Earth

Checkpoint 2

Draw a free-body force diagram for:
(a) a space-rocket
(b) its exhaust jet.

You may notice slight variations in the ways free-body diagrams are treated in different books. Some people like to draw the object realistically, showing where each force acts (as well as showing each force's direction and size). This approach is right and proper, but it complicates things unnecessarily. Unless you are actually going to consider each part separately, or calculate turning moments, it is simpler and more sensible to treat the entire body as a single point, with force arrows pointing outwards wherever possible.

Exam practice answers: pages 39-40

1 An aircraft carrier catapults a 12 000 kg jet from rest to 50 m s^{-1} in a distance of 50 m. Find the average force on the jet. (15 min)

2 An arrow is fired from a bow. When the average force on the arrow is ***F***, the arrow accelerates from rest to a top speed of 20 m s^{-1}. What speed would the arrow reach if the average force on it was doubled (to 2***F***)? (15 min)

3 A truck of mass 2 000 kg pulls a trailer of mass 3 000 kg along a level road. The truck and trailer accelerate at 1.2 m s^{-2}. Friction and air resistance act on both the truck and the trailer. The total drag force is a constant 3 000 N, with a half of this force acting on the trailer. Draw free-body force diagrams for the truck and for the trailer and work out the drive force, ***D***, and the tension, ***T***, in the tow-bar. (25 min)

Some important forces

Some forces deserve special attention, simply because we come across them so often – in life and in A-level Physics questions!

Weight

Mass is a measure of the amount of material that makes up an object. Its units are kilograms and its value does not change if the object is moved to a place where the gravitational field is different.

Weight is the common name for the *force* of gravity on an object. Weight $\boldsymbol{W}$ (in newtons) is given by:

$$\boldsymbol{W} = m\boldsymbol{g}$$

Where m is the object's mass in kilograms and $\boldsymbol{g}$ is the Earth's gravitational field strength. At the surface of the Earth, a good average value for $\boldsymbol{g}$ is:

$$\boldsymbol{g} = 9.81 \text{ N kg}^{-1}$$

Notice that gravitational field strength determines the acceleration due to gravity. Since every kilogram *weighs* 9.81 N, every kilogram *accelerates* at 9.81 ms^{-2} (by $\boldsymbol{F} = m\boldsymbol{a}$).

Links

See *gravitational fields*, pages 142–3, to find out more about what $\boldsymbol{g}$ depends on and how it varies with location.

Watch out!

Mass and weight are not the same thing. Make sure you know the difference. It is easy to mix them up.

The normal-reaction force

If anything presses against a surface, the surface pushes back with an equal and opposite force. This reaction is at right angles, 90°, to the surface and is called the normal force. Normal in this sense just means at right angles. The normal force is caused by the atoms in the surface being pushed slightly together, and therefore pushing back, like the force you feel when you press down on a compression spring.

Links

See *Newton's third law of motion*, page 20.

Watch out!

If the weight of the object is too large, then the surface will break, but for as long as two surfaces remain in contact, the action of the body on the surface equals the reaction of the surface on the body.

Normal reaction on horizontal surfaces

The diagram shows a body of weight $\boldsymbol{W}$ newtons resting on a horizontal surface. The surface will react by pushing up on the object with the normal force $\boldsymbol{N}$ newtons. The body is not moving, so the normal force $\boldsymbol{N}$ must be equal and opposite to $\boldsymbol{W}$.

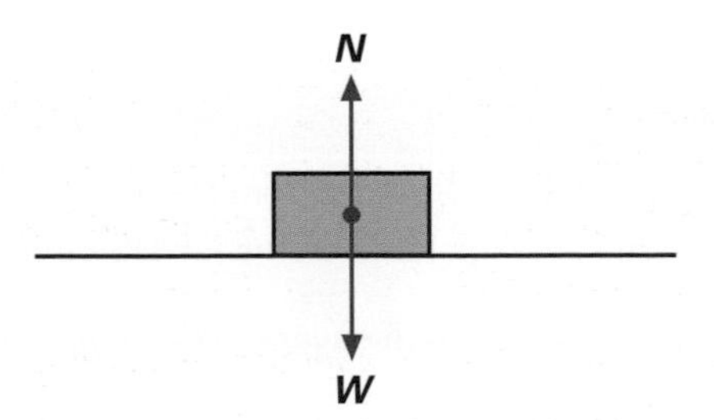

Note that $\boldsymbol{W}$ and $\boldsymbol{N}$ are not a Newton's third law pair of forces as they act on the same body and $\boldsymbol{W}$ is gravitational while $\boldsymbol{N}$ is an electrostatic contact force.

Checkpoint 1

The normal reaction is not always equal in size to a body's weight. Racing cars use aerodynamics to increase the surface contact force. What happens to the normal force as a racing car's speed increases? What are the benefits?

Watch out!

It is important that you remember that action and reaction forces always act on different bodies. Weight and the normal reaction both act on the same body, they are *not* action and reaction.

Normal reaction on sloping surfaces

Here the forces acting must be resolved into components acting along and at right angles to the surface.

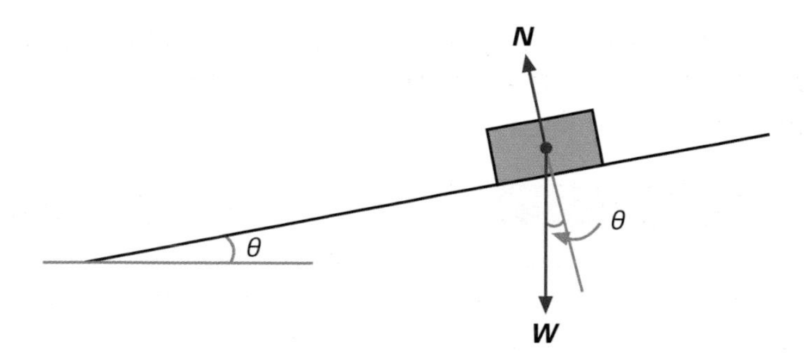

If the body is not moving:

- The forces at right angles to the slope must balance. So the normal force must be equal and opposite to the component of the body's weight at right angles to the slope: $\boldsymbol{N} = -\boldsymbol{W}\cos\theta$.
- The forces along the slope must balance. So the component of the body's weight along the surface must equal the force of friction, $\boldsymbol{F}$: $\boldsymbol{F} = -\boldsymbol{W}\sin\theta$.

Friction

Friction is caused by surface roughness on a microscopic scale (and quite complex intermolecular forces).

- Friction always opposes relative motion.
- Friction acts along the surface.
- Friction between *moving* surfaces generates thermal energy (heat).

Solving problems on forces

1. Always draw a free-body force diagram.
2. Resolve forces into components acting along and at right angles to the surface.
3. Solve the problem!

Forces perpendicular to a surface must balance (or the object will sink into or fly off the surface).

Exam practice

answers: page 40

1. A car of mass 900 kg climbs a 15° hill. If the car's engine provides a forward thrust of 8 000 N and the drag and friction on the car add up to 800 N, calculate the car's rate of acceleration. (10 min)
2. A rope is used to pull a single skier up a friction-free incline of 25°. If the skier's mass is 50 kg, calculate the tension in the rope when the skier is moving steadily up the slope. (10 min)
3. A man of mass 70 kg stands in a lift on a set of weighing scales. The scales are calibrated in newtons. Assuming a value of 10 N kg^{-1} for $\boldsymbol{g}$,
   (a) What weight will the scales register when the lift is static?
   (b) What weight will the scales register when the lift accelerates downwards at 10 m s^{-2}?
   (c) What weight will the scales register when the lift accelerates upwards at 10 m s^{-2}? (10 min)

Checkpoint 2

Car tyres do not grip the road as well on hills as they do on horizontal roads. Why not? (*Hint* think about what happens to the normal force as the hill's gradient increases.)

The jargon

Your *apparent weight* is the reaction that you feel from the surface you are standing on. If you are not accelerating, this will be equal to the pull of gravity on your mass: $\boldsymbol{R} = -m\boldsymbol{g}$ If you have an acceleration, a, (in a lift for example) then your apparent weight will be given by $m\boldsymbol{a} = \boldsymbol{R} - m\boldsymbol{g}$

Checkpoint 3

Be careful with directions. In which direction does friction act when:
(a) a block is at rest on a slope?
(b) it is being dragged up a slope?
(c) it is sliding down a slope?

Grade booster

Many questions involving friction require you to spot the *resultant* (or net) force acting.

Moving through fluids

"Eureka!"

Archimedes

Checkpoint 1

Is sea water more dense or less dense than pure water? Is warm water more dense or less dense than cold water?

The jargon

The term *fluid* covers liquids and gases and some powdered solids. Anything that *flows* is fluid.
Air resistance is another term for the drag force for an object moving relative to the air.

The jargon

A *viscometer* measures viscosity.

Checkpoint 2

Some engine oils are described as 'viscostatic'. What does this mean and why is it important?

Checkpoint 3

Sketch a velocity–time graph for a sky diver who free-falls for two minutes before opening his or her parachute. (Variations on this theme are examiners' favourites.)

So far the effect of air resistance has been ignored, but for sky divers air resistance is crucial!

Density

Density must be a very familiar quantity by now and density measurements come up regularly in practical exams.

$$\text{density} = \frac{\text{mass}}{\text{volume}} \qquad \rho = \frac{m}{V}$$ The SI unit of density is kg m^{-3}.

Density questions often involve unit conversions. These are not always easy, so take them one step at a time. For example, to convert from grams per cubic centimetre to kilograms per cubic metre, we need to work out how many kg in 1 g (10^{-3}) and how many m^3 in 1 cm^3 (10^{-6}):

$$1 \text{ g per cm}^3 = 10^{-3} \text{ kg per } 10^{-6} \text{ m}^3 = 10^3 \text{ kg m}^{-3}!$$

→ Check whether you need to know practical methods for determining the densities of (regular and irregular) solids, liquids and gases.

Drag

Whenever an object moves through a fluid, it experiences a force which opposes its motion, called the drag force. At high speeds, the **drag** force depends on the shape of the object as well as its *velocity* relative to the fluid and the fluid's *density* and *viscosity*. The drag force is caused by molecules colliding with the object.

→ Drag always opposes the relative motion of a body in a fluid.

Viscosity

Viscosity measures how *runny* a fluid is: the lower the viscosity, the runnier the fluid. The viscosity of a fluid decreases rapidly with temperature. Be careful! Density and viscosity measure different properties: water is denser than oil, but oil is more viscous than water.

Terminal velocity

Drag depends on velocity. When something moves through a fluid: the faster it goes, the larger the drag. Eventually accelerating objects reach a maximum velocity called the **terminal velocity** when the drag force equals the accelerating force. Sky divers reach their terminal velocity when the downward force of gravity is exactly balanced by the upward force of drag. They control their speed by changing their area (which is at right angles to the direction of fall). At terminal velocity, weight = drag (= a constant value). A larger area increases drag and reduces the terminal velocity.

Stokes' law

For low speeds, the value of the viscous drag force on a spherical object falling with constant velocity through a fluid is given by Stokes' law:

$$\mathbf{F} = 6\pi\eta r\mathbf{v}$$

Where η is the coefficient of viscosity of the fluid, r is the radius of the sphere and $\mathbf{v}$ is the velocity of the fluid relative to the sphere (it doesn't matter whether the sphere is falling through the fluid or the fluid is moving past the still sphere.)

Checkpoint 4

The coefficient of viscosity of water at 20°C is 1.000×10^{-3}. Use Stokes' law to work out the units of viscosity. Why is the temperature given too?

Archimedes' principle

Objects which are in a fluid appear lighter because the fluid exerts an upward force or upthrust on the object. Archimedes realised that:

→ The **upthrust** on a body immersed or floating in a fluid is equal to the weight of fluid displaced.

So upthrust = ρVg
where ρ is the fluid's density, V is the volume displaced (which also equals the submerged volume of the body) and $\mathbf{g}$ is the gravitational field strength.

Examiner's secrets

You should be able to rearrange and simplify the equation for the forces on the ball bearing to get an expression for the viscosity of the liquid.

Measuring viscosity

Stokes' law can be used to find the viscosity of a liquid, by taking measurements on a ball bearing falling through it at its terminal velocity. The three forces acting on the ball bearing are shown below.

Since the velocity is constant:

upthrust + viscous drag = weight

And the ball bearing is a sphere, so its volume = $4/3\ \pi r^3$, then

$$4/3\pi r^3\ \rho_l\ \mathbf{g} + 6\pi\eta r\mathbf{v} = 4/3\pi r^3\ \rho_s \mathbf{g}$$

Where ρ_l is the density of the liquid and ρ_s is the density of the ball bearing's material.

Links

See *Turning points in physics 1*, pages 174–5 to see how Millikan used this to find the charge of an electron.

Streamlined and turbulent flow

In streamlined flow, the fluid does not get mixed by the passing body. Every particle follows the streamline it is in. At any chosen point, every passing particle will have the same velocity. At low speeds, viscous forces dominate, fluid flow is streamlined and drag forces are small. As a body picks up speed, there will come a point where the fluid switches from streamlined to turbulent flow and drag forces become far greater.

The jargon

Laminar flow means the same as *streamlined flow*.

Exam practice answers: page 40

1 The value of the force of air resistance, F_{drag}, on an object falling at high velocity through the air, is given by the formula:

$F_{drag} = kv^2$

where k is a constant, and v is the object's velocity.

A mass 2.0 kg, falls from a hot-air balloon and eventually reaches a terminal velocity of 40 m s^{-1}.

(a) Explain why the mass eventually reaches a terminal velocity.
(b) Calculate the value of k for the mass. (gravitational field strength = 9.81 N kg^{-1})
(c) Calculate the acceleration of the mass when its velocity is 30 m s^{-1}. (10 min)

Work, energy and power

Energy is quite a difficult concept. You only really notice it when it is being transferred from one place to another, or converted from one form to another, and yet it is all around us. Energy transfer drives every process!

Links

See *binding energy and mass defect*, pages 54–5 According to Einstein's theory of special relativity, matter and energy are two forms of the same thing *mass–energy*!

Work

Work is done whenever a force is used to move something, and is given by

work done = force × distance moved $\quad W = \boldsymbol{Fs}$

- Work is an energy transfer, measured in joules.
- 1 J = 1 newton metre (N m). A joule is the amount of work done when a 1 newton force moves something 1 metre in the direction of the force.
- It is important to note that $\boldsymbol{s}$ is the distance moved in the direction that the force acts.

Watch out!

If nothing is actually moving, no work is being done – no matter how great the force involved (since no energy is being *transferred*). Work is defined by the work equation. If $\boldsymbol{s} = 0$, $W = 0$.

Work done by forces acting at angles to direction of motion

In the diagram below, the boat travels a distance $\boldsymbol{s}$ parallel to the bank. The distance moved in the direction of the force is $\boldsymbol{s}\cos\theta$, so the work done is

work = $\boldsymbol{F}\,\boldsymbol{s}\cos\theta$

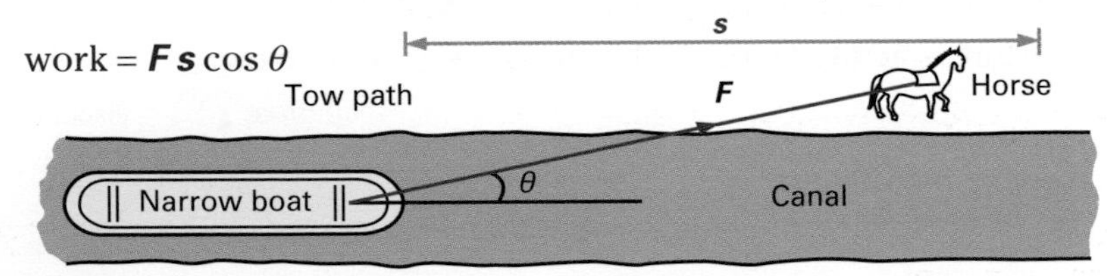

The jargon

Work is a scalar quantity (so are both energy and power), but you can have positive and negative work. *Positive work* is where the force pulls in the same direction as the movement. *Negative work* is where the force is in the opposite direction (e.g. gravity does positive work to accelerate a cyclist down a hill and negative work to decelerate him up the other side).

Work and kinetic energy

Kinetic energy E_k is the energy a body has by virtue of its motion. It is defined by the equation:

$E_k = {}^1\!/\!_2 m\boldsymbol{v}^2$

Where E_k is kinetic energy (in joules), m is mass (in kilograms) and $\boldsymbol{v}$ is speed (in metres per second).

You can do work on an object to change its kinetic energy. In fact, the work done on a body is equal to its change in kinetic energy. (This is called the *work–energy theorem*.):

$\boldsymbol{Fs} = {}^1\!/\!_2 m\boldsymbol{v}^2 - {}^1\!/\!_2 m\boldsymbol{u}^2$

- *Note* $\boldsymbol{F}$ is the resultant (net) force acting.

Don't forget

Do you know your sines and cosines? It is important here to remember that *cos 90°* = 0 and *cos 0°* = 1.

Links

See *scalars and vectors*, pages 10–11.

This equation is particularly useful in calculations of minimum stopping force or minimum stopping distance. If a body is brought to a halt, the work done to it (e.g. by the braking force) is equal to the kinetic energy lost.

- Note the dependence on $\boldsymbol{v}^2$. If you double your speed, your minimum stopping distance is quadrupled (assuming the same retarding force). Speed limits save lives!

Action point

If you enjoy algebra, try deriving the formula for kinetic energy (as the energy gained when a force does work to accelerate an object) by combining the equations $\boldsymbol{F} = m\boldsymbol{a}$, $W = \boldsymbol{Fs}$ and $\boldsymbol{v}^2 = \boldsymbol{u}^2 + 2\boldsymbol{as}$. If you don't, don't!

Gravitational potential energy

When you lift an object, you do work against the force of gravity. The energy that the object gains is called gravitational potential energy.

For a mass m raised a height Δh, the gravitational potential energy ΔE_p it gains is given by

$\Delta E_p = m\boldsymbol{g}\Delta h$

Watch out!

When a body is travelling at constant speed, the net force on it is zero. Any work done is *not* being done to the body in question.

(i.e. the work done to lift its weight mg by a height Δh against the pull of gravity).

→ Note that Δh is the vertical height raised, so it does not matter what path is taken between two points.

Conservation of mechanical energy

The term *mechanical energy* refers to the sum of a body's kinetic and (gravitational) potential energy.

$$E = E_k + E_p$$

For objects falling moving under gravity, mechanical energy is conserved - if you assume that no energy is converted to other forms. The diagram below shows how conservation of energy is applied to a roller-coaster ride:

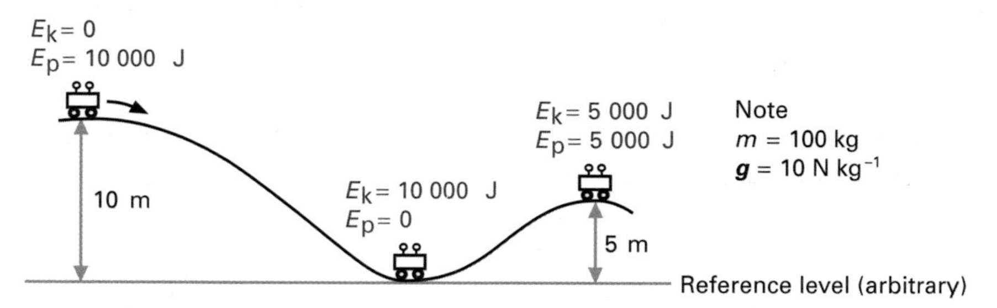

Law of conservation of energy

Energy can be neither created nor destroyed. It can only be converted from one form to others. The total amount of energy in any isolated system remains constant.

In practice, whenever energy is transferred, it does not all go into the required form – into useful energy. Some is converted to thermal energy and sound and dissipated into the surroundings. The efficiency of a system is given by

$$\text{Efficiency (\%)} = \frac{\text{useful energy out}}{\text{total energy in}} (\times 100)$$

Sankey diagrams are used to visualise energy transfers.

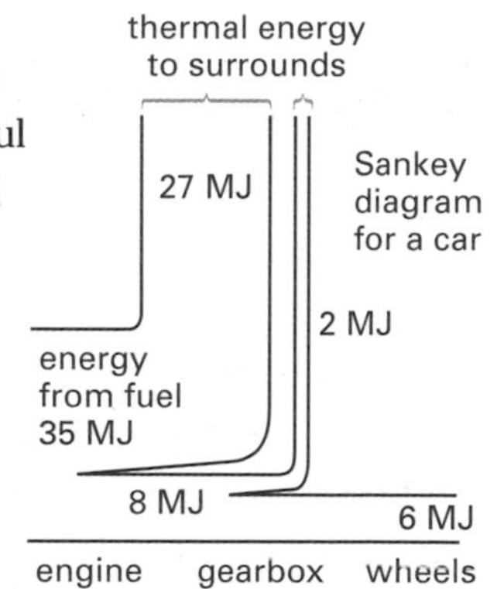

Power

Power is defined as rate of transfer of energy, or rate of doing work. It is measured in watts (W). 1 watt = 1 joule per second.

average power = work done (or energy transferred) ÷ time taken

$$P = \frac{\Delta W}{\Delta t}$$

For any vehicle travelling at speed v, power output is given by:

$P = Fv$ (F = *motive* or *drive* force and v = average speed)

Exam practice answers: page 41

A boat of mass 8 000 kg approaches a jetty at a speed of 1.2 m s^{-1}. Its engine is put in reverse and the boat comes to a halt in a distance of 6 m. (a) What is the engine's (backward) thrust? (b) What is the engine's power? Assume that drag can be ignored(!) (20 min)

Checkpoint 1

Comets follow eccentric elliptical paths around the Sun. Where along its path does a comet have the greatest potential energy?

Action point

Check other sections to find formulae and graphical methods for calculating other energy transfers (electrical, heat, elastic, nuclear, etc.).

Don't forget

Power P is rate of work

$$P = \frac{\text{force} \times \text{distance}}{\text{time}}$$

Since speed = distance/time we get

$$P = Fv$$

Watch out!

In the *motive power* equation ($P = Fv$), F is the drive force, *not* the net force acting. If the velocity is constant, all the engine's power is being used to overcome resistive forces.

Grade booster

Energy can *never* be lost or gained. It can be transferred from one form to another. Many students lose marks by stating that energy is *lost* without saying where it has gone!

Momentum and impulse

In life, we may often 'act on impulse', but in physics, impulse more often acts on us! Impulse is the product of force and time. Impulses always change the momentum of the body they act on.

Examiner's secrets

A quantity's units can be derived from the equations they appear in. Mass is measured in kg, velocity is measured in $m\,s^{-1}$, so momentum can be expressed in $kg\,m\,s^{-1}$. If you substitute F/a for m, you get an alternative unit for momentum – the *newton second*. Choose whichever unit suits your needs! (They are entirely equivalent.)

Momentum

If a body of mass m has a velocity $\mathbf{v}$, then its **momentum** $\mathbf{p}$ is:

$$\mathbf{p} = m\mathbf{v}$$

The units of momentum are $kg\,m\,s^{-1}$ or newton seconds (N s).

Law of conservation of momentum

Momentum is conserved in *all* collisions, explosions and interactions! There are no exceptions to this law.

→ The total momentum of a system before any interaction is exactly equal to the total momentum after it, provided no external forces act (external forces would allow momentum to be transferred to external bodies).

Watch out!

Momentum is a vector quantity. Don't forget to give it a direction. In problems involving motion in a straight line, you must state clearly which direction you are taking as the positive.

When two objects collide, the changes in their momenta will be equal in size, but opposite in direction. The momentum gained by one body equals the momentum lost by the other.

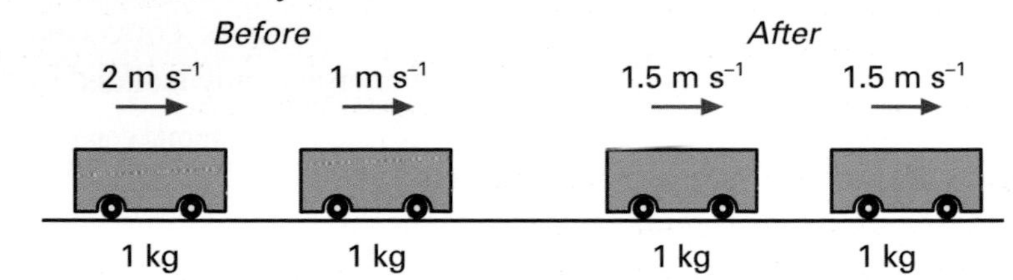

Checkpoint 1

The law of conservation of momentum is sometimes difficult to demonstrate convincingly with trolley experiments. Why? What could you do to improve things?

Tackling problems

1 Choose which direction is positive.
2 Draw before and after sketches of the objects involved.
3 Calculate every momentum you can.
4 Apply the law of conservation of momentum for collisions involving two bodies:

$$(m_1\mathbf{v}_1 + m_2\mathbf{v}_2)_{\text{before}} = (m_1\mathbf{v}_1 + m_2\mathbf{v}_2)_{\text{after}}$$

2D collisions

Grade booster

Students often lose marks through carelessness with directions and signs. Be warned!

You may have to solve problems about two-dimensional collisions, for example, when a moving subatomic particle collides with a stationary one. The law of conservation of momentum still applies but here you have to apply it to the components of momenta along the initial direction of movement and also at right angles to this direction.

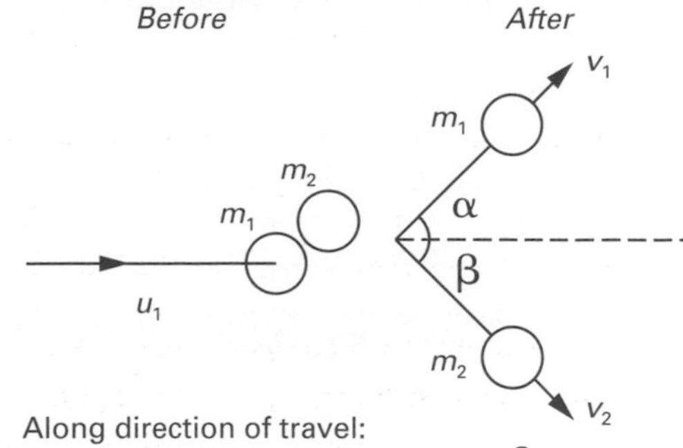

Along direction of travel:
$m_1 u_1 + 0 = m_1 v_1 \cos\alpha + m_2 v_2 \cos\beta$

At right angles to direction of travel:
$0 = m_1 v_1 \sin\alpha + m_2 v_2 \sin\beta$

Recoil and explosions

Guns and cannons *recoil* when fired because of the law of conservation of momentum. The positive momentum gained by the bullet or cannon ball is equal to the negative recoil momentum of the gun or cannon, and so the total momentum before and after the explosion is zero.

Elastic and inelastic collisions

- Kinetic energy is conserved in **elastic** collisions.
- Kinetic energy is not conserved in **inelastic** collisions.
- Momentum is conserved in *all* collisions.
- Total energy is conserved in *all* collisions.

In inelastic collisions some kinetic energy is converted to other forms of energy (usually mainly heat). Large scale collisions are inelastic: collisions between hard steel spheres are nearly elastic: some particle collisions are elastic.

Impulse and Newton's second law of motion

Most bodies have constant mass, so we normally (rightly) note that unbalanced forces cause acceleration. At a more fundamental level we can state that unbalanced forces cause a *change in momentum*. The change in momentum depends on the size and direction of the force and the period of time over which it is applied; i.e. it depends on its *impulse*.

- Impulse is the product of force and time.
- Impulse = change in momentum.
- $\boldsymbol{F}t = m\boldsymbol{v} - m\boldsymbol{u}$ (Where $m\boldsymbol{u}$ is initial and $m\boldsymbol{v}$ is final momentum).
- Force = rate of change of momentum = $(m\boldsymbol{v} - m\boldsymbol{u})/t$. This is another version of Newton's second law of motion.

Impulse is measured in either newton seconds (N s) or kg m s^{-1} (exactly the same units as momentum).

Exam practice

answers: page 41

1 A bullet of mass 40 g is fired with a horizontal velocity of 500 m s^{-1} from a rifle of mass 2.5 kg.

(a) Find: (i) the bullet's forward momentum, (ii) the bullet's kinetic energy, (iii) the speed of recoil of the rifle, (iv) the rifle's kinetic energy after the explosion.

(b) Can explosions ever be perfectly elastic? Explain your answer. (15 min)

2 What driving force is necessary to accelerate a car of mass 1 400 kg from rest to a speed of 35 m s^{-1} in 20 s? (5 min)

3 Two skaters are skating together at a steady velocity of 8 m s^{-1}. Their masses are 80 kg and 50 kg. The lighter skater is pushed forwards and accelerates to 10 m s^{-1}. Calculate the new speed of her partner. (10 min)

4 Hard snowballs bounce back off you when they hit at 90°; soft snowballs don't. Explain why these hard snowballs exert the greater force. (10 min)

Check the net

In a head on elastic collision between a moving mass and a stationary one of equal mass, the moving mass comes to rest and transfers its momentum to the other mass. If the collision is not head on, there is a right angle between the resulting paths. Try some collisions on http://galileoandeinstein.physics.virginia.edu/more_stuff/Applets/Collision/jarapplet.html

Links

See *work, energy and power*, pages 26–7. Kinetic energy, $E_k = \frac{1}{2} m\boldsymbol{v}^2$ and momentum, $\boldsymbol{p} = m\boldsymbol{v}$ are calculated from the same quantities, but they are not the same thing. *Be clear* Momentum is always conserved, whereas kinetic energy is conserved only in special cases (in perfectly elastic collisions).

The jargon

Quantitative and *qualitative* If a question asks for a quantitative answer, you must give numbers and units; if it asks for a qualitative answer, you should give general trends ('When F = 2 N, x = 8 cm' is a *quantitative* answer. 'The extension of a spring is directly proportional to the force applied – until the elastic limit is reached' is a *qualitative* answer.)

Don't forget!

If a car changes its momentum, $\Delta\boldsymbol{p}$, in a short time, Δt, then by $\boldsymbol{F} = \Delta\boldsymbol{p}/\Delta t$, this will involve a large force. This is why modern cars have crumple zones, air bags and seat belts so they take longer to stop and the forces involved are smaller.

Grade booster

The area under a force–time graph is the change in momentum (impulse). To make a tennis ball leave the racket with a faster velocity you can hit it with a bigger force *or* for a longer time.

Stress, strain and Hooke's law

Robert Hooke was a brilliant scientist who had the misfortune to be a contemporary of Isaac Newton. He had theories on light and on gravity, but so did Newton. Hooke's lasting legacy to the world of science is his law of elastic deformation: that within bounds, strain is proportional to stress.

The jargon

A substance's behaviour is *elastic* if it springs back into its original shape after a force has been applied. *Plastic* is the opposite to elastic. *Plastic deformation* occurs when intermolecular bonds are repeatedly broken and remade, changing a solid's shape (think Plasticine!).

Springs

- Solids are made of particles (atoms) that are bound together by forces. These particles behave like tiny, sticky, spongy balls. They stick together: they resist if you try pulling the atoms apart; they resist if you try pushing them together.
- Springs behave like the forces between the atoms in a solid. Understanding how springs respond to forces is a useful first step towards a more general understanding of the behaviour of solid materials under stress.

Hooke's law applied to springs

Within the elastic limits of the spring:

$$\boldsymbol{F} = -k\Delta \boldsymbol{x}$$

Where $\boldsymbol{F}$ = force applied, $\Delta \boldsymbol{x}$ = extension (increase in length) and k = the *spring constant* – the force per unit extension.

- The minus sign shows that the force acts in the opposite direction to the extension (it is a *restoring force*); it can often be ignored!

Action point

How would you determine the spring constant of the springs in the shock absorbers of a car? (*Hint* You need a metre ruler, lots of friends and a helpful car owner!)

The jargon

The term *elastic limit* is used fairly loosely to mean the limit beyond which further extension causes permanent stretching or deformation. Under compression, Hooke's law breaks down when a spring's coils touch – even without any permanent deformation.

The energy stored in a stretched spring

The energy stored in a spring = the work done in stretching the spring. For a spring that obeys Hooke's law, its force-extension graph is shown below: as the extension increases, so does the force.

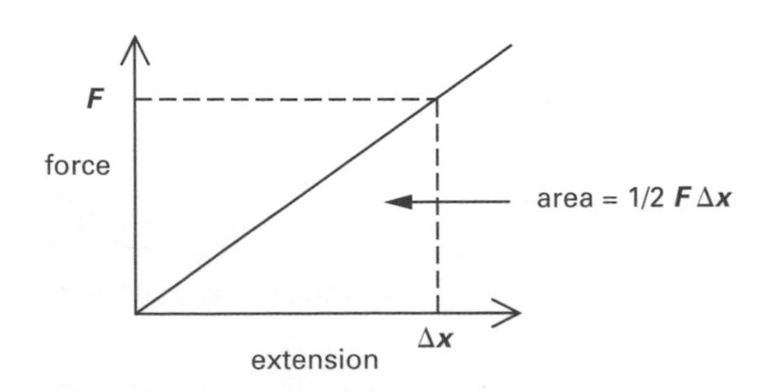

So energy stored, or elastic potential energy (strain energy), E_p = average force × extension.
When the force increases from 0 to $\boldsymbol{F}$, the average force = ½ $\boldsymbol{F}$
So $E_p = \frac{1}{2}\boldsymbol{F}\Delta\boldsymbol{x}$ which also equals $\frac{1}{2}k(\Delta\boldsymbol{x})^2$ since $\boldsymbol{F} = k\Delta\boldsymbol{x}$

- Provided no energy is lost in heating the surroundings, the work done on the spring also equals the elastic potential energy it gains.

Watch out!

If a material does not obey Hooke's law, the work done in stretching it is still the area under the force-extension graph. However the area has to be estimated by counting squares as the formula $E_p = \frac{1}{2}\boldsymbol{F}\Delta\boldsymbol{x}$ only applies for the area under a straight line.

Links

See *work, energy and power*, pages 26–7 and consider the energy conversions in a weight oscillating on a spring.

Stress and strain

The stiffness of a sample of material depends on the shape of the sample. The Young modulus measures the stiffness of a *material* and is constant for that material.

Defining stress as: stress = force / cross-sectional area $\sigma = \boldsymbol{F}/A$

Units N m^{-2} or Pa (same as the units of pressure)

And strain as: strain = extension / original length $\varepsilon = \Delta \boldsymbol{x} / l$

Units None (strain is a dimensionless ratio).

The Young modulus = stress / strain $E = \sigma / \varepsilon$

Units N m^{-2} or Pa

→ The larger the value of Young's modulus, the more rigid the material.

It is sometimes useful to expand the equation for Young's modulus to:

$$E = \frac{(\boldsymbol{F}/A)}{(\Delta \boldsymbol{x}/l)}$$

Rearranging, this becomes: $E = \dfrac{\boldsymbol{F}l}{\Delta \boldsymbol{x} A}$

→ *Tensile* refers to stretching. *Compressive* refers to squashing.
→ The *breaking strength* or *ultimate tensile stress*, UTS, is the maximum stress a material can stand in tension before breaking.
UTS = breaking force / cross-sectional area

Stress–strain graphs for different materials

→ Material **A** is stiffer than material **B**.
→ After the elastic limit or *yield point*, material **B** is plastic and has a permanent stretch. Material **A** shows no plastic behaviour – it is brittle.
→ The *limit of proportionality* is the point after which the line is no longer straight.

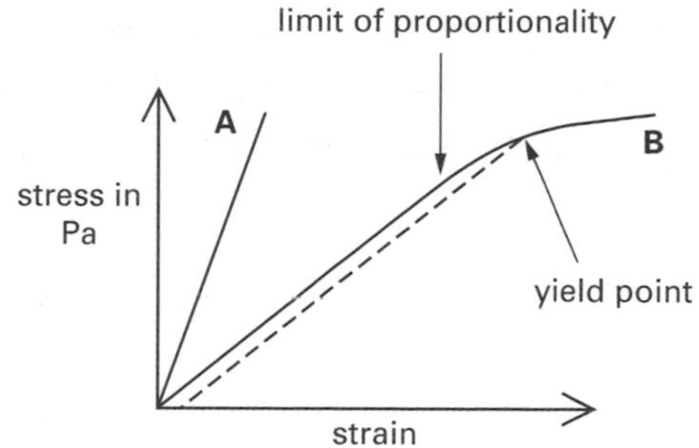

Exam practice

answers: page 41

1 A steel cable of diameter 2.1 cm and length 12 m is used on a crane. The steel has a Young's modulus of 2.0×10^{11} Pa. (a) How much does it stretch when used to lift a 1 500 kg load? (b) Repeat the calculation (same length, diameter and load) for a copper cable.
(Young's modulus for copper = 1.1×10^{11} Pa.) (15 min)

2 You are given a bag of identical springs, each with a spring constant of 15.0 N cm^{-1} and an elastic limit of 20 N. (a) Calculate the smallest number of springs required to lift a 500 N weight. (b) How are the springs arranged? (c) Calculate the extension of each spring. (d) Find the spring constant of the single spring that would behave in the same way. (10 min)

Watch out!

Strain is sometimes multiplied by 100 and written as a percentage, but it needs to be converted back to a ratio when doing calculations.

Checkpoint 1

Remember extension $\Delta \boldsymbol{x}$ is the extra length gained when an object is stretched. The definition still stands under compression (negative extension). What is the strain for an elastic band stretched to double its original length?

Watch out!

Strain depends on stress, so you would normally plot strain on the *y*-axis. However, stress–strain graphs are usually plotted the 'wrong way around' with stress on the *y*-axis, so that the gradient is the Young modulus.

Watch out!

E is used for energy and also for the Young modulus!

Checkpoint 2

A material obeys Hooke's law if strain is directly proportional to stress. Check for yourself that this general statement of Hooke's law can be used to derive $\boldsymbol{F} = k\Delta \boldsymbol{x}$ for a spring.

Watch out!

Stay in shape – don't go beyond the elastic limit. There's no going back!

Vibrations and resonance

Links

See *types of wave and their properties*, pages 112–3.

The jargon

Reciprocal This is simply 1 divided by the number. The reciprocal of 2 is ½; the reciprocal of 0.1 is 10, etc. Frequency is the reciprocal of period (and period is the reciprocal of frequency).

The jargon

A *simple pendulum* has its mass concentrated in the bob. A *compound pendulum* has a bob attached to a (significantly) heavy oscillating bar. The distribution of mass in a compound pendulum complicates its behaviour slightly.

Checkpoint 1

What length would a simple pendulum have to be in order to have a period of 1 s?

Vibrating objects can generate waves. Waves carry energy and information. Resonance allows waves to transfer this energy and information to new objects – by making them vibrate or oscillate.

Frequency and period of oscillation

→ **Frequency** f is the number of complete vibrations per unit time. Frequency is measured in *hertz* (Hz). 1 Hz = 1 vibration per second.

→ **Period** T is the time taken for one *complete oscillation* (the time between one vibration and the next). Period is measured in seconds.

Provided we stick to SI units, frequency and period are linked by the following equations:

$f = 1/T \qquad T = 1/f$

Natural frequency

Every oscillator has a **natural frequency**. If you give a swing a single push, it will swing back and forth for some time at its natural frequency. The natural frequency is determined by the **restoring forces** which tend to return the oscillator to its equilibrium position (where its displacement is zero).

Swings and springs

The **natural period** T of oscillation of a simple pendulum (or a swing) is given by: $T = 2\pi\sqrt{(l/g)}$

Where l is the length of the pendulum and g is gravitational acceleration.

The natural period of oscillation of a mass on a spring is given by: $T = 2\pi\sqrt{(m/k)}$

Where m is the mass and k is the spring constant.

In both examples, we can see that the period is defined by a set of constants; it does not vary, so oscillating pendulums and weights on springs can mark time. All clocks feature oscillating systems.

Damping

Most oscillators gradually lose energy to their surroundings – perhaps through friction or air resistance. This loss of energy **damps** the oscillation, *reducing its amplitude*. There are degrees of damping:

→ *Light damping* It takes many oscillations before the amplitude is reduced to zero.

→ *Critical damping* The displaced body just returns to zero displacement without over-shooting. Critical damping is the minimum degree of damping required to prevent oscillation.

→ *Heavy (over) damping* There is no oscillation! The displaced body slowly returns towards zero displacement.

Forced vibrations and resonance

If you push a swing just once, it will oscillate gently at its natural frequency. Its amplitude will gradually diminish because of air resistance and friction (damping). If you push it again every time it returns to you, its amplitude of oscillation will increase. It will *resonate*! You will have succeeded in transferring energy (repeatedly) to the swing.

→ **Resonance** is the strong vibration that builds up when an oscillator is driven at its natural frequency.

To find an oscillator's natural frequency, you could try driving it at different frequencies. The oscillator will normally resonate (vibrate with greatest amplitude) when driving frequency matches natural frequency. At resonance, the oscillator is one quarter of an oscillation behind the driver – it is said to lag the driver by a phase of $\pi/2$.

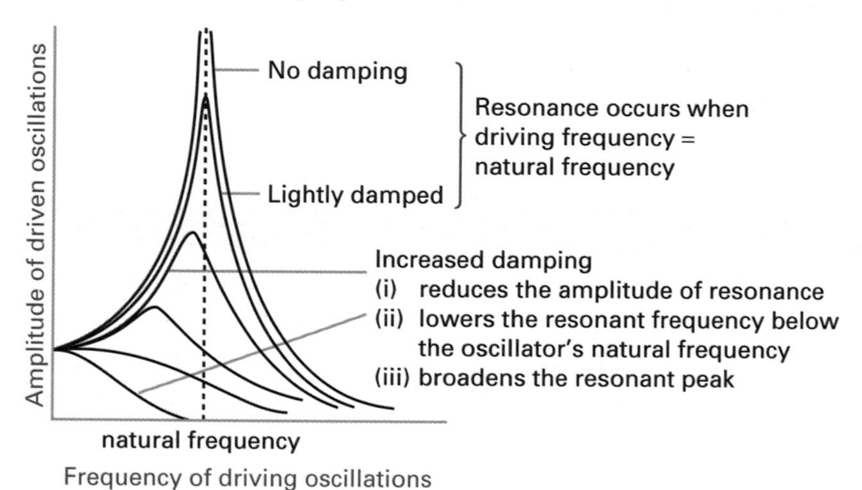

The importance of resonance

Wherever waves transfer energy, resonance plays a part. Here are just a few examples:

→ *Earthquakes* Buildings with natural frequencies close to the quake's frequency are most susceptible to damage.
→ *Microwave ovens* Interatomic bonds in water molecules have natural frequencies in the microwave region of the electromagnetic spectrum.
→ *Hearing* Different regions of the cochlea (a shell-shaped organ in the inner ear) resonate at different audible frequencies.

Car shock absorbers – applied damping

Shock absorbers are meant to cushion the car user from uncomfortable jolts and bumps, by soaking up vibrational energy peaks. The damping provided by shock absorbers should be just short of critical damping.

→ If damping is too heavy, the shock absorbers will not recover from one jolt in time to respond to the next.
→ If damping is too light, a single pot hole can set the car bouncing up and down uncomfortably.

Exam practice answers: page 41

Buildings situated close to railway lines should be constructed in a manner which minimises noise and vibrations from passing trains.

(a) Vibrations could cause parts of the building to resonate. Describe the meaning of the word resonate.

(b) Some buildings, which are subject to vibrations, are constructed on springs. Suggest how spring could prevent these buildings from vibrating.

(5 mins)

The jargon

Forced vibration Obvious forces drive the oscillator. These forces are typically repetitive and periodic. Truly *free vibrations* should be free of both driver and damping. In practice, the term is used to describe any undamped or very lightly damped vibrations at natural frequencies.

Links

See *simple harmonic motion*, pages 36-7.

Check the net

In 1940 the first Tacoma Narrows suspension bridge collapsed due to wind-induced vibrations. It had been open for traffic for only a few months. To see a film of the spectacular collapse, type in 'Tacoma Narrows video clip' into an Internet search engine. 'Millennium Bridge oscillations' should find a more recent pedestrian induced resonance.

Checkpoint 2

How can heavy damping change the resonant frequency of an oscillator? (What effect does damping have on the net restoring force acting on an oscillator?)

Checkpoint 3

Can you think of more examples of resonance which are:
(a) beneficial?
(b) harmful?

Links

See more about the ear, *medical and health physics 1*, pages 164–5.

St... page...

Circular motion

> *"People travel to wonder at the height of the mountains, at the huge waves of the seas, at the long course of the rivers, at the vast compass of the ocean, at the circular motion of the stars, and yet they pass by themselves without wondering."*
>
> St Augustine, AD400

We are bound by gravity to a spinning planet orbiting a (spinning) star on a very nearly circular path. On a big scale, true linear motion must be a rarity in a universe governed by gravity!

Angles and angular velocities

Radians

Radians are more fundamental units of angle than degrees. The definition of a degree is arbitrary: 1/360th of a full circle turn. In radians, an angle is defined by:

$$\text{angle (rad)} = \frac{\text{arc length (along a circle's circumference)}}{\text{radius}}$$

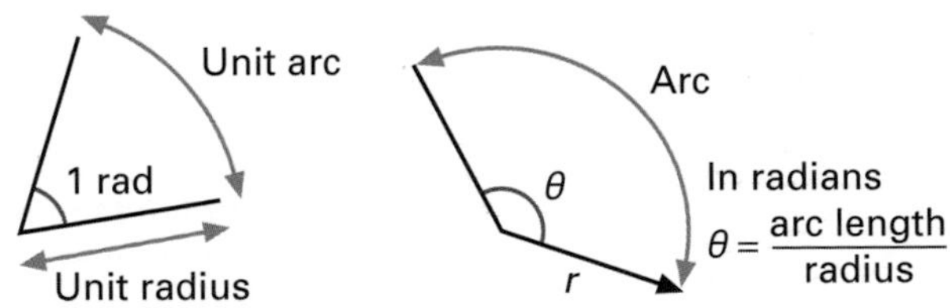

Since the circumference of a circle is $2\pi r$, there are 2π radians in one full revolution.

$$2\pi \text{ radians} = 360° = 1 \text{ revolution}$$

Measuring angles in radians has the additional benefit that *for small angles*:

$$\sin\theta \approx \tan\theta \approx \theta \text{ (in radians)}$$

Checkpoint 1

Convert 5° to radians. Find the sine and tangent of 5°. How close are the three values? (*Optional extra*: repeat the exercise, using greater angles to see how the three values diverge as the angle increases.)

Links

See *vibrations and resonance*, pages 32–3. Frequency and period are related by $f = 1/T$. If you know one of them, you can calculate the other.

Uniform circular motion and angular velocity

- **Uniform circular motion** is simply motion along a circular path, at a constant speed. If the period (time for one complete orbit) is T, speed v is given by:

$$v = 2\pi r/T$$

- **Angular velocity** is rate of change of angular displacement – a measure of rate of rotation. Angular velocity is usually denoted by the Greek letter ω. Its units are rad s^{-1}.
- To convert from a frequency f (revolutions per second), to angular velocity ω (radians per second), simply multiply by 2π:

$$\omega = 2\pi f$$

Angular velocity ω and linear speed v are related by the equation:

$$v = r\omega$$

Checkpoint 2

Write equations:
(a) giving speed of uniform circular motion v in terms of orbit radius and *frequency*
(b) giving angular velocity ω in terms of orbit *period*

Checkpoint 3

Calculate (a) the angular velocity ω and (b) the speed v of the Earth's orbit around the Sun. (The distance from the Sun to the Earth is 1.50×10^8 km.)

Centripetal acceleration and force

Nothing will follow a circular path unless it is forced to.

- **Centripetal force** is the force required to keep a body in uniform circular motion.

Links

Newton's laws of motion, 20–1

→ Centripetal force and acceleration are always directed towards the centre of the circular path.

If you need to convince yourself of the direction of the centripetal force, consider a conker swinging around on a string. The string can only pull the conker (you can't push something with a string!) and it pulls it inwards.

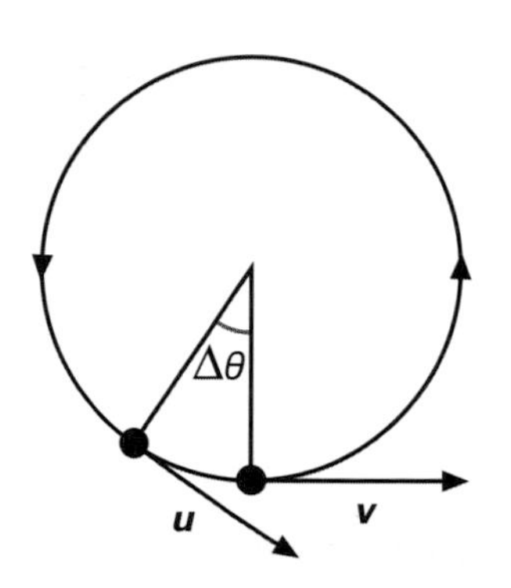

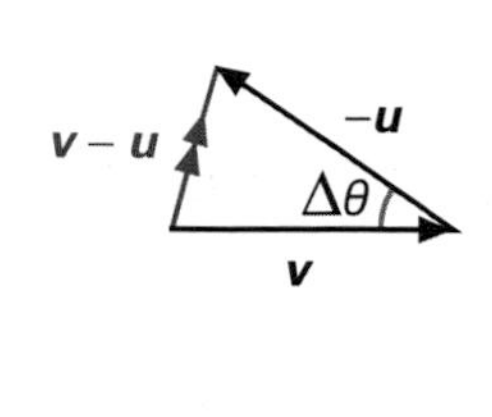

Consider the conker at two places on its path having moved through a small angle, θ. The conker is travelling at a constant speed, but its velocity has changed from u to v because its direction has changed. The change in velocity is towards the centre of the circle as shown by the vector diagram. You don't need to know how to derive the equation for the centripetal acceleration given below but you will need to use it:

$a = v^2/r$,

or in terms of the angular velocity,

$a = \omega^2 r$

Using $F = ma$, centripetal force given by:

$F = mv^2/r$

or

$F = m\omega^2 r$

Banking

Aircraft steer by banking. A component of the lift force acting on the wings is directed horizontally into the turn. Banking questions test your ability to resolve forces *and* your understanding of circular motion – making them firm favourites with examiners.

Links

See *scalars and vectors*, pages 10–11 for rules of vector subtraction.

Grade booster

To get you to really think, how would you explain that an object can be accelerating *and* moving at constant speed?

Checkpoint 4

(a) Calculate the centripetal acceleration of the Earth in its orbit around the Sun.
(b) Calculate the centripetal acceleration of the Moon around the Earth.
(Period of Moon's orbit = 27.3 days, Earth–Moon separation = 3.8×10^8 m.)

Checkpoint 5

Calculate the centripetal force needed to keep a person of mass 80 kg on the surface of the Earth from flying off into space (due to tangential motion):
(a) at the equator
(b) at the North pole.
(Assume a value of 6 400 km for the Earth's radius.)

Exam practice answers: page 42

1 A bobsleigh corners on a frictionless, horizontal banked track.
(a) Draw a free-body force diagram for the bobsleigh.
(b) By resolving forces, prove that the banking angle θ (the angle of the track relative to the horizontal) is given by $\tan\theta = v^2/(rg)$, where v is the bobsleigh's speed, r is the radius of its turning. (20 min)

2 The radius of a CD is 0.06 m and it rotates at 3.5 rev/s when playing music at the outer edge. Find the maximum tangential speed of the disk. (5 min)

Simple harmonic motion

Understanding simple harmonic motion is the first step towards understanding any mechanical oscillation, but be warned – SHM is not as simple as its title suggests!

Conditions for SHM

A body will oscillate with **simple harmonic motion** if the restoring force acting on it (pulling the body back towards a rest position) is directly proportional to the body's displacement. A restoring force results in an acceleration which is in the opposite direction to the body's displacement.

→ The conditions for simple harmonic motion are summarized by:

$$\boldsymbol{a} \propto -\boldsymbol{x}$$

Where $\boldsymbol{a}$ is acceleration and $\boldsymbol{x}$ is displacement. The minus sign shows that acceleration is in the opposite direction to the displacement vector.

SHM and circular motion

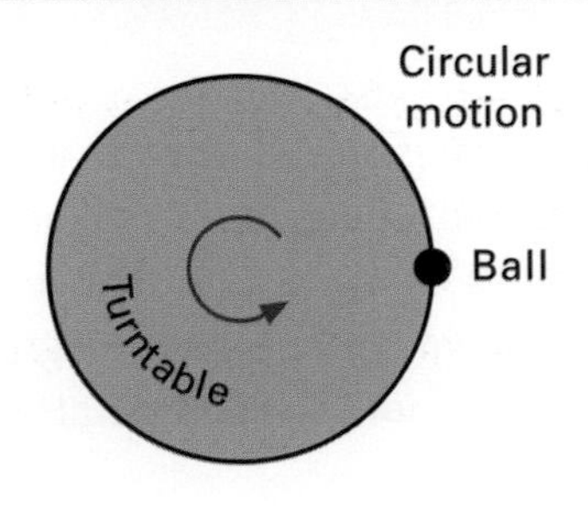

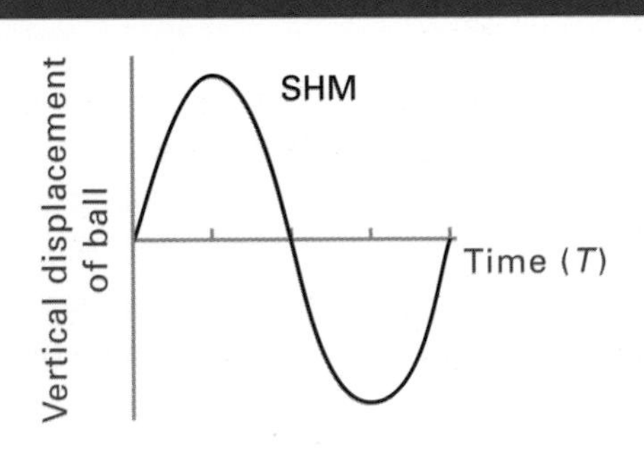

An object on a turntable rotating with constant speed will have a constant angular velocity, ω. When viewed from the side, the object moves in a straight line with simple harmonic motion. ω is also used for simple harmonic motion where:

$$\omega = 2\pi/T = 2\pi f$$

T is the time for one oscillation and f is the frequency or number of oscillations per second.

Phase and phase difference

Different points on the rim of a spinning wheel are said to be out of step or *out of phase*. The phase difference between any two such points is the angle between them (subtended at the wheel's centre). When the term phase is applied to waves and oscillations more generally, one complete oscillation is taken to be equivalent to one rotation or 2π radians and phase differences (usually denoted by the Greek letter epsilon, ε) are often measured in radians.

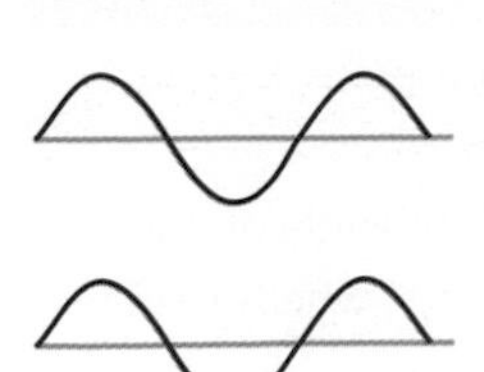

In phase
phase difference = 0

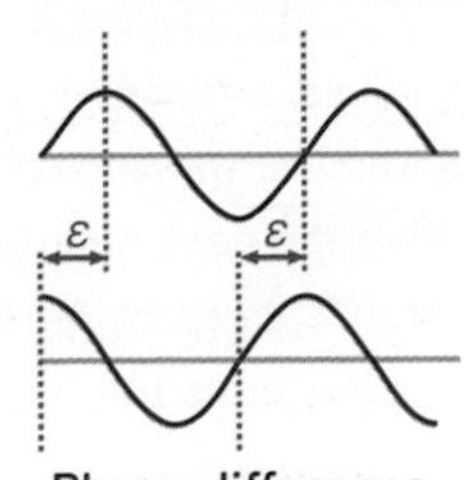

Phase difference
= $\frac{1}{4}$ cycle
= $\frac{\pi}{2}$ radians

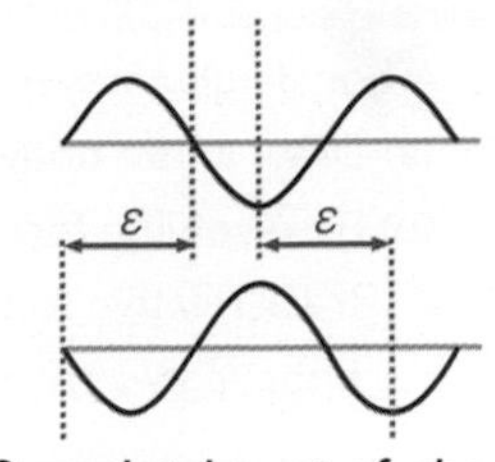

Completely out of phase
Phase difference
= $\frac{1}{2}$ cycle
= π radians

Links

Pendulums (oscillating through a small angle, eg <5° to the vertical) and weights on springs are both examples of simple harmonic oscillators that you should be aware of. See *vibrations and resonance*, pages 32–3 for equations.

Grade booster

Time period is independent of amplitude for objects that oscillate with SHM, but this does *not* define SHM.

The jargon

The term *angular velocity* is sometimes dropped in favour of *omega* when used to describe SHM. (Some books use the term *angular frequency*, which makes sense given that ω is measured in radians per second.)

Checkpoint 1

What are the angular velocities of:

(a) a spoke on a bicycle wheel of radius 20 cm if the bike is travelling at 12 m s^{-1}

(b) a loudspeaker cone vibrating at 300 Hz?

Don't forget

Time period, $\boldsymbol{T}$, is the reciprocal of frequency, $\boldsymbol{f}$.

Grade booster

Many students try to use the wave equation ($\boldsymbol{v} = \boldsymbol{f}\lambda$) in questions about oscillations because there is a link with wave motion. Don't even go there!

Graphs and equations for SHM

The equations and graphs below work for continuous, undamped SHM.

- Damping complicates things by reducing the amplitude A over time.
- The oscillator must already be oscillating before time $t = 0$.

Displacement against time

The displacement $\boldsymbol{x}$ of a particle vibrating with SHM is given by:

$$\boldsymbol{x} = A \sin \omega t$$

Where A = amplitude (i.e. maximum displacement), ω is angular velocity and t is time. ωt is measured in radians.

Velocity against time

Velocity is rate of increase in displacement. At any instant it is the gradient of a displacement–time graph:

$$\boldsymbol{v} = A\omega \cos \omega t$$

Maximum velocity depends on both amplitude and angular velocity and since the maximum value a cosine can have is 1:

$$\boldsymbol{v}_{max} = A\omega$$

Maximum velocity is achieved when displacement $\boldsymbol{x} = 0$, that is when the oscillator is at its mean position.

Acceleration against time

Acceleration is rate of increase in velocity. At any instant it is the gradient of a velocity–time graph:

$$\boldsymbol{a} = -A\omega^2 \sin \omega t, \text{ but } \boldsymbol{x} = A \sin \omega t, \text{ so } \boldsymbol{a} = -\omega^2 \boldsymbol{x}$$

Maximum acceleration is achieved at maximum displacement; i.e. the displacement is equal to the amplitude and the velocity is zero.

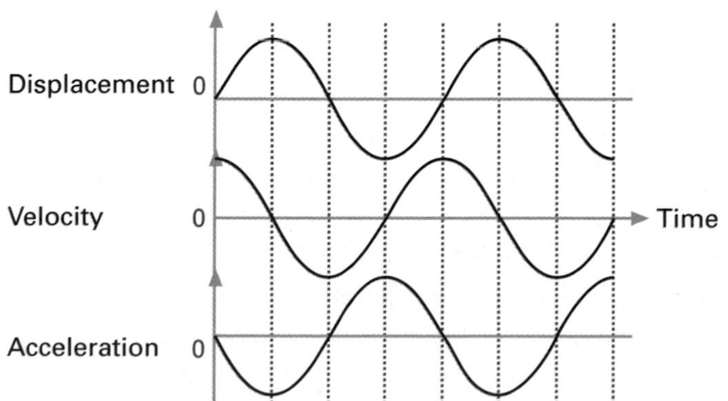

Velocity is the gradient of the displacement–time graph.

Acceleration is the gradient of the velocity–time graph

Watch out!

The displacement equation assumes that when $t = 0$, the oscillator's displacement is zero. This is a bit dodgy, because zero initial displacement implies zero initial restoring force and therefore no oscillation! Some exam boards prefer to use the more proper, but mathematically tricky: $\boldsymbol{x} = A \cos \omega t = A \cos 2\pi f t$ (same shape, but the oscillation starts at maximum displacement), in which case: $\boldsymbol{v} = -A\omega \sin \omega t = \pm 2\pi f \sqrt{(A^2 - \boldsymbol{x}^2)}$ and $\boldsymbol{a} = -A\omega^2 \cos \omega t = -(2\pi f)^2 \boldsymbol{x}$.
Check your syllabus!

Watch out!

Take care with trigonometric functions. Make sure your calculator knows that the numbers you are typing in are radians, not degrees (a common mistake)!

Action point

SHM involves the continuous transfer between kinetic energy and potential energy. The *total* amount of energy is constant. On the same axes, show how the KE and PE changes with displacement of a pendulum.

Examiner's secrets

SHM is a good topic to use to check your understanding of the links between displacement, velocity and acceleration. Given a graph of displacement against time, you must be able to say what is happening to velocity and acceleration.

Exam practice answers: page 42

1 Define simple harmonic motion. (5 min)

2 A fairground attraction has a vibrating floor. It oscillates with simple harmonic motion, with an amplitude of 1 m. (a) Find the minimum frequency for customers to just lift off the ground. (b) What would happen to the minimum frequency if the amplitude were reduced? (15 min)

Answers
Mechanics

Scalars and vectors

Checkpoints

1 *Scalar* mass, temperature, energy;
vector weight, acceleration.

2 If your ruler is accurate to the nearest mm, a 1 cm measurement has a 10% error whilst a 10 cm measurement has a 1% error. (If absolute errors are the same, increasing the scale reduces relative errors.)

3 $\sin\theta$ = opp/hyp = $\mathbf{v}_y/\mathbf{v}$, so $\mathbf{v}_y = \mathbf{v}\sin\theta$
$\cos\theta$ = adj/hyp = $\mathbf{v}_x/\mathbf{v}$, so $\mathbf{v}_x = \mathbf{v}\cos\theta$

Action point

7.55 kg or 7 550 g.

Exam practice

1 (a) Magnitude: $R^2 = (120)^2 + (50)^2 = 14\,400 + 2\,500 = 16\,900$
$R = 130$ N

(b) Direction: $\tan\theta = 50 \div 120 = 0.4167$
$\theta = \tan^{-1}(0.4167) = 22.6°$

2

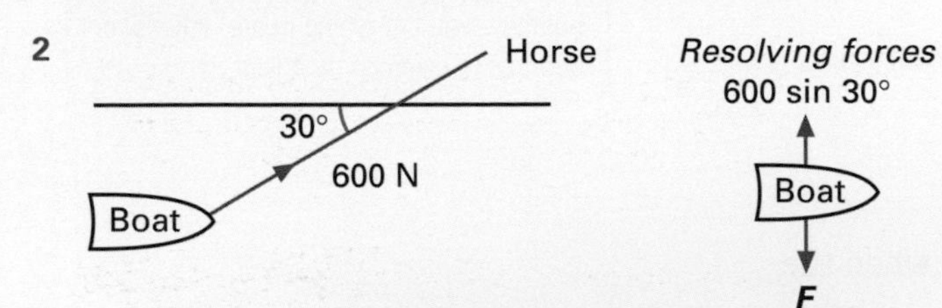

Resolving the tension in the rope into its components, the component towards the bank must be balanced by an equal and opposite force (**F** say) from the rudder and keel.
$\mathbf{F} = 600 \sin 30° = 300$ N away from the bank.

Forces and moments in equilibrium

Checkpoints

1

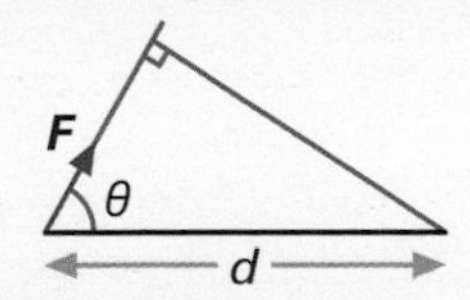

Using moment = perpendicular force × distance;
perpendicular force = $\mathbf{F}\sin\theta$, so moment = $\mathbf{F}d\sin\theta$
Using moment = force × perpendicular distance;
perpendicular distance = $d\sin\theta$, so moment = $\mathbf{F}d\sin\theta$

Action point

Wide bases and low centres of gravity both mean that the object has to be tilted through a larger angle before its centre of gravity acts to the side of its base.

Exam practice

1 $\tan\theta = 7.5/0.6 = 12.5$ $\theta = 85°$
Total upward force on walker = $2T\cos\theta$
Equilibrium ⇒ $2T\cos\theta = 500$ N
$T = 500/2\cos\theta = 2.9 \times 10^3$ N (to 2 sig. fig.)

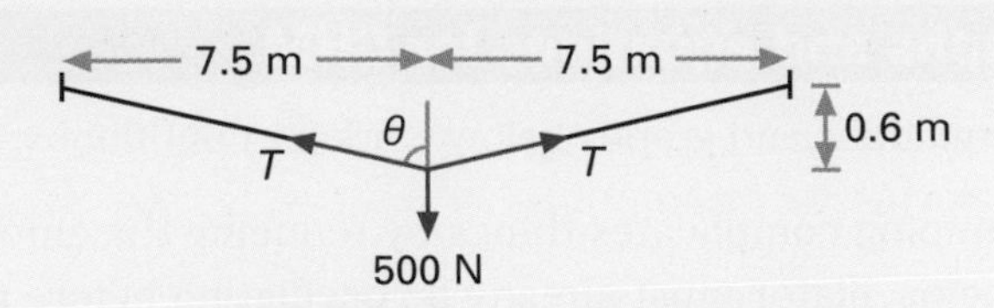

2 Vertical equilibrium ⇒ $\mathbf{F}_A + \mathbf{F}_B = 2.0 \times 10^5$ N
Rotational equilibrium ⇒ moments about any point are balanced.
Moments about point of support on pillar A
$2.0 \times 10^5 \times 100 = \mathbf{F}_B \times 140$
$\therefore \mathbf{F}_B = 1.43 \times 10^5$ N
Moments about point of support on pillar B
$\mathbf{F}_A \times 140 = 200\,000 \times 40$
$\therefore \mathbf{F}_A = 0.57 \times 10^5$ N
Check $\mathbf{F}_A + \mathbf{F}_B = 2.0 \times 10^5$ N

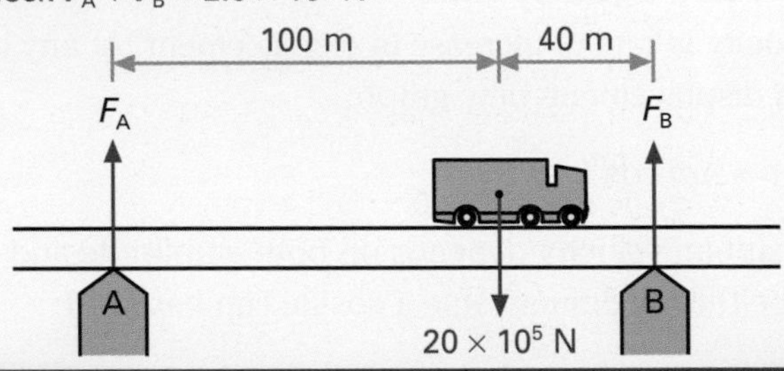

Ways of describing motion

Checkpoints

1 Average velocity = displacement/time taken. The start and finish of a lap are in the same place – zero displacement.

2

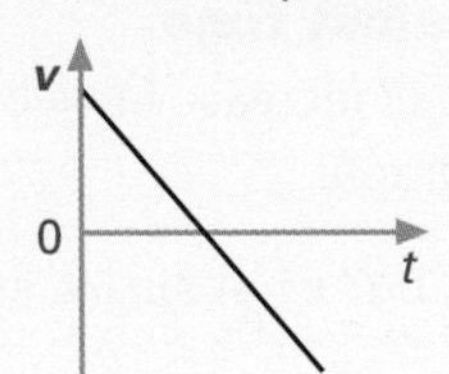

Displacement is the area under the graph.

Exam practice

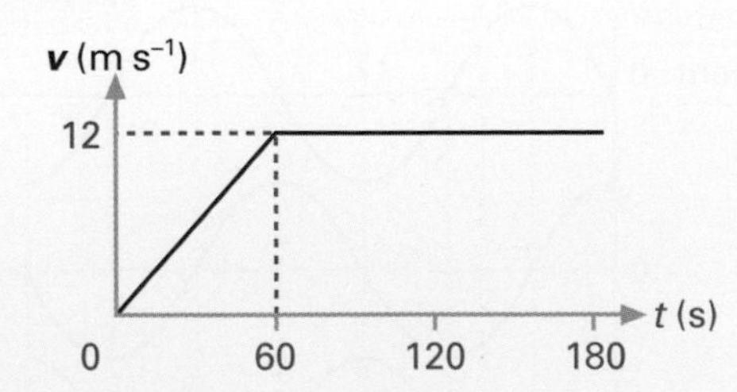

Top speed = $0.2 \times 60 = 12$ m s^{-1}
Distance travelled = area under graph
$= (\frac{1}{2} \times 60 \times 12) + (120 \times 12) = 1\,800$ m

Equations of motion

Checkpoints

1 *Equation 4* $\mathbf{v} = \mathbf{u} + \mathbf{a}t \Rightarrow \mathbf{u} = \mathbf{v} - \mathbf{a}t$
Substituting into $\mathbf{s} = \mathbf{u}t + \frac{1}{2}\mathbf{a}t^2$ we get $\mathbf{s} = (\mathbf{v} - \mathbf{a}t)t + \frac{1}{2}\mathbf{a}t^2$
which simplifies to $\mathbf{s} = \mathbf{v}t - \frac{1}{2}\mathbf{a}t^2$
Equation 5 Average velocity = $(\mathbf{v} + \mathbf{u})/2$;
Displacement $\mathbf{s}$ = average velocity × time taken = $(\mathbf{v} + \mathbf{u})t/2$

2 $a = (v - u)/t = (18 - 10)/4 = 2\ m\,s^{-2}$; distance travelled = 56 m

3 $u = 10\ m\,s^{-1}$, $v = 0\ m\,s^{-1}$, $a = -10\ m\,s^{-2}$

Using $v^2 = u^2 + 2as$, $0 = 100 - 20\,s \Rightarrow s = 5$ m

4 Rearrange the second equation to get an expression for v and substitute into the first.

Exam practice

1 Use $v^2 = u^2 + 2as$

$100 = 0 + 2 \times a \times 20$

$\therefore a = 2.5\ m\,s^{-2}$

2 (a) Use $v = u + at$

$v = 0 + 2.5 \times 4$

$\therefore v = 10\ m\,s^{-1}$

(b) Average speed = distance/time

Time taken for accelerating phase = 4 s

Distance travelled in accelerating phase:

$s = ut + \frac{1}{2}at^2 = \frac{1}{2} \times 2.5 \times 16 = 20$ m

$\therefore$ Remaining 80 m were covered at $10\ m\,s^{-1}$

Time for constant speed phase = 8 s

Total time = 12 s

Average speed = $100/12 = 8.3\ m\,s^{-1}$

3 Use $v^2 = u^2 + 2as$

Top of flight, $v = 0$, so

$0 = u^2 + 2 \times (-10) \times 8$

$u = \sqrt{160} = 13\ m\,s^{-1}$

Time taken to reach peak:

$v = u + at$

$t = (v - u)/a = (0 - 13)/(-10) = 1.3$ s

Total time taken = 2.6 s

4 Thinking distance = $20 \times 0.5 = 10$ m

Braking distance: using $v^2 = u^2 + 2as$

$0 = 400 - 20 \times s$

so $s = 20$ m

Total distance travelled by car = 30 m

It stops 15 m past the traffic lights!

Projectiles

Exam practice

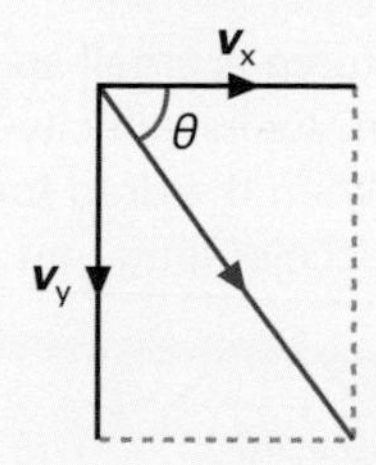

1

(a) Horizontally, distance = $8 \times 3 = 24$ m

(b) Vertically, $s = ut + \frac{1}{2}at^2$

$a = 9.81\ m\,s^{-2}$

$s = 0 + \frac{1}{2} \times 9.81 \times 3^2 = 44.1$ m

Height of cliff = 44.1 m

(c)

$\tan\theta = v_y/v_x$

Vertical velocity on impact, $v_y = u + at = 29.4\ m\,s^{-1}$

$\tan\theta = 29.4/8 = 3.675$ so $\theta = 74.8°$

2 Resolve initial velocity v into horizontal and vertical components v_x and v_y. $v_x = v\cos\theta$; $v_y = v\sin\theta$.

Range = $v_x \times$ time of flight.

Vertically, time taken to reach maximum height is given by $v = u + at$, which gives $0 = 20 \times \sin 45° - 9.81t$.

So $t = 1.44$ s.

Range = $20 \times \cos 45° \times 2.88 = 40.7$ m

3 (a) Vertically, $s = ut + \frac{1}{2}at^2$

$500 = 0 + \frac{1}{2} \times 9.81 \times t^2$

$t = \sqrt{(1\,000/9.81)} = 10.1$ s

(b) Horizontal distance travelled = $50 \times 10.1 = 505$ m

(c) Speed is the resultant of horizontal and vertical velocities. Vertical velocity as the package hits the ground = $9.81 \times 10.1 = 99.1\ m\,s^{-1}$. Using Pythagoras' theorem, speed $v = \sqrt{(50^2 + 99.1^2)} = 111\ m\,s^{-1}$

Newton's laws of motion

Checkpoints

1 (a) The pull force of the floor on the person.

(b) The pull of the person's gravity on the Earth!

(c) The force of the ball on the bat.

2 (a)

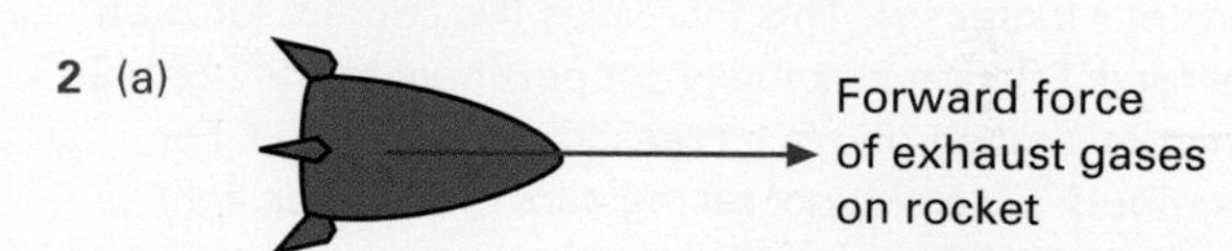

(b)

Backward force of rocket on exhaust gases

Exam practice

1 Use $v^2 = u^2 + 2as$ to find acceleration.

$50^2 = 0 + 2 \times a \times 50$ so $a = 25\ m\,s^{-2}$

$F = 12\,000 \times 25 = 300\,000$ N

2 This question is designed to make you think. It is not explained in enough detail for an answer that will satisfy everyone. You have to make some dodgy assumptions!

Doubling the average force doubles the average rate of acceleration. But it also increases the distance. Doubling the force on a spring usually doubles its extension. The arrow's top speed v is given by: $v^2 = 2as$ ($u = 0$). If both a and s are doubled, v^2 will be four times as big, and v will be twice as big.

Top speed is doubled by doubling the force. Try tackling the problem from an energy conservation viewpoint – see *work, energy and power*, pages 26–7, and *stress, strain and Hooke's law*, pages 30–1. (Real bows are not usually quite so simple in their behaviour!)

3 Treating truck and trailer as one unit:

Trailer and truck together

net force = $5\,000 \times 1.2 = 6\,000$ N

drive – drag = 6 000 N

drive force = 9 000 N

Trailer and truck separately

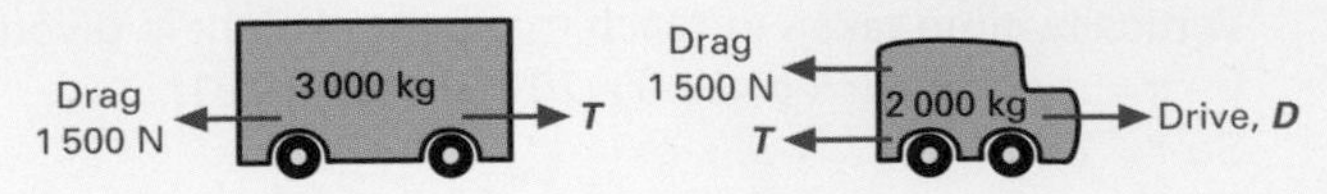

For the trailer:

T – drag = 3 000 × 1.2

T = 3 600 + 1 500 = 5 100 N

For the truck:

D – 1 500 – T = 2 000 × 1.2

T = 9 000 – 1 500 – 2 400 = 5 100 N

Note It is worth doing calculations 'from both sides' to check for mistakes. If you got a different answer for T from the truck calculation, you would know something was wrong.

Some important forces

Checkpoints

1 As a racing car speeds up, the downward force on the aerofoils increases. This increases the contact force on the road. Friction is actually proportional to the normal reaction, and so grip is increased at high speed. For maximum acceleration, racing cars must be as light as possible. Aerodynamic lift (upside down!) compensates for the lack of weight and keeps the contact force high.

2

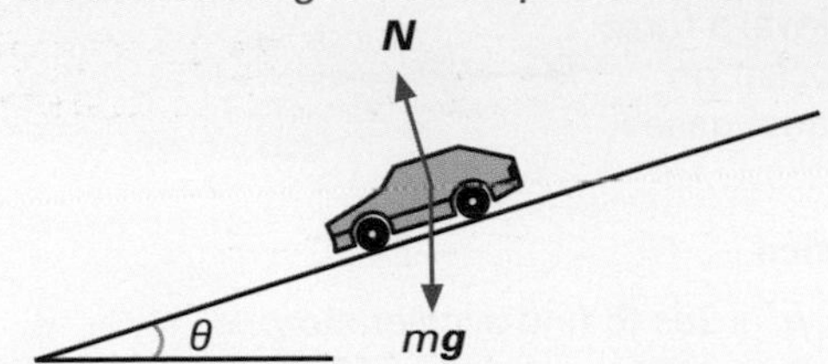

On a hill, the normal reaction is usually given by $mg\cos\theta$, where θ is the slope of the hill; this has a maximum value (of mg) when $\theta = 0$ (horizontal road).

3 Remember that friction opposes motion!
(a) up the slope (b) down the slope (c) up the slope.

Exam practice

1

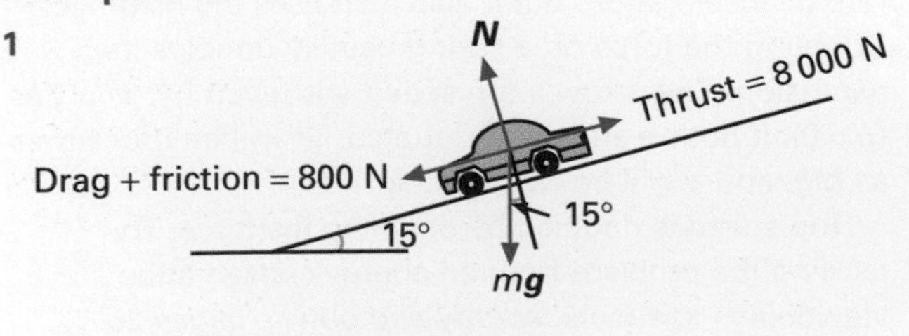

Resolve weight into components normal to the surface and along the surface.

Normal forces balance ($mg\cos\theta = N$).

Along the plane, forward force = 8 000 N;

Backward force = 800 + $mg\sin\theta$

= 800 + 900 × 9.81 × sin 15° = 3 085 N

Resultant = 8 000 – 3 085 = 4 915 N

a = 4 915/900 = 5.46 m s^{-2}

2 Resolve weight into components along and normal to the slope (normal forces balance). Forces along slope balance ∴ $T = mg\sin\theta = 50 \times 9.81 \times \sin 25°$ so T = 207 N

3 (a) The scales register the man's apparent weight. Taking down as positive, and g = 10 ms^{-1}, R = –700 N: the scales register 700 N
(b) 70 × 10 = R + 70 × 10, so R = 0: the man is in free fall so will be weightless
(c) 70 × –10 = R + 70 × 10, so R = –1 400 N
(Note These kind of problems helped Einstein to develop his theory of general relativity.)

Moving through fluids

Checkpoints

1 Sea water is more dense than pure water. Warm water is *generally* less dense than cold water, but strange things happen close to freezing point. Ice is less dense than water. Water's density hits a maximum at 4 °C. As a pond cools, the cold water sinks. This continues until 4 °C is reached, then the pond freezes from the top down. Fishes etc. are usually safe in the dense 4 °C water below!

2 Viscostatic means the viscosity does not change (very much) with temperature: required so that the viscosity of the oil does not change too much when the engine is run and heats up.

3

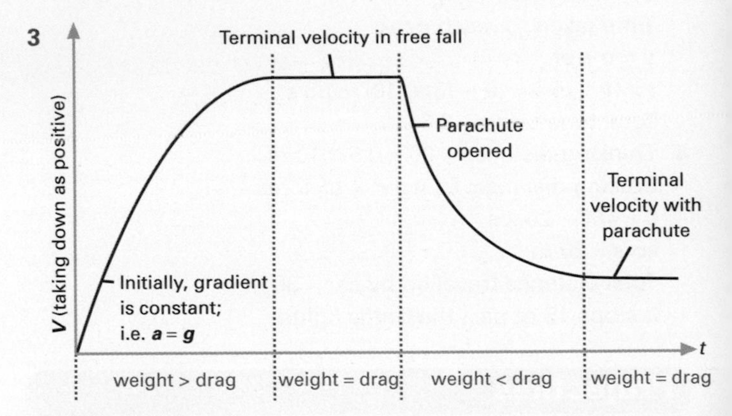

Exam practice

1 (a) At first the drag force is small so the object accelerates down. As its velocity increases so does the drag force. When the drag force equals its weight, it stops accelerating and reaches its terminal velocity.

(b) weight = $F_{drag} = kv^2$

$2.0 \times 9.81 = k \times 40 \times 40$

k = 0.0122 N s^{2}m^{-2}

(c) $ma = mg - kv^2$

$2.0 \times a = 2.0 \times 9.81 - 0.0122 \times 30 \times 30$

a = 4.32 m s^{-2}

Work, energy and power

Checkpoint

1 A comet has its greatest gravitational potential energy when it is furthest from the Sun.

Exam practice

If the engine supplies the only significant force slowing the boat down, the work it must do is equal to the kinetic energy it must lose.

$E_k = \frac{1}{2}mv^2 = \frac{1}{2} \times 8\,000 \times 1.2^2 = 5\,760$ J

work done $= F \times 6.0 \quad \therefore F = 960$ N

power $= F \times v$, where v = average speed

$\therefore$ power $= 960 \times 0.6 = 580$ W (to 2 sig. fig.)

Note You could also tackle this using equations of motion which gives exactly the same answer.

Momentum and impulse

Checkpoint

1 Momentum is always conserved, but to demonstrate it, you need to be able to measure the masses and velocities of all bodies involved. Friction links trolleys to the Earth. We can't detect any changes in the Earth's velocity, and so it is hard to prove momentum conservation. Better bearings, use of an air-track, etc., improve the situation by reducing the transfer of momentum to the Earth. A friction-compensated track could also be used.

Exam practice

1 (a) (i) $\boldsymbol{p} = m\boldsymbol{v} = 0.04 \times 500 = 20$ kg m s^{-1}.
(ii) $E_k = \frac{1}{2}m\boldsymbol{v}^2 = 5\,000$ J. (iii) recoil speed = 20/2.5 = 8.0 m s^{-1}. (iv) rifle's $E_k = 80$ J.

(b) No. *Collisions* can sometimes be perfectly elastic (same total amount of kinetic energy before and after), but in an *explosion*, energy is converted from some other form into kinetic energy. If you don't have any kinetic energy before the explosion, and you do after, there has to have been an increase in the system's kinetic energy!

2 $\boldsymbol{a} = 35/20 = 1.75$ m s^{-2}
$\boldsymbol{F} = 1\,400 \times 1.75 = 2\,450$ N

3 Initial momentum of the two skaters is 130×8 = 1040 kg m s^{-1}. Final momentum of 50 kg skater is 500 kg m s^{-1}, so final momentum of 80 kg skater = 540 kg m s^{-1}. Final velocity = 540/80 = 6.75 m s^{-1}.

4 The hard snowball has a greater change in momentum than the soft snowball. The impulse on it is greater and so the impulse on the person is greater (Newton's third law). The hard snowball could impart an impulse of up to twice that imparted by the soft snowball. If the acceleration period is the same for both cases, the force exerted by the hard snowball will be correspondingly greater. In fact, the contact time is generally shorter when hard objects collide than when soft objects collide. This increases the impact.

Stress, strain and Hooke's law

Checkpoints

1 Strain = $\Delta x/l$, here $\Delta x = l$ so strain = 1

2 Stress $\propto$ strain ==> $F/A \propto \Delta x/l$ ==> $F \propto \Delta x$ if A and l are constants.

Exam practice

1 (a) $d = 2.1 \times 10^{-2}$ m, so cross-sectional area
$A = 3.46 \times 10^{-4}$ m^2
weight of load = 14 700 N
$\therefore$stress = 4.25×10^7 Pa
E = stress/strain
$\therefore$ strain = $4.25 \times 10^7/2.0 \times 10^{11} = 2.13 \times 10^{-4}$
$\therefore \Delta x = 12 \times 2.13 \times 10^{-4} = 2.6 \times 10^{-3}$ m (2.6 mm)
(to 2 sig. fig.)

(b) $\Delta x = 4.6$ mm

2 (a) 25 springs (500/20)
(b) In parallel
(c) For each spring, F is exactly 20 N
so $\Delta x = 20/15 = 1.33$ cm.
(d) The total load = 500 N, $\Delta x = 1.33$ cm,
so $k = 376$ N cm^{-1}.

Vibrations and resonance

Checkpoints

1 $T = 2\pi\sqrt{(l/g)} \Rightarrow T^2 = 4\pi^2 l/g \Rightarrow l = gT^2/4\pi^2$
For $T = 1$ s, $l = 0.248$ m

2 Damping reduces the net restoring force acting on an oscillator, increasing its oscillation period and reducing its resonant frequency.

3 (a) *Beneficial* Grills and radiant heaters cause surface molecules to resonate and absorb heat rays. Lasers at certain frequencies can destroy the colour in certain tattoo dyes by resonance. Radio receivers are tuned to resonate at certain frequencies. Musical instruments make use of resonance.

(b) *Harmful* Too much energy can always be harmful. Regular gusts of wind can cause havoc at sea – to sailing boats with resonant frequencies equal to the gust frequency. Car wheels can develop a dangerous wobble at certain speeds due to resonance (generally cured by balancing the wheel).

Exam practice

(a) Resonance is when an oscillator is forced to vibrate at its natural frequency which causes the amplitude of the oscillation to build up to a maximum.

(b) If the springs have a different natural frequency to that of the driver frequency then the building will not resonate

Circular motion

Checkpoints

1 $5° = 0.087\,3$ rad; $\sin 5° = 0.087\,2$; $\tan 5° = 0.087\,5$ (very little difference between θ in radians, $\sin\theta$ and $\tan\theta$).
$10° = 0.174\,5$ rad; $\sin 10° = 0.173\,6$; $\tan 10° = 0.176\,3$ (some divergence, but still *approximately* equal – OK for rough calculations).
$20° = 0.349\,1$ rad; $\sin 20° = 0.342\,0$; $\tan 20° = 0.364\,0$ (significant errors would arise if you tried assuming $\theta = \sin\theta = \tan\theta$).

2 (a) $v = r\omega$; $\omega = 2\pi f$ $\therefore$ $v = 2\pi rf$
(b) $\omega = 2\pi f$; $f = 1/T$ $\therefore$ $\omega = 2\pi/T$

3 (a) The Earth's period of orbit is 1 year. The easy answer is $\omega = 2\pi$ radians per year. To convert to radians per second: 1 year = $365 \times 24 \times 60 \times 60$ = 31 536 000 s
$\therefore$ $\omega = 2\pi/31\,536\,000 = 1.992 \times 10^{-7}$ rad s^{-1}
(b) Earth's speed in orbit around the Sun, $v = r\omega$
$= 1.50 \times 10^8 \times 1.992 \times 10^{-7}$. $v = 29.9$ km s^{-1}

4 (a) Centripetal acceleration of Earth around the Sun; use either $\mathbf{a} = \mathbf{v}^2/r$ or $\mathbf{a} = r\omega^2$.
$\mathbf{a} = 5.95 \times 10^{-3}$ m s^2. (*Note* convert $\mathbf{v}$ to m s^{-1} and r to m before calculating.)
(b) Centripetal acceleration of Moon around Earth = $r\omega^2$.
$T = 27.3$ days $= 27.3 \times 24 \times 3\,600$ s. $T = 2.3 \times 10^6$ s
$\omega = 2\pi/T = 2.664 \times 10^{-6}$ rad s^{-1}; $r = 3.8 \times 10^8$
$\therefore$ $\mathbf{a} = 2.69 \times 10^{-3}$ m s^{-2} (to 3 sig. fig.)

5 $F = mr\omega^2$. $m = 80$ kg, $T = 24$ hours $= 8.64 \times 10^4$ s
$\omega = 2\pi/T = 7.27 \times 10^{-5}$ rad s^{-1}
At the equator, $r = 6.4 \times 10^6$ m, so $F = 2.71$ N (not much danger of being thrown off!). At the poles, $r = 0$ and the force needed to keep you from flying off = 0.

Exam practice

1 (a)

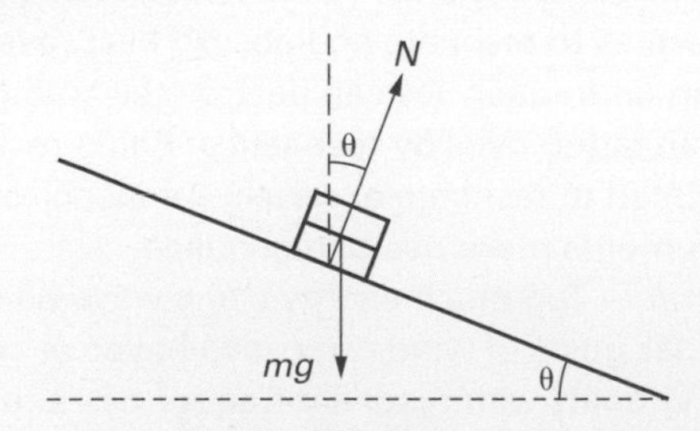

(b) θ is the angle of the bobsleigh relative to the normal. N is the normal reaction of the surface on the bobsleigh. Resolving forces:

horizontally $N\sin\theta = \dfrac{mv^2}{r}$ (the centripetal force)

vertically $N\cos\theta = mg$

Dividing the two equations gives

$$\frac{\sin\theta}{\cos\theta} = \frac{v^2}{gr} = \tan\theta$$

2 $\omega = 2\pi \times 3.5 = 22$ rad s^{-1}
$v = r\omega = 0.06 \times 22 = 1.32$ m s^{-1}

Simple harmonic motion

Checkpoints

1 (a) $\boldsymbol{\omega} = v/r = 12/0.2 = 60$ rad s^{-1}
(b) $\boldsymbol{\omega} = 2\pi f = 1\,880$ rad s^{-1}

Exam practice

1 SHM is oscillatory motion where acceleration is proportional in size, but opposite in direction, to displacement; the simplest definition is $\mathbf{a} \propto -\mathbf{x}$.
$\mathbf{a} = -A\omega^2 \sin \omega t$ would do!

2 (a) The problem is this: What is the frequency when the floor's maximum downward acceleration just equals $\mathbf{g}$?
$\mathbf{a} = -A\omega^2 \sin \omega t$
When the floor is at its greatest positive displacement, $\sin\omega t = +1$ and $\mathbf{a} = -A\omega^2$ (i.e. maximum downward acceleration = $A\omega^2$)
($\mathbf{a} \propto -\mathbf{x}$. When $\mathbf{x} = +A$, $\omega t = \pi/2$)
$\omega = 2\pi f$, so when the maximum downward acceleration is $\mathbf{g}$, $A(2\pi f)^2 = \mathbf{g}$ $f = \sqrt{(9.81/4\pi^2)}$ $f = 0.498$ Hz
At any greater frequency, the customers' feet will leave the ground.
(b) If you reduce A, the necessary frequency increases ($f \propto 1/\sqrt{A}$).

Radioactivity, Nuclear and Particle Physics

The discoveries of radioactivity and atomic structure just over 100 years ago heralded a new era for physics – an era dominated by the newly discovered laws of quantum physics and relativity. You will get a flavour of both of these pillars of modern physics in this chapter. About 50 years later, dramatic new developments led to an appreciation of the fundamental building blocks of matter, and to theories that are still being tested at facilities such as CERN.

Exam themes

- *Scientific evidence* Your ability to remember, analyse and interpret the results from key experiments can (and will) be tested in this section of the course. What is the evidence for:
  - the nuclear structure of the atom
  - the nature of α-, β- and γ-emissions
  - the size and spacing of atoms
  - the size of nuclei
  - the existence of quarks?
- *Applications* How can radioactivity be put to use? Examiners like to see if you can see the benefits and limitations of new techniques, especially in the How Science Works context. Nuclear power (fission and fusion), medicine, measurement techniques and radioactive dating such important applications that examiners find difficult to ignore!
- *Knowledge* of nuclear structure, including quarks and leptons, the notation and the rules governing nuclear reactions.
- *Radioactive decay* Awareness of randomness and the need for a statistical approach.
- *Ability to interpret graphical data* N v Z for stable nuclides; mass defect per nucleon.
- Particle accelerator design principles

Topic checklist

	Edexcel		AQA/A		AQA/B		OCR/A		OCR/B		WJEC		CCEA	
	AS	A2	AS	A2	AS	A2	AS	A2	AS	A2	AS	A2	AS	A2
The atom and its nucleus		●	○		○			●		●	○			●
Elements and isotopes		●	○		○			●		●	○	●		●
Nuclear instability		●	○		○			●	○	●				●
Properties of ionizing radiation		●		●	○			●		●	○	●		●
Radioactive decay		●		●	○			●		●	○	●		●
Binding energy and mass defect		●		●		●		●		●		●		●
Nuclear fission and fusion		●		●		●		●		●				●
Other applications of radioactivity	○	●		●		●		●		●		●		●
Probing matter	○	●	○		○			●		●		●		●
Particles – production and patterns		●	○		○			●		●	○			●
More about leptons and quarks		●	○		○			●		●	○			●
Forces/interactions and conservation laws		●	○		○			●		●	○			●

The atom and its nucleus

"When it comes to atoms, language can be used only as in poetry."

Niels Bohr

The idea that everything is made of tiny particles is deep seated in physics. Until the start of the 20th century, atoms were believed to be the tiny indivisible ("fundamental") particles everything else was made from, but then the electron was discovered and it became clear that atoms were themselves made of still smaller parts.

Thomson's 'plum-pudding' model of the atom

Atoms are not charged. Since electrons are negatively charged constituents of an atom, the rest of the atom must be positively charged. Thomson's model of the atom shown below consists of a positively charged 'pudding', with negatively charged 'plums' or 'raisins' (electrons) embedded in it.

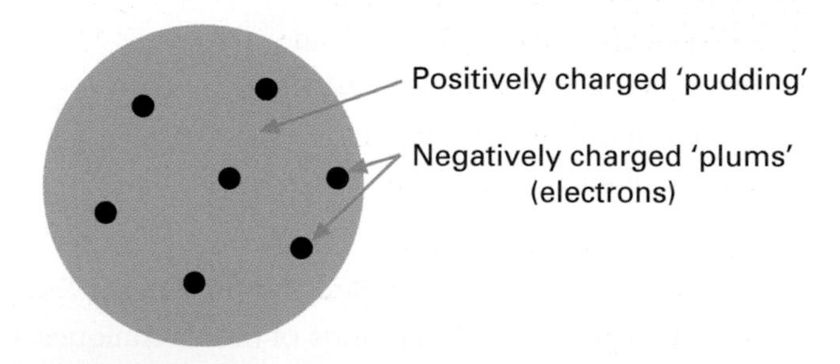

Checkpoint 1

Why did Thomson think the positive and negative charges should be evenly distributed within the atom?

Discovery of the nucleus

In 1899, Ernest *Rutherford* discriminated between three types of ionizing radiation emitted by unstable nuclides: alpha, beta and gamma radiation. In 1909, researchers in his group were using *alpha particles* to probe gold atoms. *Geiger* and *Marsden* fired alpha particles at a thin gold foil. They used a zinc sulphide screen (a simple *scintillation detector*) to see where the alpha particles ended up.

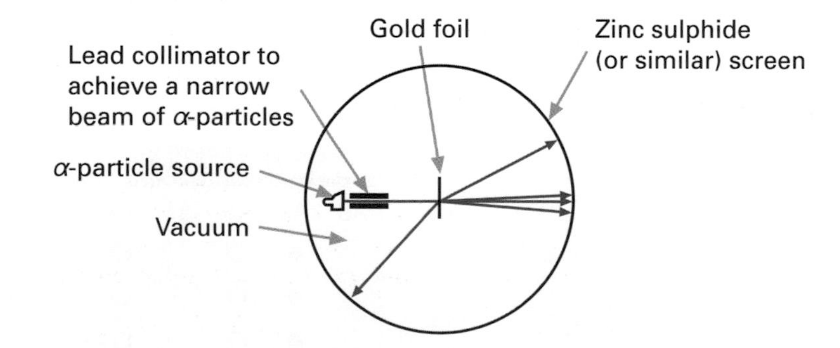

Geiger and Marsden discovered that:

- Most of the alpha particles went straight through the gold foil without any change in direction and without any loss of energy.
- Some alpha particles were deflected and some even rebounded!

Checkpoint 2

Alpha particles have about 1/50th the mass of a gold atom and they are positively charged. If a gold atom was a pudding of evenly distributed mass and charge, what do you think would happen to the alpha particles?

- Would they be slowed down?
- Would they be deflected?

Links

A modern equivalent of Rutherford scattering uses fast moving electrons to probe deep into protons. So-called *deep inelastic scattering* resulted in the discovery of quarks and other exotic particles. See *Particle physics* section on pages 62–7.

Grade booster

You should be able to draw the paths of alpha particles scattered by a nucleus. The closer the path to the nucleus, the greater the deflection.

Grade booster

This is a classic experiment. Learn the set-up and the reasons for:

- vacuum
- collimated source and detector
- thin foil of gold
- why gold, in a thin foil?

Rutherford set about interpreting the data. The most obvious conclusions to be drawn from these experiments were:

- ➜ most of an atom is empty space
- ➜ the positive charge is concentrated in the nucleus
- ➜ the mass is concentrated in the nucleus

Rutherford was able to determine the size of the nucleus by working out the force needed to give the necessary alpha deflections, using Coulomb's law (an inverse-square law for the force between two charges). He found that while an *atom's diameter* might be around 10^{-10} m, the *diameter of the nucleus* is typically around 10^{-15} m – 100 000 times smaller! If you imagine that the atom is the size of a huge sports stadium, the nucleus is the size of a 10 pence piece.

Problems for 'classical physics'

Rutherford's new model posed serious problems that classical physics could not answer. The electrons had to be in orbit, since static electrons would be captured by the nucleus, but

- ➜ orbiting electrons are being constantly accelerated
- ➜ according to classical physics, accelerated charges should emit radiation (they don't!)

The evidence for the nuclear structure of the atom was so strong that it was classical physics that had to give way. Niels Bohr used atomic line spectra to show that electron orbits had specific energy levels and Louis de Broglie provided a theoretical mechanism for these states by linking particle and wave properties!

The nuclear atom

Moseley's suggestion that positive charge should be quantized and Chadwick's discovery of the neutron completed the basic picture. The approximate radius of an atom = 10^{-10} m, and the approximate radius of the nucleus = 10^{-15} m.

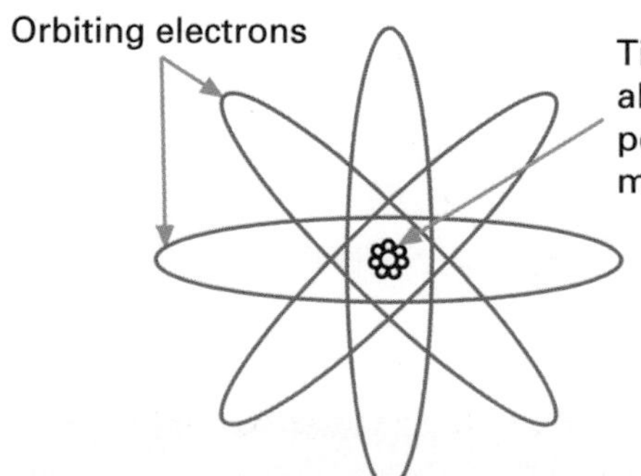

"It was almost as incredible as if you had fired a fifteen inch shell at a piece of tissue and it came back and hit you."

Ernest Rutherford

Links

See *Coulomb's law*, page 144.
A version is $F = k\,q_1q_2/r^2$
Where k is a constant, q_1 and q_2 are charges and r represents the separation of the charges. Check for yourself that halving r quadruples the force between the charges.

The jargon

Classical physics is the old stuff. Galileo, Newton and even Maxwell were classical physicists. Classical physicists believed in determinism: perfect laws of nature from which everything could ultimately be predicted. Modern physicists accept uncertainty!

Checkpoint 3

What would happen if electrons constantly lost energy by electromagnetic radiation?

Links

See *probing matter*, page 60 and *standing waves*, pages 126–7. According to quantum theory, particles moving at speed behave like waves. Standing waves don't transfer energy. In certain orbits, electrons behave as standing waves!

Exam practice answers: page 68

1 What were the key results of Geiger and Marsden's α-particle scattering experiments and what conclusions were drawn about the structure of the atom? (15 min)

Elements and isotopes

"'Excellent!' Dr Watson cried. 'Elementary', said Holmes."

Sir Arthur Conan Doyle

Watch out!

Isotopes of an element have the same number of electrons, so they have the same *chemical* properties. They have different *physical* properties due to their different number of neutrons.

Don't forget

Nucleon number (A) is sometimes called *atomic mass number* and the proton number (Z) is also known as the *atomic number*.

Checkpoint 1

Notice the distinction between *atomic mass* and A. The mass number A is always a whole number (an integer). Atomic mass is not. Explain.

Grade booster

Generally, physics questions are about physical processes. Try not to wander off into chemistry – easily done when thinking about electrons.

Elements are the substances that contain the same type of atom. They were first identified by their chemical behaviour. Measurements of atomic mass were then used to put the elements in order and some underlying chemical patterns began to emerge, but it was hard to make much sense of the periodic table until the discovery of the basic nuclear structure of the atom.

Proton number or atomic number Z

→ Z completely defines the *element*.

All carbon atoms have six protons, all nitrogen atoms have seven, etc. The number of electrons in an atom equals the number of protons (protons and electrons have equal and opposite charge), which is why Z determines an atom's chemical behaviour.

Nucleon number, or mass number, A

→ A defines the particular *isotope*.

A is the total number of *nucleons* (protons and neutrons) in the atom. Different **isotopes** of an element have different numbers of neutrons, but the same number of protons. All carbon atoms have six protons; most also have six neutrons (so for "normal" carbon, $A = 12$), but there is an isotope of carbon which has eight neutrons ($A = 14$). C-14 atoms are heavier than C-12 atoms, but they react chemically in exactly the same way as "normal" carbon. (This isotope also happens to be radioactive.) Hydrogen has three isotopes.

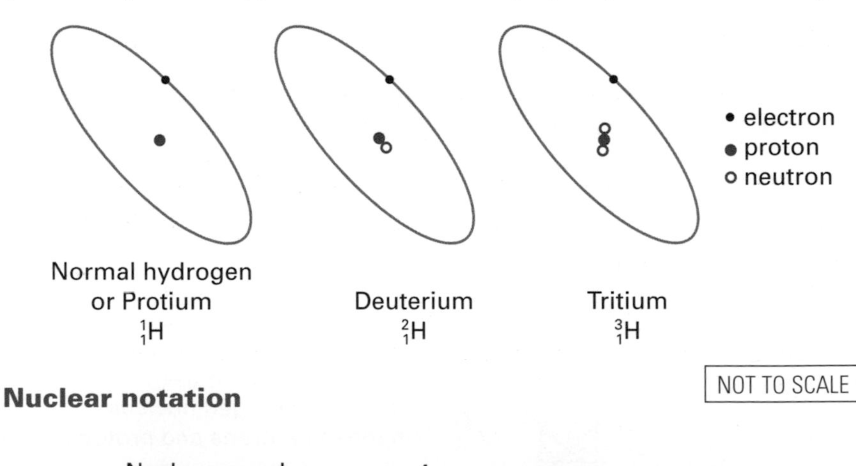

Nuclear notation

Nucleon number → A

Proton number → Z

X ← Chemical symbol

Variation in nuclear size with nucleon number

Variation in nuclear volumes and radii with nucleon number (data from high-energy electron diffraction experiments) show that volume is roughly proportional to nucleon number A. This tells us that nucleons are incompressible; nuclear density is constant. The radius, r, of a nucleus is given by:

$$r = r_0 A^{1/3}$$

Where r_0, the constant of proportionality, equals 1.2×10^{-15} m. r_0 is the radius of a single nucleon.

Atomic mass units u

Measuring the mass of atoms in kilograms gives tiny numbers. A suitably small alternative unit is needed. The mass of a proton or neutron would be a sensible choice, and this is very close to what an **atomic mass unit** (symbol u) is, but u is actually defined as one-twelfth of the mass of a normal carbon atom, i.e. carbon-12. The conversion factor is:

$$1 \text{ u} = 1.660\,566 \times 10^{-27} \text{ kg}$$

Particle	*Z*	*A*	*Charge (C)*	*Mass (u)*
Electron	−1	0	-1.6×10^{-19}	0.000 55
Proton	1	1	$+1.6 \times 10^{-19}$	1.007 28
Neutron	0	1	0	1.008 67

Nuclear reactions

In *nuclear reactions* (and *only* in nuclear reactions), elements can **transmute**, i.e. change into new elements.

The rules

- Proton number Z is conserved.
- Nucleon number A is conserved.

Some examples of nuclear reactions

- In 1919 *Rutherford* bombarded nitrogen gas with alpha particles

$$^{4}_{2}\text{He} + ^{14}_{7}\text{N} \Rightarrow ^{17}_{8}\text{O} + ^{1}_{1}\text{H}$$

(α-particle) (nitrogen) (oxygen) (proton)

and observed transmutation for the first time. Notice that an alpha particle is identical to a helium nucleus and a proton is identical to a hydrogen nucleus.

Exam practice answers: page 68

1 A caesium atom has an atomic mass number of 137 and an atomic number of 55.
(a) Give the number of:
(i) neutrons
(ii) protons
(iii) electrons.
(b) Explain why all atoms are neutral.

2 A material is known to be an isotope of tin. Given only this information, can you specify:
(a) its proton number
(b) its neutron number
(c) its nucleon number?
Explain. (10 min)

3 Uranium-238 has a proton number of 92. Write a balanced equation for its decay by alpha particle emission (to an isotope of thorium, Th). (5 min)

Grade booster

Generally your answers should be quoted to 2–4 significant figures. In questions involving atomic mass units, you are allowed to use 6, but don't make a habit of it!

Checkpoint 2

Values of atomic mass for most elements are close to whole number multiples of the mass of one hydrogen atom, but some are not (e.g. chlorine has an atomic mass of 35.5). Why is this?

Checkpoint 3

Chlorine's proton number is 17. How many neutrons are there in each of its isotopes: Cl-35 and Cl-37? Can you work out which isotope is more abundant? By how much?

Links

See *properties of ionizing radiation*, pages 50–1 to find out more about alpha particles.

Examiner's secrets

Nuclear equations appear with great regularity in exam papers. They generally aim to test whether you:
(a) can balance a nuclear equation
(b) understand how *Z* and *A* define the nuclide.

Nuclear instability

Whether a particular nuclide will be stable or not depends upon the balance (or imbalance) that exists between the forces tending to push it apart and the forces tending to hold it together.

The jargon

β-decay posed problems for nuclear physicists. β-decay seemed to break the law of conservation of energy (and that won't do!). Wolfgang Pauli solved the problem by proposing the existence of a new, difficult to detect, particle – the *neutrino* – which escapes with all the missing energy. (The new particle needed a new force to eject it: the weak nuclear force.) *Neutrinos* have since been detected and are thought to exist in huge numbers throughout the universe. (See page 63.)

Fundamental forces

There are only four known fundamental forces: **gravity** and the **electromagnetic force** account for every push or pull encountered outside the nucleus, but inside the nucleus, two more forces exist. These are called, rather unimaginatively, the **strong nuclear force** and the **weak nuclear force** (or interaction). The strong force is the main focus of this section; it binds the nuclides together. The weak nuclear force plays a role in β-decay. The two forces that battle it out in the nucleus and determine whether or not it will be stable are the electrostatic repulsion between protons and the strong nuclear force.

The strong nuclear force

- Is a force of attraction between neighbouring nucleons (though it turns to repulsion when the nucleons squeeze too close!).
- Does not depend on charge (the attraction between two protons is about the same as that between two neutrons, or between a proton and a neutron).
- Has an *extremely* short range (10^{-15} m).

The strong nuclear force can be thought of as a contact force. Each nucleon is strongly attached to its immediate neighbours (only). *The binding effect of each nucleon is extremely localized.* If it helps you, think of the nucleons as tiny balls with Velcro surfaces!

Checkpoint 1

How do we know that:

(a) The nuclear force is stronger than the electromagnetic force at close range?
(b) The strong force acts on both protons and neutrons?

The electromagnetic force

Although every nucleon is not necessarily attracted to every other nucleon by the strong nuclear force, *every proton in the nucleus is repelled by every other proton.* The electromagnetic force:

- is a force of repulsion within the nucleus (between protons, due to their similar charge)
- obeys an inverse-square law and acts over a far greater range than the strong nuclear force

Links

See *Coulomb's inverse-square law*, page 144.

The strong nuclear force dominates at short distances. This explains why atoms of low atomic mass tend to be stable (and why the helium nucleus is a particularly stable unit). In larger atoms, extra protons tend to exert a destabilizing influence. Bismuth ($Z = 83$) is the stable nuclide with the highest proton number.

The stable nuclides

The word **nuclide** describes a unique combination of particles in nucleus. Any particular nuclide is defined by its mix of protons and neutrons. The stable nuclides are shown below. There are about 270 of them. More than 2 500 unstable nuclides (not shown) have also been discovered (or artificially induced). They tend to decay, leaving more stable 'daughters' (i.e. nuclides lying closer to the trend line).

Notice that stable nuclides with the highest atomic numbers tend to have the greatest excess of neutrons over protons. No stable nucleus has more than 83 protons.

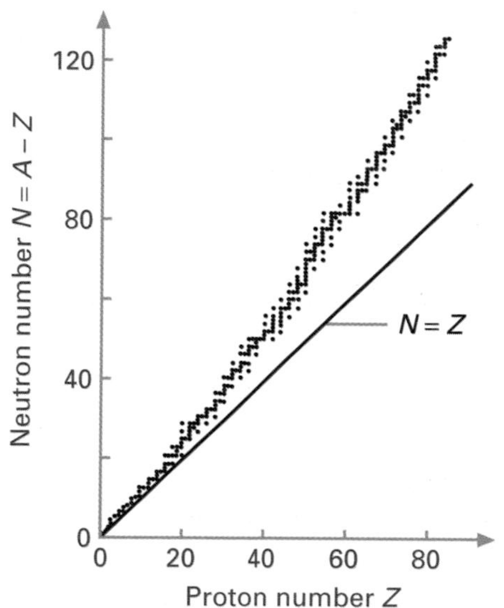

The jargon

Get used to words like *nuclide, radionuclide, isotope, radioisotope,* but don't get scared by them. The context often gives away the meaning. Physicists tend to talk about *nuclides,* because the nucleus is the physicist's domain! The definition of a nuclide ignores the existence of electrons (and therefore of chemistry!) completely, but the words nuclide and isotope are usually interchangeable. The *radio* prefix is used as in *radioactive* and just means connected with rays or radiation.

Important

→ $Z_{max} = 83$ There is a limit to the number of protons a nucleus can hold before their mutual repulsion makes the nucleus unstable.

→ *For* $Z < 20$, $Z \approx N$ At low atomic number, the strong nuclear force dominates – to such an extent that nucleon charge makes very little difference to stability.

→ *For* $Z > 20$, $N > Z$ At high atomic numbers, extra neutrons increase stability. Extra neutrons add a bit of extra nuclear binding (at least locally) and increase the average proton separation (which reduces their inclination to blow the nucleus apart).

Links

A radionuclide above the stability line decays by beta emission – see *properties of ionizing radiation*, pages 50–1.

Grade booster

When you are asked a question concerning nuclear stability, always mention *both* forces involved.

Exam practice answers: page 68

1 'The discovery of the nuclear structure of the atom was also the discovery of the third fundamental force – the strong nuclear force, which binds the nucleus.' True or false? Give your reasons. (15 min)

2 (a) Krypton's atomic number is 36.
(i) Plot the position of Kr-89 on the figure above.
(ii) It decays by β^--emission. Plot its daughter's position.
(b) Radon has an atomic number of 86.
(i) Plot the position of Rn-222 on the figure above.
(ii) It decays by α-emission. Plot the position of its daughter.
(c) What in general terms is the effect of each type of decay on stability? (20 min)

Properties of ionizing radiation

Radioactive decay releases ionizing radiation, i.e. radiation capable of knocking electrons out of atoms and molecules.

The jargon

The word *ray* can be used to describe radiated particles as well as photons of electromagnetic radiation. (A ray is just something which has been radiated!)

Checkpoint 1

Beta particles are fast moving electrons that come from the nucleus, but the nucleus does *not* contain any electrons! Can you explain this?

Checkpoint 2

What is the link between the degree of ionization each type of ray causes and its range?

Grade booster

Never write just radiation (which includes light and anything else that is radiated) when you really mean *ionizing radiation*. The ability of α-, β- and γ-rays to cause ionization is the thing that makes them special!

The jargon

X-rays are high-energy photons, like γ-rays. The only difference between the two is the way in which they are generated. X-rays are emitted by excited atoms when electrons make big jumps in energy level, but γ-rays are emitted by the nuclear quantum energy leaps associated with radioactive decay!

Three types of ionizing radiation

Three types of **ionizing radiation** are produced by naturally occurring radioisotopes. They are named after the first three letters of the Greek alphabet (alpha α, beta β and gamma γ), in order of increasing ability to penetrate matter.

- *α-particles* are easily stopped (e.g. by a sheet of paper, or a few centimetres of air). They cause the most ionization in the shortest distance.
- *β-particles* are stopped by a thin sheet of lead, or a few tens of centimetres of air. They cause less local ionization than α-particles, but more than γ-rays.
- *γ-rays* are far more penetrating and may pass through a layer of lead several centimetres thick. High-energy γ-rays may pass through entire buildings!

Radiation	Description	Z	A
α	High-energy helium nucleus	2	4
β^-	High-energy electron	−1	0
β^+	High-energy positron	+1	0
γ	High-energy photon	0	0

Detecting radiation

Photographic film was the first detector of ionizing radiation ever used (by Henri Becquerel in 1896). There are now better detectors for most applications, though film still has its uses. Detectors that count individual events include:

- Geiger–Müller tubes
- scintillation detectors
- semiconductor devices

Summary of properties

The properties are best presented in the form of a table:

	Alpha	Beta minus	Gamma
Nature	2 protons + 2 neutrons	High energy electron	High energy photon of electromagnetic radiation
Charge	+ 2	-1	neutral
Mass in u	4	approx. $\frac{1}{2000}$	zero
Stopped by	skin or paper	few cm of Al	30 cm of Pb or 1 m of concrete
Range in air	few cm	1m or so	several m

Deflected by magnetic fields?	yes	yes	no
Deflected by electric fields?	yes	yes	no
Ionising ability	high	moderate	low

Deflection by electric and magnetic fields

Moving charged particles are deflected by electric and magnetic fields. The degree of deflection depends on charge, mass and speed; the direction of deflection depends on whether the charge is positive or negative. The drawing below shows the effects of a perpendicular magnetic field on α-, β- and γ-rays.

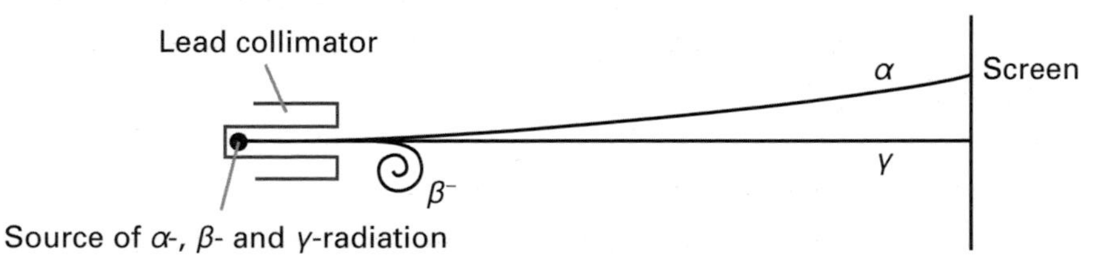

γ-rays are not deflected by a magnetic field, proving they are uncharged. α- and β-rays are deflected in opposite directions by the field, proving they are oppositely charged. β-rays are deflected very much more than α-rays, proving they have a significantly higher *charge/mass ratio.*

Watch out!

Many people think that γ-rays are the most dangerous because they are the most penetrating. In fact, α-particles can do you the most damage. Can you explain why?

Links

See *Risk = probability × consequences,* page 171.

Checkpoint 3

Use Fleming's left-hand rule to work out the direction of the magnetic field used in the figure opposite. Does it go into or out of the page?

Checkpoint 4

Explain why β-particles have a higher charge-to-mass ratio than α-particles. How much higher is it?

Exam practice answers: pages 68–69

1 Which type of radiation could be used to detect: (a) cracks in steel girders; (b) smoke; (c) under-filled cereal packs? Give reasons. (10 min)

2 (a) Describe experiments which could be used to determine the types of ionizing radiation emitted by an unknown source.
(b) How could you compare the energies of: (i) α-particles and (ii) β-particles from different sources? Explain your method(s). (20 min)

Radioactive decay

Unstable nuclei have too much energy locked up inside them for comfort. Sooner or later, they have to release some of this energy – by radioactive decay – in order to achieve a more stable (less energetic) state. Nuclear parents are less stable than their offspring!

Checkpoint 1

Why do we have to measure background radiation levels *every time* we carry out quantitative experiments with ionizing radiation?

Background radiation

If you turn on a Geiger counter well away from any known radioactive sources, it will start to click away gently and randomly. It is recording **background radiation**. Sources of background radiation include:

- the *Sun* (particularly during periods of sun-spot activity)
- *cosmic rays* (from far off stellar explosions etc.)
- the *Earth* (radioactive rocks and gases)

Background radiation has always existed; you can't escape it.
It contributes to our radiation dose and it provides a constant background of radiation, which must be subtracted from any other radiation measurements we might be interested in. Background radiation levels vary with location and they vary over time.

Checkpoint 2

If the *absolute error* in a radiation count N is $\pm\sqrt{N}$, what is the *relative error* in the same count? Calculate absolute and percentage errors in radiation counts of 100 and 10 000.
Note $\sqrt{N}/N$ simplifies to $1/\sqrt{N}$.

Randomness

- Radioactive decay is unaffected by temperature and pressure (at least within the normal earthly range of conditions!).
- Radioactive decay is a *random* process. You may know that a particular nucleus *is* unstable, but you cannot know exactly *when* it will decay.

You can hear this randomness in the uneven clicking of a Geiger counter measuring low-level radiation. You can see it in the fluctuating count rate from any radioactive source. We have to use large radiation counts if we want to minimize the effects of these random fluctuations.

Activity

Activity is the rate of decay of a source – its output, measured in *Becquerels* (Bq).

$$1 \text{ Bq} = 1 \text{ decay event per second} = 1 \text{s}^{-1}$$

The activity of a source depends on the number of unstable atoms it contains. If a source contains N unstable atoms, then we can write:

$$A \propto -N$$

Where A is the activity of the source. The minus sign is necessary because each decay event *reduces* N (by one). This becomes:

$$A = -\lambda N \qquad [1]$$

λ is called the **decay constant**. λ is closely related to (but not equal to) the probability that an atom will decay in any given interval. It is characteristic of a given isotope. The most stable radionuclides have the smallest decay constants (and vice versa).

Grade booster

Calculus notation You should be familiar with the notation used in calculus even though you are not required to understand how to do calculus (integration and differentiation of graphs). In calculus notation, activity A is written 'dN/dt'; which means the rate of change in N (the gradient of a graph of N against t).

The exponential decay equation

If we integrate equation 1, we get the *exponential decay equation*:

$$N = N_0\, e^{-\lambda t} \qquad [2]$$

Where N_0 is the number of unstable parent atoms at the start (when $t = 0$), N is the number of unstable parent atoms remaining after time t, λ is the decay constant and e is the special number, loved by mathematicians, which is the base of natural logarithms ($e \approx 2.72$).

→ Since a source's activity is directly proportional to the number of unstable atoms remaining, we can substitute initial and final activity or decay rates for N_0 and N and the equation will still hold.

Thus $A = A_0\, e^{-\lambda t}$

The graph on the left shows raw data for exponential decay and the right-hand graph shows the best-fit curve from this data used to determine the half-life.

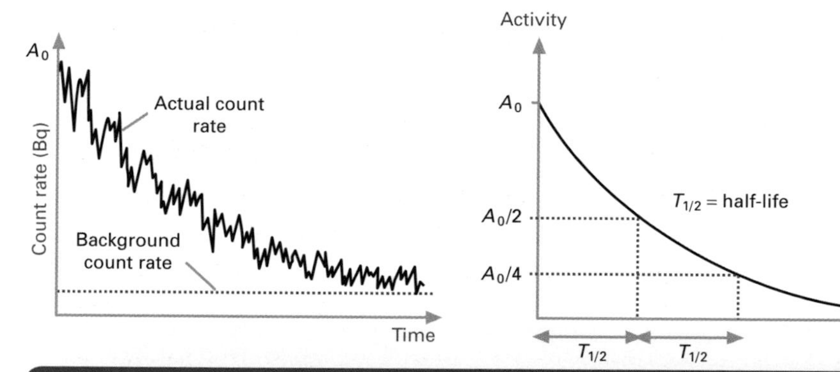

Half-life $T_{1/2}$

The radioactive **half-life** of a source is the average time it takes for the source's activity to drop by a half. Half-life can normally be found from the decay curve. Remember to:

→ subtract background radiation
→ draw a best-fit curve to smooth out the random variations
→ work out half-life from at least two starting points and take an average

Half-life and the decay constant

The radioactive half-life of a source is linked to its decay constant by the equation:

$$\lambda T_{1/2} = \ln 2$$

ln 2 is the natural logarithm of 2, which has the value 0.693. This equation can be derived from the exponential decay equation by substituting $T_{1/2}$ for t and $N_0/2$ for N (see action point).

Exam practice answers: page 69

1 A radionuclide contains 400 000 nuclei. Its decay constant is 0.30 s^{-1}. What is the initial activity? (2 min)

2 Two samples A and B of the same radionuclide have different activities. Give the reason for this. (5 min)

3 The half-life of thorium-228 is 1.91 years. Calculate its decay constant and hence find the time needed for its activity to fall by a factor of ten. (*Note* If $e^x = y$, then $x = \ln(y)$.) (15 min)

The jargon

Mathematically speaking, an *exponent* is a power, and so an equation with powers in can be called an *exponential equation.*

Checkpoint 3

Rearrange equation 2 to make $e^{-\lambda t}$ the subject. Write down in your own words what the new version of the equation means.

Checkpoint 4

If a source has a half-life of 1 day, how many days will it take for its activity to fall below 10% of its initial activity?

Examiner's secrets

Make sure you learn that N equals the number of unstable nuclei *remaining*. It is easy to forget this because the number remaining is equal to the number that have decayed after the first half-life.
This is the *only* time this is true.

Watch out!

The units used for time and for decay constant must cancel. If time is measured in seconds, λ must be given in s^{-1}; if time is measured in years, λ must be in y^{-1}.

Action point

You can work out for yourself that $e^{-\lambda t}$ represents the fraction of unstable parent atoms (N/N_0) remaining after time t. So when $t = T_{1/2}$, this fraction must equal 1/2. So:

$e^{-\lambda T_{1/2}} = {}^1/_2$

or $e^{\lambda T_{1/2}} = 2$

Taking natural logarithms gives us:

$\lambda T_{1/2} = \ln 2$

Binding energy and mass defect

With the exception of hydrogen, all atoms weigh slightly less than the sum of their constituents! They have a mass defect. The law of conservation of mass is broken and must be replaced with the more general law of conservation of mass–energy discovered by Albert Einstein. In this section, we get to use the most famous equation in physics: $E = mc^2$!

The jargon

Special relativity is based on the premise that the speed of light is constant for all observers (regardless of their motion).

Binding energy

Nucleons are bound together by the *strong nuclear force*. You would have to do work against this binding force to pull the nucleons apart. The amount of work you would have to do to separate all the constituent nucleons from a nucleus is called the **binding energy**.

Mass–energy

Einstein showed in his work on relativity that mass and energy are two forms of the same quantity, more properly called **mass–energy**. Matter and energy can be interconverted according to the equation:

$$E = mc^2$$

Where E is energy, m is mass and c is the speed of light ($c = 3.00 \times 10^8\ \text{m s}^{-1}$).
You should see from the equation that the annihilation of a small amount of matter yields a lot of energy. (1 kg $\rightarrow 9 \times 10^{16}$ J!)

Checkpoint 1

How long could you run a 1 000 MW power station on 1 kg of matter (assuming *all* of it is converted into useful energy)?

Mass defect

The **mass defect**, Δm, of a nucleus is the difference between the summed mass of its constituent nucleons and electrons and its actual mass.

$$\Delta m = (m_{\text{protons}} + m_{\text{neutrons}} + m_{\text{electrons}}) - m_{\text{atom}}$$

It is the mass equivalent of the atom's binding energy (the work you would have to do to separate each nucleon from the nucleus increases its mass!)

$$\Delta E = \Delta m\, c^2$$

binding energy = mass defect × speed of light squared

Links

Refer to *more about leptons and quarks*, pages 64–5 to read about the formation and annihilation of matter and antimatter.

Total binding energy increases with *nucleon number*, but this does *not* mean that high nucleon number nuclides are the most stable. (With more nucleons, the glue is spread more thinly.)

Links

See the section on the strong nuclear force in *nuclear instability*, pages 48–9.

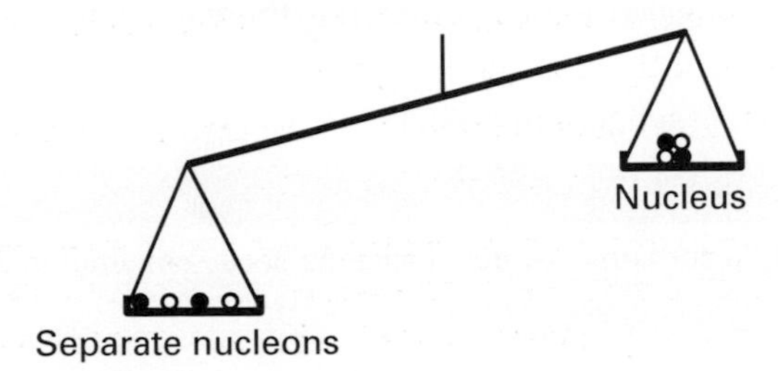

An atom or nucleus always weighs less than the particles it is made of.

Mass defect and binding energy per nucleon

If you divide mass defect or binding energy by the number of nucleons, you get a useful measure of nuclear stability. The greater the mass defect (and binding energy) per nucleon, the more stable the nucleus.

- 1 eV (electron volt) = 1.6×10^{-19} J
- 1 u = 1.661×10^{-27} kg = 931.5 MeV

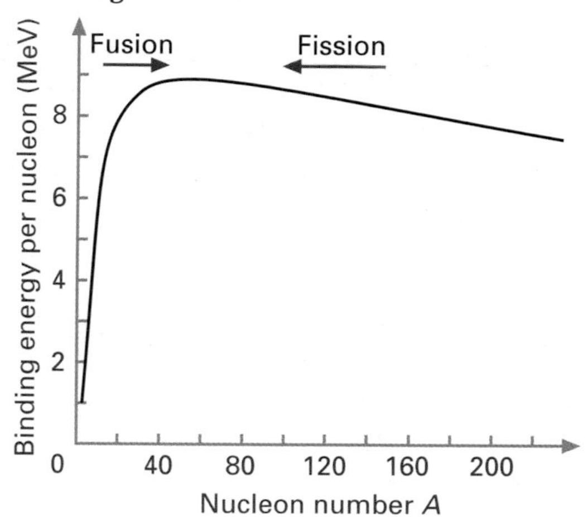

Some features of the graph

- Binding energy per nucleon of normal hydrogen is zero, because there is only one nucleon, so the strong nuclear force is not involved, and so no work need be done to overcome it!
- Binding energy per nucleon rises rapidly with A, peaking between $A = 50$ and $A = 80$. Iron ($A = 56$) is the most stable atom.
- Binding energy per nucleon falls as A increases beyond this point.

Energy release from fusion and fission

- **Fusion** is sticking together, **fission** is breaking apart.
- Fusion powers the Sun and the stars. Fission powers nuclear power stations.
- Both processes release energy. (How is this possible?)

The graph above holds the key to understanding energy release by fusion and by fission. Any process which increases mass defect *per nucleon* releases energy. Looking at the graph:

- fusion can be interpreted as a shift to the right (increasing A)
- fission is a shift to the left (decreasing A)
- fusion of low atomic mass nuclei (only) increases mass defect per nucleon and therefore releases energy
- fission of high atomic mass nuclei (only) also increases mass defect per nucleon and therefore releases energy

Grade booster

Make sure you are happy with the definitions of mass defect and binding energy and with the idea of mass–energy conversions.

The jargon

Binding energy per nucleon can be thought of as an energy *well*. The deeper the well, the more tightly bound the nucleus.

Grade booster

This is a key graph. Study it closely. Energy release leads to greater stability. Fusion of low A nuclei and fission of high A nuclei release energy.

Watch out!

Binding energy is sometimes considered a negative quantity, since it is an energy deficit – so you might find the figure on the right in an inverted form (a reflection in the x-axis), in which case iron sits at the bottom of the binding energy trough.

Action point

Notice from the graph that the binding energy per nucleon is typically 8 MeV. Find out which element has the most stable nucleus.

Watch out!

Once you reach iron in nuclear reactions, it's all over. No more energy available. Iron is cold and dead in terms of available nuclear energy.

Exam practice answers: page 69

Two deuterium (^2H) nuclei (mass = 2.0141 u) fuse to produce one ^3He nucleus (mass = 3.0160 u) and a neutron.

(i) Write the nuclear equation for this reaction.

(ii) Calculate the total mass defects before and after the fusion.

(iii) Find the energy released in MeV. (20 min)

Nuclear fission and fusion

A major application of radioactivity at the beginning of the 21st century is in nuclear power stations, where the energy released by processes in the nucleus is converted to electrical energy. As was explained in the previous section, in theory, it is possible to use two processes (fission or fusion), though at present only reactors that release energy via fission are commercially viable. However, research is continuing into making controlled fusion feasible – trying to create in a reactor what happens continuously in the Sun.

Nuclear fission (splitting)

Uranium-235 is used as an energy source in nuclear power stations. U-235 is an α-emitter, but it also decays by *spontaneous fission* into two large parts (plus a few neutrons), releasing a great deal of energy. The neutrons released can trigger further fission, causing a *chain reaction* which can release huge amounts of energy. Controlling this chain reaction is the key to nuclear power generation.

Checkpoint 1

99.3% of uranium is of the isotope U-238; only 0.7% of mined uranium is of the isotope U-235 which provides the power in a normal fission reactor. Why can't U-235 be separated chemically from U-238?

neutrons released
thermal neutron
uranium 235 nucleus splits
further fission events

Ejected neutrons travel very fast and interact weakly with most matter – so they tend to escape. A *moderator* must be used to slow them down (they are then known as *thermal* neutrons) and keep them from escaping. Thermal neutrons are very effective at inducing further fission. Hydrogen-rich substances readily absorb neutron energy, but they tend to also absorb the neutrons themselves (as hydrogen is converted to deuterium) which spoils things. Pure graphite (carbon) is usually used as a moderator.

Watch out!

Be sure you understand the distinction between absorbing neutrons and absorbing their energy. The moderator should absorb only the energy; the control rods must mop up the particles themselves.

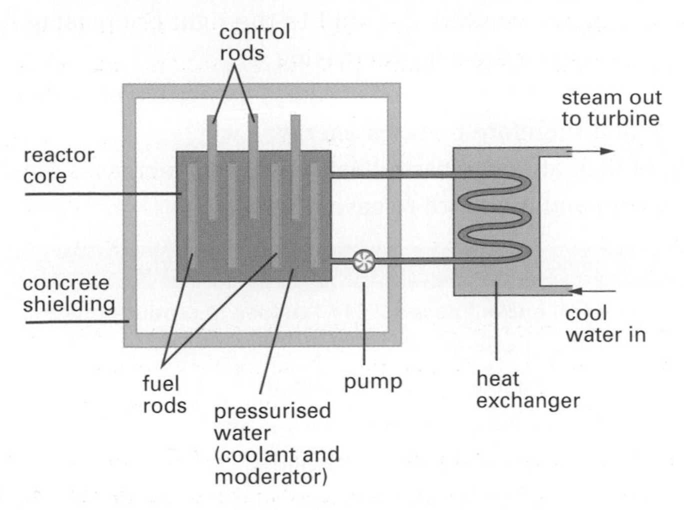

Check the net

www.british-energy.com/pagetemplate.php?pid=314

Checkpoint 2

In PWRs (pressurised water reactors), the water has two functions. What are they?

If more than one neutron per fission goes on to cause further fission, the reaction rate will rapidly rise. *Control rods* made of cadmium or boron (good neutron absorbers) are used to make sure the chain reaction does not get out of hand.

The thermal energy is taken away by a *coolant* that is circulated past the fuel, before being pumped to a *heat exchanger* to produce steam, that drives a *turbine* and generator.

Nuclear fusion (joining together)

In this process, light (small) nuclei are joined together to form a heavier (larger) nucleus, with the release of energy. The nuclei (which are positively charged) have to be moving extremely quickly to achieve this, in order to overcome the force of repulsion between them, and get close enough for the (attractive) *strong* force to hold them together. The speeds required are so big that huge temperatures (at least 100 million K) are necessary, which makes the engineering of a fusion reactor really challenging.

Checkpoint 3

What is the most plentiful natural source of deuterium?

Practical fusion reactors

The most promising reaction for commercial reactors is that between deuterium (${}_1H^2$) and tritium (${}_1H^3$), forming helium:

$${}_1H^2 + {}_1H^3 \rightarrow {}_2He^4 + {}_0n^1 + \text{energy}$$

There is a lot of naturally occurring deuterium and tritium can be produced in the reactor itself by surrounding the core with lithium and allowing it to absorb the neutrons released in the reaction.

At such high temperatures, matter exists in a form known as a *plasma*, in which the electrons are free to move, leaving behind positively charged ions. The main problem is in keeping this plasma in one place long enough for fusion to take place. It cannot be allowed to touch anything else, as it would quickly cool. Three methods are available:

- gravitational confinement – works in the Sun and other stars, but requires huge masses
- inertial confinement – changes take place so rapidly that particles do not have time to move away. They are held by their own inertia.
- magnetic confinement – works because the plasma is charged, and can be contained in a *torus* shape (like a ring doughnut) by magnets. This is probably the most promising design and is used in the JET (Joint European Torus) fusion project at Culham, near Oxford.

Check the net

For more about JET, see www.jet.efda.org

Links

See *forces/interactions and conservation laws*, pages 66–7.

Exam practice answers: pages 69–70

1. In a fission reactor, the processes of *moderation, control* and *cooling* are required. Explain what these are and how they are achieved in practice. (10 min)
2. A uranium nucleus releases about 200 MeV of energy when fission takes place. If a reactor is 40% efficient and has an electrical power output of 1500 MW, how many fission events take place each second? (10 min)
3. Where does fusion occur continuously in nature? Explain how the conditions needed for fusion are created in these circumstances. (8 min)

Other applications of radioactivity

Other applications of radioactivity and nuclear physics range from treating cancer to smoke detection. Nuclear physics has changed the world!

Checkpoint 1

Carbon-14 dating is only accurate(ish) for organic objects between 200 and 10 000 years old. Why?

Checkpoint 2

The age of the Earth All uranium on Earth is older than the Earth itself. U-238 has a half-life of around 4.5 billion years; U-235 has a (much shorter) half-life of 7.1×10^8 years. Is it any surprise that U-238 is the more abundant isotope? (If you assume they were equally abundant when formed, you can calculate how long ago that was! The answer comes to around 6 billion years, which is okay as an upper limit.)

Links

Penetration (see *properties of ionizing radiation*, pages 50–1) and half-life (see *radioactive decay*, page 53) are key properties in any application.

Radioactive dating

Radioactive decay takes time. If you know a source's initial activity and its half-life, its present activity tells you the time (give or take all the uncertainties involved) since it started to decay. Radioactive dating techniques have to rely on certain assumptions about initial conditions.

Carbon dating

Atmospheric levels of carbon-14 are thought to have been fairly steady for thousands of years. C-14 is constantly being produced in the atmosphere by the action of cosmic rays. It decays by β^--emission, with a half-life of 5730 years. Over the millennia, an equilibrium has been set up so that the rate of decay is balanced by the rate of formation. Every living thing contains C-14 in the same proportion as occurs in the atmosphere. When organisms die, the proportion of C-14 begins to fall, so the specific activity (activity per kg) of C-14 in any organic matter tells us how long ago it died.

Non-destructive measurement and detection

- *α-particles* are used to detect smoke (open up your smoke detector and look for the radiation danger sticker).
- *β-particles* are used to check the thickness of paper and card.
- *γ-rays* are used to check aircraft wings and oil rigs for cracks. They are used in the coal mining industry to monitor the rate of production automatically (on the output conveyor belt). Y-rays have found an enormous range of uses. They come in a range of energies, making them suitable for measuring a wide range of objects! Dual-energy beams can be used to distinguish between different materials (e.g. coal and ash or steel and plastic).
- The position of a leak from an underground or inaccessible pipe can be found by adding a small volume of a radioactive isotope to the liquid that flows through the pipe. The count rate in the area close to the leak will be increased and the penetrating nature of the emission means that the position can be identified from some distance away.

In all of these applications, the half life and activity of the radioactive nuclide must be chosen to take into account such factors as dose uptake to humans and the feasibility and economics of regular renewal of the source.

Killing microbes

- γ-rays are used to sterilize surgical equipment (γ-rays can penetrate packaging without damaging it, which offers obvious benefits). The process is quick, clean and simple.
- γ-rays can also be used to sterilize tinned foods and (more controversially) to increase the shelf life of fresh fruit and vegetables.

Checkpoint 3

What other methods could be used for sterilizing surgical equipment? What are their disadvantages (compared to γ-ray sterilization)? Which method would you support?

Nuclear medicine

Tracer techniques

Radioactive tracers are used to aid diagnosis. A (fairly) short half-life gamma emitter is injected and a gamma-camera is used to map its movement and distribution within the body. The tracer can be attached to sugars to highlight sugar-greedy cancer growth, or to a pharmaceutical which will accumulate in the site of interest.

Checkpoint 4

The range and penetration of γ-rays depends upon their energy. How would the ideal source energy for treating a child differ from that used for an adult?

Treatment of cancer

Radiation has proved a most effective treatment for many types of cancer. Cancer cells divide and grow more rapidly and are more susceptible to radiation damage than normal cells.

- *External treatment* makes use of gamma-ray beams which target the tumour from different directions so that the tumour (and only the tumour) receives a fatal radiation dose.
- *Internal treatment* can involve inserting a sealed β-source for short time periods or injecting a radiopharmaceutical which is designed to be as site specific as possible.

Links

See page 168 for more about medical applications.

Exam practice answers: page 70

The specific activity of carbon in living organisms is always 0.23 Bq per gram. A fossilized bone is discovered to have an activity of 0.0052 ± 0.0006 Bq per gram of carbon.

(a) Calculate the range of values the bone's age lies within.

(b) Explain the source of uncertainty and state any measures which could be taken to reduce it. (15 min)

Probing matter

Atoms were supposed to be the fundamental particles everything is made of, but it turns out that it's not quite that simple! Nuclear physicists continue to probe and smash atoms and their nuclei in the hopes of finding evidence for a grand unified theory – a theory of everything.

Diffraction patterns

Light *diffraction patterns* can be used to measure the spacing of slits in a *diffraction grating*. X-ray diffraction patterns can be used to measure the spacing of atoms in a crystal.

Links

See *diffraction*, pages 120–1.

→ You can't distinguish between two points closer than one wavelength apart. Fine *resolution* requires short wavelengths.

The Bragg equation gives the link between wavelength λ, spacing s and displacement angle θ for the nth constructive fringe.

$$n\lambda = s \sin \theta$$

Note The wavelength must be smaller than the spacing being measured or you get no *interference fringes* (see checkpoint). The number of *constructive interference* fringes you get is equal to s/λ, rounded down to the nearest whole number.

Checkpoint 1

If $\lambda = s$, the Bragg equation becomes $n = \sin \theta$. For $n = 1$, $\sin \theta = 1$ and θ is 90°. The first order constructive fringe grazes off at 90° to the target. What is the smallest spacing that can be measured by diffraction of light (wavelength 0.4 to 0.7 μm)?

Particle diffraction

Particles sometimes behave as waves (and vice versa). The wavelength of a particle is related to its momentum by the de Broglie equation:

$$p = h/\lambda$$

Where p is momentum (the product of mass and velocity), h is Planck's constant and λ is wavelength. The greater the particle's momentum, the shorter its wavelength.

Grade booster

Many examinations now include questions that ask you to recall details of standard laboratory experiments. You should keep concise and up-to-date records of class experiments and demonstrations.

Low-energy electron diffraction

Louis de Broglie's theory of wave–particle duality was first confirmed in 1927, when George Thomson (son of J. J. Thomson, who discovered electrons) noticed diffraction patterns produced by a beam of electrons he had fired at gold foil. The spacing of the fringes allowed calculation of atom separations.

Checkpoint 2

What wavelength (roughly) would you choose to measure:

(a) atom sizes and separations?

(b) nucleon sizes and separations? Give your reasons.

(c) High energy X-rays may have wavelengths as short as 10^{-11} m.

What are the implications for probing matter?

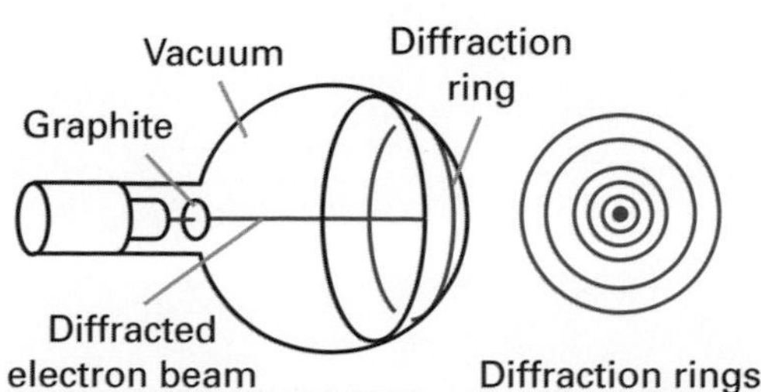

Diffraction patterns produced by a graphite sheet in a cathode-ray tube are shown above. Typical values: tube voltage ≈ 5 000 V, electron speed ≈ 4.2×10^7 m s^{-1}, wavelength ≈ 1.7×10^{-11} m, carbon atom separation ≈ 1.2×10^{-10} m.

High-energy electron diffraction

With a high enough voltage in a good enough vacuum, you can give electrons enough momentum to produce diffraction patterns when

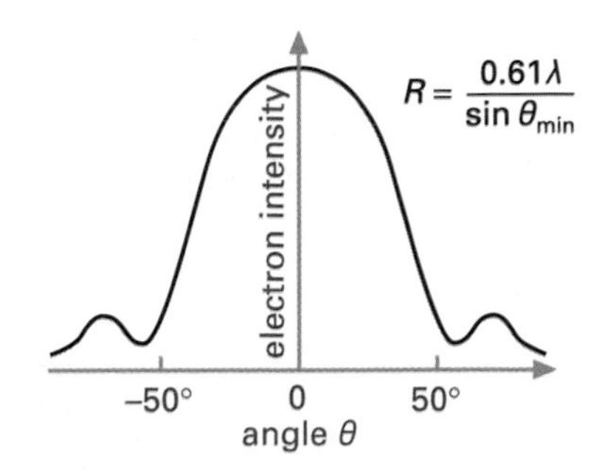

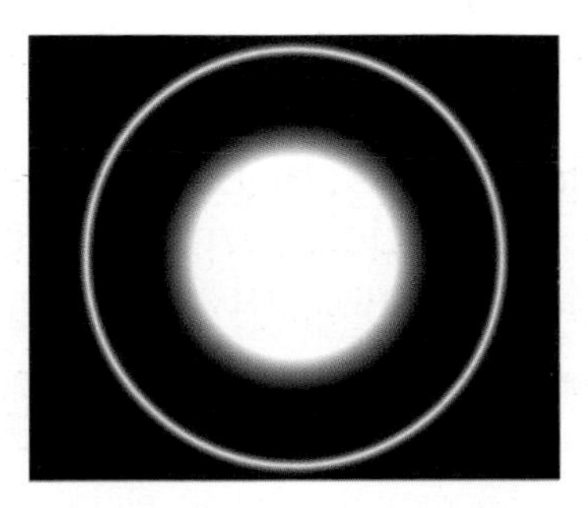

they are scattered elastically by nuclei. Being *leptons*, electrons have the advantage that they are not subject to the strong nuclear force. The high-energy (420 MeV) electron diffraction by carbon nuclei is shown above. Note that the geometry is different here. The nuclei are treated as individual tiny balls and the angular displacement of the first minimum is used to calculate the size.

Neutron diffraction

Just about every type of particle has been used to probe the atom. Charged particles such as electrons, protons and alpha particles interact with the charge held within the nucleus. Neutrons have the advantage of no charge, and so neutron diffraction can be used to give information on the distribution of the strong nuclear force.

Deep inelastic scattering

Rutherford's α-particle scattering experiment (see *The atom and its nucleus*, pages 44–5) that gave us evidence of the existence of the nucleus was followed 50 years later by another classical investigation. Here, physicists at SLAC (Stanford Linear Accelerator Center) in California probed nucleons by firing high energy *electrons* at them. They found that the electrons were scattered in many directions and concluded that nucleons themselves have structure – they are not fundamental and are made up of smaller particles. This was the first direct evidence for *quarks*.

Links

More about leptons on page 63.

Watch out!

Don't try to work out the momentum of a high-speed particle using its rest mass. Mass increases significantly as you approach light speed. (One of the effects explained by Einstein's theory of special relativity.)

The jargon

Relativistic is a term used to describe particles moving so fast that relativity must be accounted for. You can generally ignore relativity below speeds of 10^8 m s^{-1}.

Checkpoint 3

The main drawback with neutrons is that because they have no charge, there is no way to artificially accelerate them. Calculate the wavelength of a neutron ejected from a nucleus at a speed of 10^7 m s^{-1} ($h = 6.63 \times 10^{-34}$ J s).

Links

Quarks are explained in more detail on page 63.

Exam practice answers: page 70

1 The incident particles in Rutherford's α-particle scattering experiment were α-particles. What were they in the deep inelastic scattering experiment? What was the target in each case? Write a short paragraph describing the conclusion of each experiment.

Particles – production and patterns

Atoms were originally thought to be fundamental particles (the word atom comes from the Greek word *atomos* meaning indivisible) until the electron was discovered in 1897. Since then, particle physicists have used increasingly sophisticated accelerators to produce particles and investigate theories. Currently the electron is thought to be one of the 12 fundamental particles which, with their 12 antiparticles, form the standard model of particle physics.

Check the net

The web site of the European Laboratory for Particle Physics (and the home of the world wide web) has information about its particle accelerators at www.cern.ch/

Particle accelerators

The basic concept used in accelerators is that energy can be converted into mass, and the more energy that is available (via bigger particles moving at higher speeds), the more massive the "new" particles that can be produced. These decay very rapidly after they have been produced, requiring extremely responsive detectors to detect their fleeting presence. The principles of operation of the three main types of accelerator are listed in the next section.

Linear accelerator ("linac")

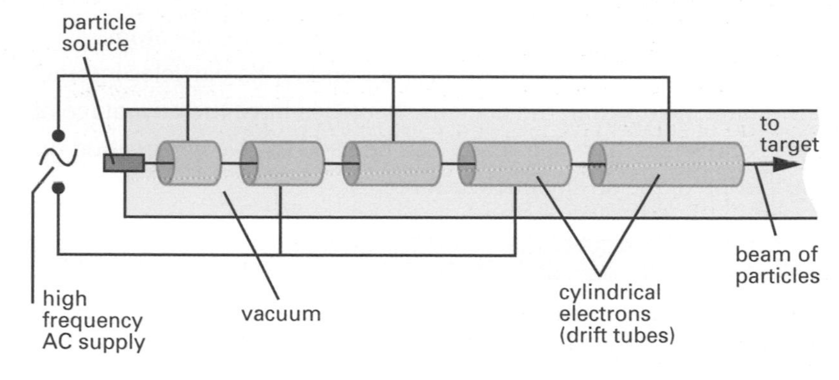

Checkpoint 1

Explain why the tubes that the charged particles travel along must be evacuated.

- Charged particles travel along a straight evacuated tube.
- Cylindrical electrodes connected to a high-frequency AC supply change polarity when the particles reach the end of an electrode.
- The particles are accelerated across the gap.
- The electrodes increase in length as the particles' speed increases.
- To get high-energy particles, the accelerators have to be very long.

Cyclotron

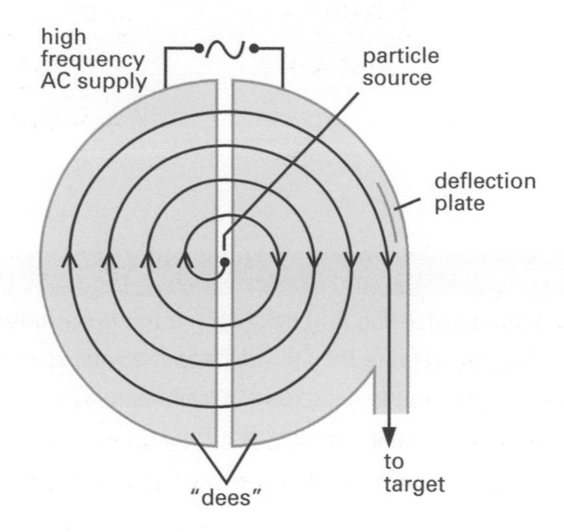

- A more compact method uses a magnetic field to make the particles move in a circle.
- Instead of moving particles colliding into a stationary target, oppositely charged particles can be accelerated in the opposite direction increasing the energy of the collision.
- The particles spiral outwards between two semicircular metal D-shaped electrodes ("dees") connected to a high-frequency AC supply.
- The frequency of the supply is given by $f = Bq/2\pi m$.
- The energy of the particles is limited by the theory of relativity – as they approach the speed of light, their mass increases so they would no longer reach the gaps at the moment that the PD changes.

Synchrotron

- Here previously accelerated particles travel around the same circular path.
- As they are accelerated more, their mass increases.
- The strength of the magnetic field in the deflecting magnets is increased to keep the radius of their path the same.

Fundamental particles and the standard model

The particles that have been detected over the last 80 years can be arranged in groups, in what is known as the *standard model*. The 12 fundamental (not made up of smaller) particles form two groups called **leptons** and **quarks**. Each group consists of three pairs of particles, known as the three *generations* of particles.

Leptons

1st generation	*2nd generation*	*3rd generation*	*Charge*
electron e^-	muon μ^-	tau τ^-	$-e$
electron neutrino ν_e	muon neutrino ν_μ	tauon neutrino ν_τ	zero

Quarks

1st generation	*2nd generation*	*3rd generation*	*Charge*
up u	charm c	top t	$+^2/_3e$
down d	strange s	bottom b	$-^1/_3e$

→ Order of increasing mass

Checkpoint 2

In a cyclotron, the particles do not get faster as they move through the 'dees'. Why is this? Where do they increase their speed?

Check the net

You'll find the particle adventure on www.cpepweb.org/

The jargon

Down, charm, etc are also called quark *flavours*.

Exam practice answers: page 70

Circular accelerators can accelerate charged particles to greater energies than linear accelerators.

(a) List the principles of physics used in a cyclotron.

(b) A proton of mass m and charge e is moving in a circle of radius *r* in a cyclotron. Work out the time taken to complete one semi-circle. (10 min)

More about leptons and quarks

"If I could remember the names of all these particles, I'd be a botanist."

Enrico Fermi

These pages give you more detailed information about the fundamental particles and introduce you to the concept of antimatter, extending the list of 12 particles on the previous page by another 12 – their antiparticles. The consequences of a particle meeting its antiparticle are considered, and also the formation of pairs of particles from energy.

Action point

Draw up two tables like the ones above listing the antiparticles with their corresponding charges.

Matter and antimatter

Try not to be uneasy about the perhaps disturbing feel of the word "antimatter"! Just accept that each particle has a corresponding antimatter particle or antiparticle. An antiparticle is identical to its particle except that it has the opposite charge. So an antielectron has the same mass as an electron but it has a charge of $+e$. Antiparticles are usually denoted by having a line over the symbol, so an antielectron would be $\bar{e}$. An antielectron also has its own name - the **positron**, and is sometimes written as e^+.

Leptons

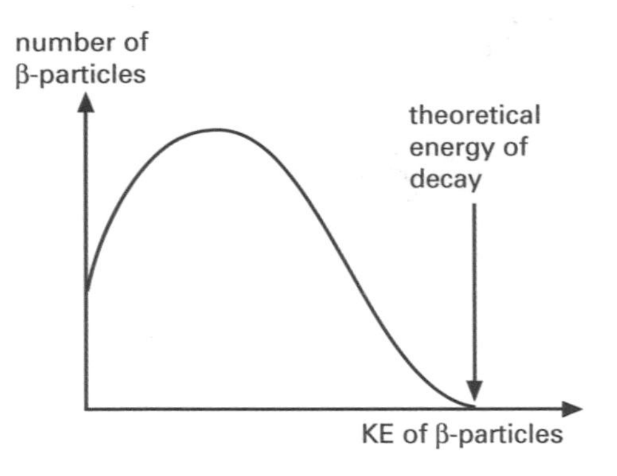

These fundamental particles all have one property in common – they only interact via the weak interaction and do not feel the strong interaction. The charged leptons can also interact via the electromagnetic interaction.

Neutrinos are really difficult to detect and were originally introduced to try to explain a mystery associated with β-emission, which is that the β particles are emitted with a range of energies, as shown in the diagram below. In theory, all the β particles from decay of the same nuclide should have the same energy (α particles do!). Dirac proposed in 1930 that the "missing" energy from the β particles was being carried off by uncharged particles of zero rest mass that he called neutrinos. They were eventually detected over 20 years later, as were their antiparticles (antineutrinos).

Hadrons, baryons and mesons

Whereas leptons can be found singly, quarks can only exist combined with other quarks. These combinations of quarks are called **hadrons**. A **baryon** is a hadron made from three quarks. For example, a proton consists of two up quarks and a down quark (uud) and a neutron is one up quark and two downs (udd).

Mesons are hadrons consisting of a quark and an antiquark.

Because hadrons are made up of quarks, they feel the strong interaction.

Checkpoint 1

Show that the combinations of quarks for the proton and neutron give the correct amount of charge.

Stability of hadrons

- All free hadrons (except the proton) are thought to be unstable and decay into other particles.
- Neutrons and protons inside the nucleus are relatively stable.
- A free neutron has a half-life of about 15 minutes.
- Free protons are much more stable and their half-life is thought to be of the order of 10^{32} years.

Beta decay

Neutrons decay to produce a proton, an electron and an antineutrino. This can happen in the nucleus of an atom that has a high neutron to proton ratio. It is an example of the weak force causing beta decay as the electron is emitted from the nucleus as a beta particle (β^-). An antineutrino is also emitted while the proton remains in the nucleus.

Using the quark model, this means that a neutron (udd) has become a proton (uud) so a down quark has changed into an up quark – the quark *flavour* has changed.

A different kind of beta decay occurs in nuclei that have too few neutrons for stability. Positron decay is the mirror image of beta decay. Here a proton becomes a neutron while a positron (a β^+) and a neutrino are emitted.

Checkpoint 2

Use the quark model to describe the changes that must take place when positron decay occurs.

The jargon

Beta capture is when one of the orbiting electrons is captured by the nucleus and combines with a proton to produce a neutron.

Annihilation

When a particle meets (interacts with) its own antiparticle, *annihilation* is said to take place and the mass of both particles is converted into energy, which appears as γ photons. If you are given the mass of the particles in u, you can use the conversion factor of 1 u = 930 MeV (approximately) to fine the energy of the photons.

Grade booster

It helps if you can spell "annihilation" correctly

Pair production

This is just the opposite of annihilation. A γ photon with sufficient energy (it needs a lot!) can produce two particles, but these *must* be a particle-antiparticle *pair*. The most common pair is an electron and a positron, since these are of relatively low mass. They usually look like this in photographs from detectors, and the tracks curve in opposite directions because the particles are oppositely charges and a magnetic field applied in the detector produces a force in opposite directions.

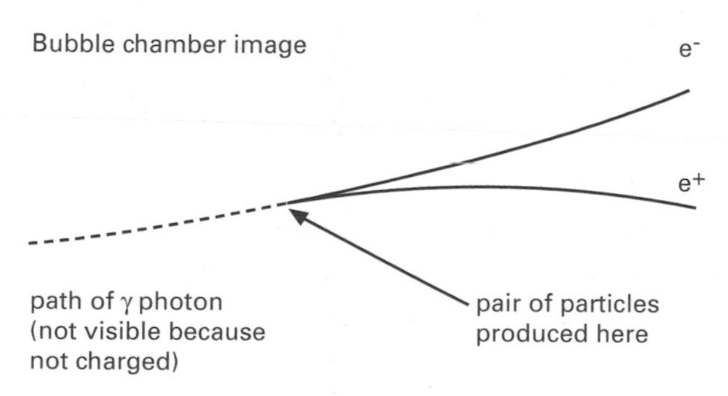

Checkpoint 3

What energy of photon is required to create an electron–positron pair?

Action point

Try to work out the direction of the field in the bubble chamber from the curvature of the tracks

Exam practice answers: page 71

(a) Explain why measurements of the energies of the β- particles emitted during nuclear decay led to the suggestion that an additional undetected particle must also be emitted during the same process.

It is possible to produce antihydrogen.

(b) Is an atom of antihydrogen positive, negative or neutral?

(c) What quarks does an antiproton contain?

(d) What other particle must be present in then atom of anthydrogen?

(e) What type of particle is this? (10 min)

Forces/interactions and conservation laws

You will learn a lot about many different types of forces in the course of A-level Physics, but these pages include ideas about the four *fundamental* forces that produce all of the others. These fundamental forces can be explained using the idea of particles that are exchanged to make the forces "work" during interactions, which themselves are only possible if certain laws apply.

Forces

There are four fundamental forces (or *interactions*). The first two weaken with distance, but extend to infinity.

- Gravity forms planets, stars and galaxies. It has no real relevance in particle physics because the particles involved have such small masses.
- The electromagnetic force keeps atoms together.

The next two have a range within the size of the atomic nucleus.

- The weak force is responsible for radioactive decay. If the quark flavour changes, the interaction must be weak.
- The strong force acts between quarks and keeps protons inside the nucleus in spite of their electrostatic repulsion.

Links

See *gravitational fields*, pages 142–3; *electric forces and fields*, pages 144–5; *nuclear instability*, pages 48–9.

Exchange particles

- The action of forces can be explained by the theory of exchange particles, sometimes called **gauge bosons**.
- There are three exchange particles that are associated with the weak interaction – W^+, W^- or Z^0.
- The exchange particle for the gravitational force, the **graviton**, has yet to be discovered.

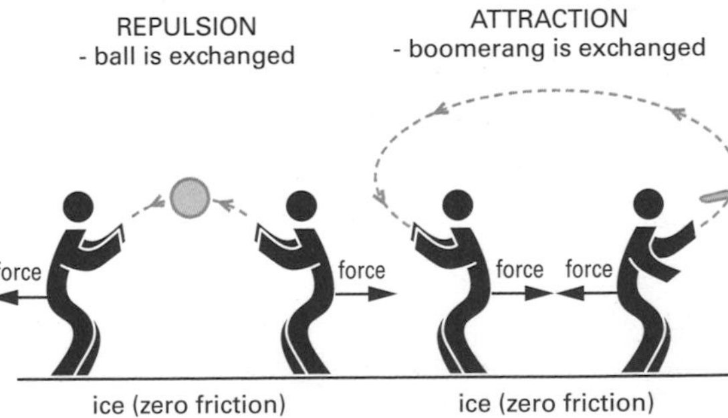

Summary table:

Force	Particles affected	Exchange particle	Range	Relative strength
Strong	quarks	gluons	10^{-15} m	1
Weak	quarks and leptons	W and Z particles	10^{-17} m	~10^{-5}
Electromagnetic	all particles with charge	photons	infinite	~10^{-2}
Gravitational	all particles with mass	gravitons	infinite	~10^{-39}

For particles that contain quarks (hadrons), the exchange particles are pions (mesons).

Checkpoint 1

Explain what the following groups of particles are and give an example of each one: hadrons, mesons, baryons, leptons, quarks, antiparticles, bosons.

Action point

Think about the momentum of the person throwing/receiving the ball/boomerang.

Examiner's secrets

Be able to match up the exchange particle with the interaction - it's a very common question.

Feynman diagrams

These diagrams were devised by the American physicist Richard P. Feynman in the 1960s in order to make the interpretation of particle interactions more straightforward. The particles are drawn as lines – straight for "ordinary" particles and wavy for exchange particles. Here is a Feynman diagram representing β^- decay:
There are some rules that apply:

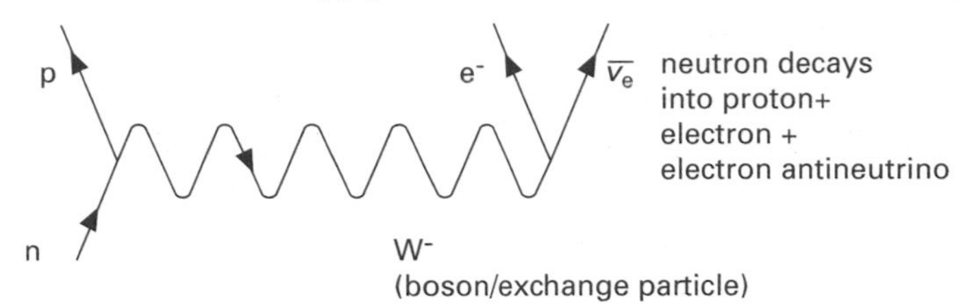

- incoming particles (the past) come in from the bottom
- outgoing particles (the future) leave at the top
- baryons (or quarks) stay on one side, leptons on the other side
- charge is conserved (see below) at each vertex. This allows you to decide which of the exchange particles is responsible for a particular weak interaction, form instance.

Conservation laws

You have already met conservation laws (or principles) in mechanics (momentum is conserved in collisions) and electricity (charge is conserved at junctions in circuits. Similar laws apply in particle physics and allow you to decide whether any particular interaction is possible, though it does not necessarily tell you that it *will* happen. The conservation laws are:

- charge is always conserved
- lepton number (+1 for leptons, –1 for antileptons, 0 for all other particles) is always conserved
- baryon number (+1 for baryons, –1 for antibaryons, 0 for all other particles) is always conserved
- strangeness is conserved in strong and electromagnetic interactions and in *some* weak interactions.

The maths is very easy – just add up the numbers on each side of the equation representing the interaction, and if they balance, the interaction is possible.

The jargon

'Conserved' just means the same *after* as *before*, – as with energy, momentum, charge, etc.

Checkpoint 2

What do the relative strengths of the strong and gravitational forces suggest about what dominates in the nucleus?

Exam practice answers: page 71

An electron and a positron can annihilate by either of the following mechanisms:

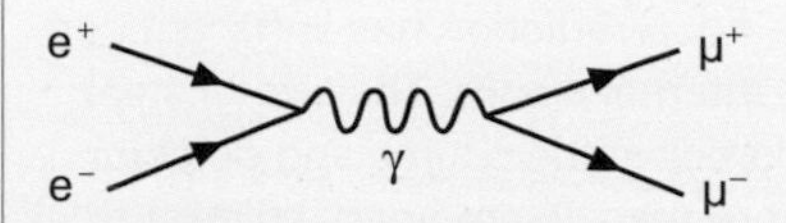

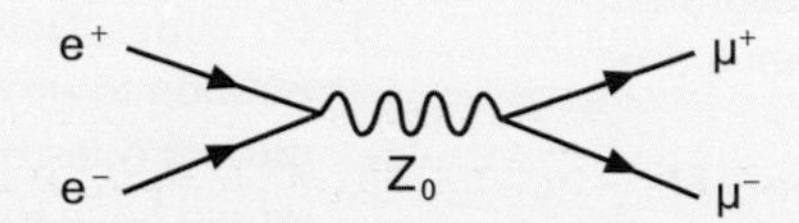

(a) Which of the fundamental interactions is represented by each figure?

(b) Draw another diagram to illustrate the exchange of a π^+ between a neutron and a proton. (5 min)

Answers

Radioactivity and the structure of the atom

The atom and its nucleus

Checkpoints

1 Because opposite charges attract.
2 They would be slowed down, but they would not be deflected (imagine firing pellets through a real pudding).
3 They would slow down and fall into the nucleus.

Exam practice

1 Key results of Geiger and Marsden's experiments and their implications:
(i) Most of the α-particles passed straight through, without loss of energy, showing that most of the atom is empty space.
(ii) Some of the α-particles are deflected and a few even rebounded, showing that mass and charge must be concentrated in the nucleus (in a tiny space).

Grade booster

Hints and tips The basic implications are easily learned. A-grade students should explain that the law of conservation of momentum requires that the things the α-particles rebounded off must have greater mass than the α-particles themselves and that electrostatic forces obey an inverse-square law; every doubling of separation quarters the size of the force. The force needed to reverse the α-particle's momentum can be calculated and from this, you can work out how close the α-particle must get to the repulsive nucleus!

Grade booster

Make your labels clear on such diagrams

Elements and isotopes

Checkpoints

1 A = number of protons + number of neutrons = an integer. Atomic mass = mass of these protons and neutrons. The mass of a proton is not quite the same as the mass of a neutron, so even in atomic mass units, atomic mass is not a whole number.
2 Cl-35 has 18 neutrons (35 – 17). Cl-37 has 20 neutrons. Average atomic mass = $35.5 = A \times 35 + B \times 37$, where A and B are the proportions of each isotope (as fractions). $A + B = 1$.
3 $35.5 = 35A + (1 - A) \times 37$; $A = 0.75$. So 75% Cl-35 and 25% Cl-37.

Exam practice

1 (a) (i) $137 - 55 = 82$ (ii) 55 (iii) 82
(b) There is an equal number of electrons and protons.
2 (a) Yes. All isotopes of any particular element have the same proton number.
(b) No. Different isotopes have different numbers of neutrons.
(c) No. Nucleon number incorporates the (unspecified) neutron number.
3 ${}^{238}_{92}U \rightarrow {}^{234}_{90}Th + {}^{4}_{2}He$

Nuclear instability

Checkpoints

1 (a) Electrostatic repulsion would tear the nucleus apart.
(b) There has to be a force to bind all nucleons. Neutrons are not repelled electrostatically, but they would tend to drift about everywhere if not held tightly in the nucleus by the strong force.

Exam practice

1 True. The nucleus is made of protons and neutrons. Protons are all positively charged and are therefore mutually repulsive. There has to be a binding force strong enough (at least at short range) to overcome this repulsion. (The new strong force must also bind the uncharged neutrons.)

Grade booster

You might argue that the discovery of the nuclear structure was *not* also the discovery of the strong force; you can still get credit if your case is good enough! (Rutherford's group discovered the nucleus before anyone knew it was made of protons and neutrons. If you treat each nucleus as a tiny blob of concentrated mass and charge, there may be no need for a new force, but this approach explains nothing!)

2 (a) (i) Kr-89: $Z = 36$, $N = 53$ plotted correctly.
(ii) Daughter after β-decay: $Z = 37$, $N = 52$.
(b) (i) Rn-222: $Z = 86$, $N = 136$ plotted correctly.
(ii) Daughter after α-decay: $Z = 84$, $N = 134$.
(c) In each case, the daughter lies closer to the line of stability; decay increases stability.

Properties of ionizing radiation

Checkpoints

1 A neutron converts into a proton and an electron. The electron cannot be held in the nucleus and so it is ejected as a beta particle. In beta$^+$ decay, a proton converts into a neutron and an anti-electron (positron).
2 The emissions lose their energy by ionising. Hence those that are most ionising lose their energy over a small distance, and have the shortest range.
3 Into the page
4 α-particles: $m = 4$ u, $q = +2e$; $q/m = 0.5e\,u^{-1}$; β-particles: $m \approx 1/2\,000$ u, $q = -e$; $q/m \approx -2\,000e\,u^{-1}$. Charge-to-mass ratio of a β-particle is about 4 000 times that of an α-particle!

Exam practice

1 (a) γ-rays. Neither α- nor β-radiation has sufficient penetration to probe for internal cracks. The size of steel girders rules out α-particles (the source and detector would have to be more than 10 cm apart, so α-particles wouldn't ever make the journey). β-particles could penetrate the air, so they could be used to detect holes (but so could light etc.). Only γ-rays have sufficient penetration to get through steel to probe for internal faults.

(b) α-particles. Only α-particles will be significantly attenuated by smoke.

(c) β-particles. α-particles would never get through the cardboard pack – whatever its contents. The intensity of a beam of γ-rays would not be sufficiently affected by the presence or absence of cereal in a small box. β-particles should be able to get through an empty cardboard box, and would be significantly attenuated by any cereal contained.

Grade booster

Most A-level students would get the right type of radiation for each job, but only the best students give full explanations of their choices. For any application, two key questions are:

(i) Is the radiation sufficiently penetrating for the job?

(ii) Is the radiation going to be measurably affected by the change it is trying to detect?

2 (a) Credit is given for any methods that work. You could exploit differences in charge, charge-to mass ratio, penetration or ionizing ability.

To test for γ-radiation, you could place a lead shield (say 1 cm thick) in front of the source. If you still detect significant radiation, then the source must be a γ-emitter. α- and β-emitters could be identified from their cloud chamber traces. α-particles cause the most ionization and the clearest tracks; α-tracks are straight and end abruptly (because α-particles have high momentum and are mono-energetic). β-particle tracks are fainter and wander from the straight and narrow. You could use an arrangement similar to that given in the drawing on page 51 to exploit differences in charge (and charge-to-mass ratio).

Essentials

(i) Methods show knowledge of distinct characteristics of α-, β- and γ-radiation.

(ii) Methods would work. Suitable detectors and clear explanations of outcomes.

(b) (i) *α-particle energies* Use cloud chamber track. The longer the track, the greater the energy.

(ii) *β-particle energies* Compare ability to penetrate a suitable absorber ('suitable' means sufficiently thick and dense to absorb some, but not all, β-particles).

More penetrating ⇒ greater energy. Greater energy ⇒ more work must be done to stop the particles ⇒ greater range.

Note cloud chamber track-lengths do not provide an easy method of comparing β-particle energies. Many types of detector can measure energy directly. Some credit can be given for suggesting their use. For full credit, details of *how* they discriminate between energies is needed.

Radioactive decay

Checkpoints

1 Background levels vary – over time as well as with location. Surface activity on the Sun and stellar explosions in outer space all affect background levels of ionizing radiation.

2 If $N = 100$, absolute error = 10 and percentage error = 10%. If $N = 10\,000$, absolute error = 100 and percentage error = 1%

3 $e^{-\lambda t} = N/N_0$, i.e. the fraction of the original activity that still remains after time $t = e^{-\lambda t}$.

4 4 days.

Exam practice

1 $A = \lambda N = 0.30 \times 400\,000 = 120\,000\ s^{-1}$ (Bq)

2 They must have different masses (different number of nuclei)

3 $\lambda = \ln(2)/T_{1/2} = 0.693/1.91 = 0.363\ y^{-1}$

$A = A_0 e^{-\lambda t}$

Problem is to find t when $A = A_0/10$

$A_0/10 = A_0 e^{-\lambda t}$

$e^{\lambda t} = 10$

$\lambda t = \ln(10)$

$t = \ln(10)/\lambda = 2.303/0.363 = 6.35$ y

Binding energy and mass defect

Checkpoints

1 Energy equivalence of 1 kg of matter $= 1 \times (3.00 \times 10^8)^2 = 9.00 \times 10^{16}$ J. $E = P \times t$, so $t = 9.00 \times 10^{16}/10^9 = 9.00 \times 10^7$ s = 25 hours (≈ 1 day)

Action point

Iron-56 at 8.79 MeV.

Exam practice

(i) ${}^2_1H + {}^2_1H \rightarrow {}^3_2He + {}^1_0n$

(ii) Mass of nucleons = $(2 \times 1.0078 + 2 \times 1.0087)$ u = 4.0330 u

Mass of nuclei before reaction = (2×2.0141) u = 4.0282 u

Total mass defect before = (4.0330 – 4.0282) u = 0.0048 u

Mass of nuclei after reaction = (3.0161 + 1.0087) u = 4.0248 u

Total mass defect after = (4.0330 – 4.0248) u = 0.0082 u

Change in mass defect = 0.0034 u

(iii) Change in mass defect in kg

$= 0.0034 \times 1.661 \times 10^{-27} = 5.65 \times 10^{-30}$ kg

$\Delta E = \Delta mc^2$

$= 5.65 \times 10^{-30} \times (3.00 \times 10^8)^2 = 5.08 \times 10^{-13}$ J

1 eV = 1.6×10^{-19} J, so 1 MeV = 1.6×10^{-13} J

1 J = $1/(1.60 \times 10^{-13})$ MeV

∴ energy released = 3.17 MeV.

Nuclear fission and fusion

Checkpoints

1 U-235 and U-238 are both uranium – same proton number, same electron structure, same chemistry!
2 Moderator and coolant.
3 Sea water.

Exam practice

1 Moderation: slowing down of neutrons to allow them to cause further fission; achieved by moderator, which consists of low mass atoms, with which the neutrons collide and lose some KE.
Control: to prevent the chain reaction from developing too rapidly; achieved by control rods that absorb neutrons.
Cooling: removal of thermal energy from the core of the reactor; achieved by circulating coolant which passes to heat exchanger.
2 Energy released = 200×10^6 MeV
$= 200 \times 10^6 \times 1.6 \times 10^{-19}$ J $= 3.2 \times 10^{-11}$ J
Energy input = 1500 MW/0.4 = 3750 MW
$= 3750 \times 10^6$ W $= 3750 \times 10^6$ Js^{-1}
Hence number of fissions $= 3750 \times 10^6$ Js^{-1}/3.2×10^{-11} J
$= 1.2 \times 10^{20}$ per second
3 In the Sun; gravitational collapse of the outer regions is converted to thermal (internal) energy that causes the temperature to rise sufficiently.

Other applications of radioactivity

Checkpoints

1 Radioactive dating works best if the time period being measured is roughly the same as the half-life of the radioisotope being used.
(i) Activity of C-14 in living matter is low (0.23 Bq g^{-1}).
(ii) Enough time must have passed for a significant change in activity to have occurred (after 200 y the activity of C-14 only falls by about 2%; shorter time scales are impossible to measure without very big samples).
(iii) If too much time has passed, the residual activity will be hard to measure (after 10 000 y, residual activity is around 30% initial activity; because initial activity is low, it is hard to get accurate C-14 datings beyond this sort of time scale).
2 No! U-238's longer half-life means the more time that passes, the greater its abundance relative to U-235's etc. will be.
3 Steam-sterilization in an autoclave; chemical sterilization (followed by rinsing).
4 Ideal source for use on a child would have a lower energy and therefore a higher attenuation coefficient.

Exam practice

$A_0 = 0.23$ Bq g^{-1}
(a) For minimum age, $A = 0.0058$; for maximum age, $A = 0.0046$ Bq g^{-1}
$A = A_0 e^{-\lambda t}$
$\lambda = \ln 2/t_{1/2} = 0.693/5\,730$ y^{-1} $= 1.21 \times 10^{-4}$ y^{-1}
Min age, $t_{min} = [\ln(0.23/0.000\,58)]/1.29 \times 10^{-4} = 28\,500$ y
Max age, $t_{max} = [\ln(0.23/0.004\,6)]/1.29 \times 10^{-4} = 30\,300$ y
(b) The main source of uncertainty is the randomness of radioactive decay.
To minimize uncertainty, you need the biggest counts possible, so
(i) count over a long period
(ii) use as large a sample as possible
(iii) take background readings over a long period, under similar conditions
(iv) use a detector with a high capture efficiency

Probing matter

Checkpoints

1 For $\lambda = 0.4$ μm, spacing must be > 0.4 μm for any constructive fringes to appear. For $\lambda = 0.7$ μm, s must be > 0.7 μm.
2 (a) and (b) The ideal wavelength must be at least as small as the particle being measured, but not too much smaller. $10^{-11} - 10^{-10}$ m for an atom, $10^{-15} - 10^{-14}$ m for a nucleus. You don't get diffraction patterns with longer wavelengths and the fringes become too closely spaced with much shorter wavelengths. (c) 10^{-11} m is not even nearly short enough for probing the nucleus, but it's ideal for measuring atom separations.
3 Momentum, $p = mv$. $m = 1.0087 \times 1.6651 \times 10^{-27}$ kg (see *elements and isotopes*, pages 42–3)
$\therefore m = 1.6796 \times 10^{-27}$ kg $\Rightarrow p = 1.6796 \times 10^{-20}$ kg m s^{-1}.
$\lambda = h/p = 6.63 \times 10^{-34}/1.68 \times 10^{-27} = 3.95 \times 10^{-7}$ m.

Exam practice

1 Electrons
Gold foil/atoms (Rutherford); nucleons
Rutherford: atom is mainly "space"; contains very small nucleus where most of mass is concentrated.
DIS: nucleons have structure/are made up of smaller particles; called quarks.

Particles – production and patterns

Checkpoints

1 The particles would collide with any gas particles in the tube and lose their energy.
2 The force (Fleming's Left Hand Rule) is always at right angles to the direction of motion, so only causes a change in direction. The particles get faster as they cross the gap between the dees.

Exam practice

(a) Charged particles experience (unbalanced) force between "dees"; hence accelerated; moving charge constitutes current; magnetic field produces force on moving charge (refer to BQv or Fleming); force is perpendicular to field and motion, hence results in circular motion.

(b) Force on moving charge provides centripetal force

$BQv = mv^2/r$

Hence $v = BQr/m$

Distance = πr

Time = distance/speed = $\pi m/BQ$

More about leptons and quarks

Checkpoints

1 uud = $+\frac{2}{3} + \frac{2}{3} - \frac{1}{3} = +1$. udd = $+\frac{2}{3} - \frac{1}{3} - \frac{1}{3} = 0$

2 An up quark has changed into a down quark.

3 $E = mc^2 = 2 \times 9.11 \times 10^{-31} \times (3 \times 10^8)^2 = 1.64 \times 10^{-13}$ J.

Exam practice

(a) Energy of decay is constant; β- particles emitted with range of energies; must be another particle (neutrino) taking away missing energy.

(b) Neutral.

(c) Anti-up, anti-up, anti-down.

(d) Positron.

(e) Lepton.

Forces/interactions and conservation laws

Checkpoints

1 *Hadrons* particles made from combinations of quarks, e.g. pion, proton.

Meson made from 2 quarks, e.g. pion.

Baryon made from 3 quarks, e.g. proton.

Lepton fundamental particle, e.g. electron.

Quark fundamental particle, e.g. up quark.

Antiparticle antimatter particle with opposite charge to its fundamental particle, e.g. positron.

Bosons particles which are exchanged when forces act, e.g. photon.

2 It is the strong force that dominates in the nucleus – the gravitational force is important only when masses are huge, as in the universe.

Exam practice

(a) Left figure: electromagnetic. Right figure: weak.

(b)

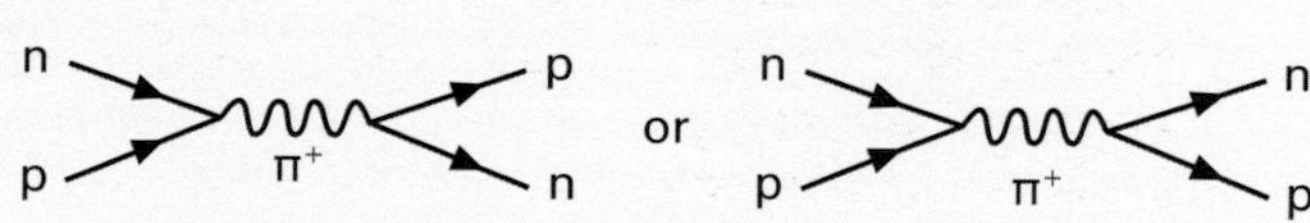

Electricity and electromagnetism

Electricity comes from the Greek word *elektron*, meaning amber. Around 600 BC, a Greek philosopher, Thales, discovered that when he rubbed amber with fur, it attracted bits of straw. The amber is said to be electrically charged. 2 300 years passed before electricity really captured the interest of the scientific world. One particularly bizarre experiment was performed by Luigi Galvani in 1786. He hung the legs of a dead frog from a railing during a thunderstorm to discover whether lightning would make them twitch. The frog's legs twitched even before the lightning arrived. This was the inspiration for Mary Shelley's horror story *Frankenstein*.

Exam themes

- *Understanding* For instance, why does a capacitor initially discharge quickly?
- *Modelling* Use of formulae such as $C = Q/V$ to model physical phenomena.
- *Mathematical competence* Such as using equations to solve circuit problems.
- *Recall* Remembering, for example, the distinction that is made between EMF and PD.
- *Applications* Such as why do car lights dim if a driver tries to start a car with the lights on?
- *Graphical analysis* For example, interpreting a graph of terminal PD against current to find E and r.
- *Links* Building an overview of physics, e.g. the recurring appearance of exponential changes.
- *Practical work* For instance, obtaining IV characteristics.

Topic checklist

	Edexcel		AQA/A		AQA/B		OCR/A		OCR/B		WJEC		CCEA	
	AS	A2	AS	A2	AS	A2	AS	A2	AS	A2	AS	A2	AS	A2
Current as a flow of charge	○		○		○		○		○		○		○	
Current, PD and resistance	○		○		○		○		○		○		○	
Resistors and resistivity	○		○		○		○		○		○		○	
Electrical energy and power	○		○		○		○		○		○		○	
Kirchhoff's laws	○		○		○		○				○		○	
Potential dividers and their uses	○		○		○		○		○		○			
EMF and internal resistance	○		○		○		○		○		○			
Capacitors		●		●		●		●	○			●	○	
Electromagnetism		●		●		●		●		●		●		●
Electromagnetic induction		●		●	○	●		●		●		●	○	
Alternating currents		●	○	●				●		●		●		●

Current as a flow of charge

Links

See page 45 to remind yourself about atomic structure.

Checkpoint 1

Electricity can be subdivided into two smaller topics – static and current electricity. Explain the difference(s) between them.

The jargon

A *cell* is a component that converts chemical energy into electrical energy. More than one cell is referred to as a battery. Remember a cell means one, like one room in a prison. A battery means more than one, like a battery of soldiers.

Action point

Can you show that 1 ampere is a flow of 6.25×10^{18} electrons every second?

Checkpoint 2

Calculate how much charge passes a point if 12 A flows for 3 hours.

Why are birds sitting on live electrical wires not electrocuted? It seems that to get an electric shock, electrons have to flow through the victim and go into the ground. There is an old saying that 'volts jolt but mills kill'. Just 26 milliamps (0.026 A) flowing through your heart may well kill you, but not many volunteers have tried to find out!

Electric current

Electrostatics experiments (like the ones that Thales did) were responsible for showing that there were two types of charge. These are now called positive and negative, and like charges repel while unlike charges attract. It is now understood that insulators which become positive when rubbed, have electrons *removed* by friction. Negatively charged insulators *gain* extra electrons from the material they are rubbed with.

The unit of charge

Charge is measured in coulombs (C). Each electron carries just -1.6×10^{-19} C so it takes nearly 10^{19} electrons to carry 1 C of charge!

Electric current

Metals are not insulators but conductors of electricity because they contain some electrons that are free to move.

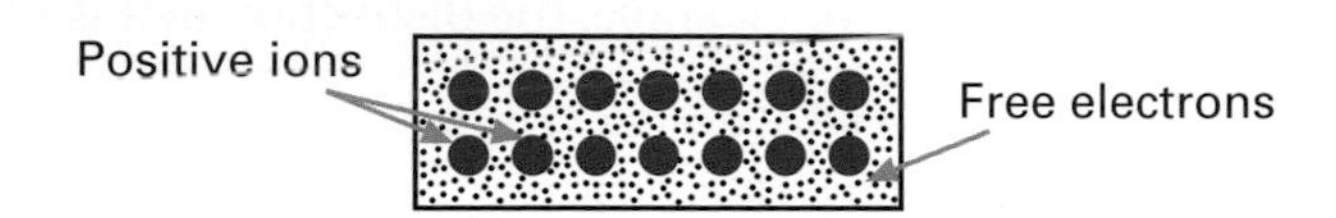

This diagram represents the internal structure of a metal. A regular lattice (framework) of atoms that have lost an electron (called positive ions) is surrounded by a sea of free electrons.

When you switch on a light bulb, it lights up almost instantly. This does not mean that electrons carry energy from the power station at the speed of light. The electrons were already in the metal filament of the bulb and start drifting along when the circuit is complete!

An **electric current** is a net (overall) movement of charged particles in a certain direction. Current is defined as the rate of flow of charge so

$I = \Delta Q/\Delta t$ where ΔQ is the charge that passes a point in the circuit in Δt seconds. From this, the coulomb is defined as the amount of charge transferred by a current of 1 A in 1 s.

Charge carried equals the area under an *I*–*t* (current against time) graph. The rule applies whether current is steady or varying.

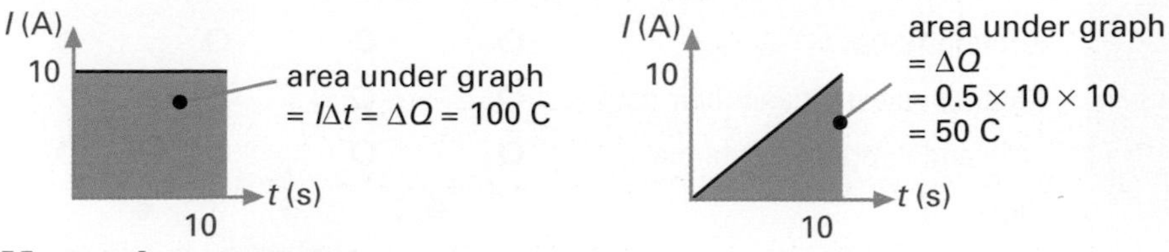

Measuring current

Points to remember when using ammeters to measure current include:

- ➜ connect your ammeter in series
- ➜ ammeters can be delicate – start on the least sensitive setting, then

move on to a more sensitive scale when you know that the current will not exceed the upper limit of that scale

- an ideal ammeter would have zero resistance

Checkpoint 3

Why would an ideal ammeter have zero resistance?

Conventional current and electron flow

Think of a number! You almost certainly did not think of a minus number. When scientists were trying to decide what would move when current flowed, they assumed that it would be a positive quantity. (The electron had not been discovered then!) A positive charge would be repelled from the positive terminal of a battery and attracted to the negative. This is the direction of *conventional current*. In metals, the charged particles that can be persuaded to move by completing the circuit with an energy source (e.g. a battery) are electrons, which move in the opposite direction to the conventional current.

In electrolysis, the chemical solution contains both positive and negative ions. The positive ions flow in the same direction as the conventional current and the negative direction in the opposite direction.

Grade booster

All arrows representing current should point in the direction of conventional current. This rule extends to include all circuit symbols, e.g. in transistors.

+ charges would be repelled from the + terminal of the cell and attracted towards the – terminal

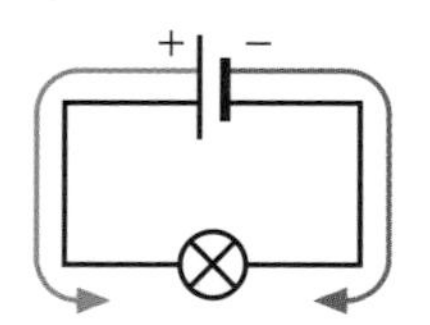

– charges are repelled from the – terminal of the cell and attracted to the + terminal

This is the direction of conventional current.

This is the actual direction of electron flow.

Drift velocity

The current I through a metal, e.g. a wire, depends upon:

- the number of free electrons per unit volume n (metals are good conductors because they have many free electrons)
- the cross-sectional area of the wire A (thick wires allow the electrons to move more easily than thin wires)
- the charge carried by each free electron e
- the drift speed with which the electrons move v (typically about 1 mm s^{-1})

Current is therefore given by:

$$I = nAev$$

But current can flow through materials other than metals; e.g. ions can flow through liquids in electrolysis experiments. We need a more general equation:

$$I = nAqv$$

Where n is the number of charge carriers per unit volume and q is the charge on the charge carriers.

The jargon

Electrons normally move around constantly in a haphazard way with no overall movement in any one direction. But they can be persuaded to drift in a certain direction by an electric cell. A gradual drift towards the positive plate is superimposed on their normal haphazard motion. This is their *drift speed* (or *drift velocity* if we mention direction too).

Grade booster

Many students use clumsy expressions when describing physical processes. Current flows *through* components. It can help you visualize what is happening if you learn this kind of phrase.

Exam practice answers: page 96

(a) If 26 mA flows through your heart, how long would it take for 4.68 C to pass?

(b) Describe the difference between semiconductors and conductors in terms of the number of charge carriers they contain.

(c) Calculate the drift speed of free electrons in a copper wire of diameter 1 mm, carrying a current of 7 A. Take n to be 1.0×10^{29} and the charge on each electron to be 1.6×10^{-19} C. (15 min)

Current, p.d. and resistance

"Truth is ever to be found in the simplicity, and not in the multiplicity and confusion of things."

Sir Isaac Newton

Current is a flow of charged particles. Electrons pass along the electrical circuits in our houses, entering appliances via the live wire and leaving along the neutral wire. So what do we buy when we pay our electricity bills? We are not paying for electrons as they enter and leave. We pay for the energy they deliver.

Energy transfers

Energy is stored as chemical energy in cells. When it is transferred to electrons it is known as electrical energy, or even kinetic energy as the electrons are moving. In the circuit below, electrons 'pick up' energy at the cell and deliver it to the bulbs. If the electrons were still carrying any energy when they returned to the cell, the bulbs in this circuit would get brighter and brighter. As this does not happen, we know that all the energy picked up at the cell is delivered to the circuit.

For **series circuits** $V = V_1 + V_2 + \ldots$

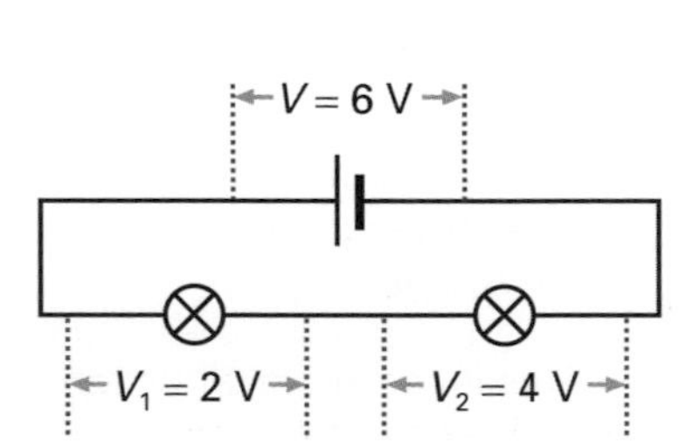

Imagine if the circuit above could still work if it was 'opened up' as shown on the right. A graph showing energy changes would look like this.

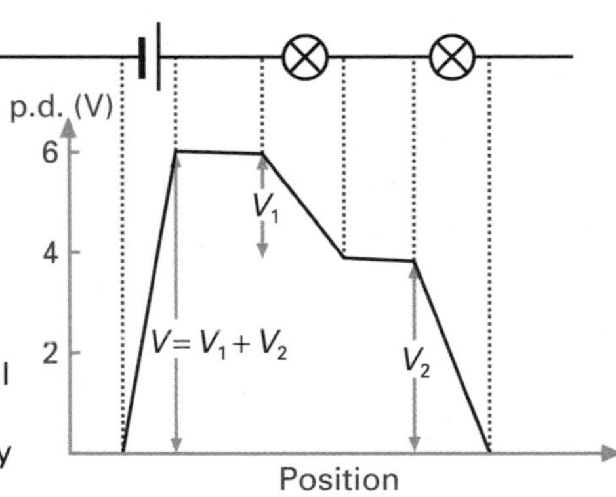

Electrons give up all of their energy before they return to the cell. There are many electrons distributed all the way around the circuit, all carrying the same (negative) charge, all repelling one another. Electrons leaving the second bulb may have given up all their energy but there are many more coming behind to push them forward. Remember that:

- the **potential difference** (p.d.) between two points is the work done (or energy transferred) in moving +1 C of charge from one point to another
- p.d. = $\frac{\text{energy (work done)}}{\text{charge}}$ or $V = \frac{W}{Q}$
- 1 volt = 1 joule per coulomb ($1\ \text{V} = 1\ \text{J}\,\text{C}^{-1}$)

Measuring potential difference

As potential difference is used to compare two points in a circuit, say before and after a resistor, voltmeters are connected in parallel with the component. To avoid changing what we set out to measure, the voltmeter should have a very high resistance to avoid diverting electrons away from their original route through the resistor.

Checkpoint 1

State and explain three assumptions that have been made in this section.

Grade booster

It is often helpful to visualize electrons picking up and dropping off energy. However, remember that they are not living things with minds of their own!

Check the net

For brief introductory notes on electricity go to http://www.abdn.ac.uk/physics/bio/fi20www/index.htm

Examiner's secrets

Don't annoy the examiner! Voltage does *not* go through things. Voltage, or p.d., is measured *across* components.

Checkpoint 2

What would be the resistance of an ideal voltmeter?

Resistance

As free electrons move through a wire, they collide with positive ions and with one another, slowing down as they do so. Some of their (kinetic) energy is transferred to the ions making the ions vibrate more so that the temperature of the wire increases. This is **resistance**.

→ $\text{resistance} = \dfrac{\text{potential difference}}{\text{current}}$ or $R = \dfrac{V}{I}$

Resistance is measured in ohms (Ω or omega for short). 1 Ω is defined as the resistance of a conductor in which a current of 1 A is produced by a p.d. of 1 V across its terminals.

Series and parallel circuits

Series circuits have only one path for the electrons to follow.

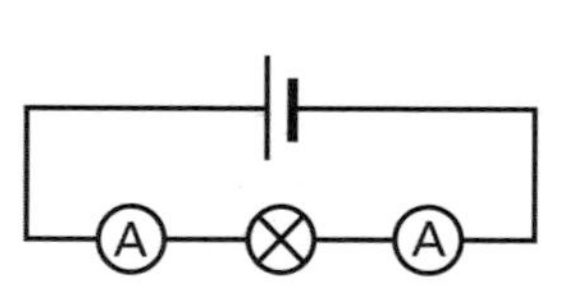

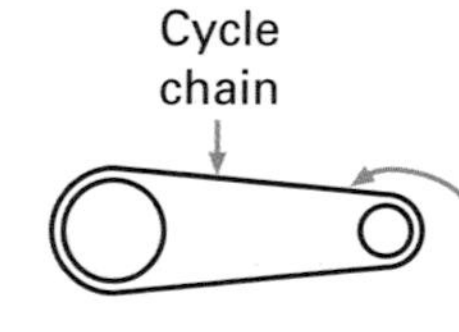

Count how many chain links go past every second

- → The current is the same at every point in a series circuit.
- → 1 A is defined as 1 C of charge flowing past a point every second.
- → As each electron carries a charge of -1.6×10^{-19} C, when 1 A flows, 6.25×10^{18} electrons go past every second!
- → The links in a bicycle chain are like electrons. There is only one path for them to take, and as they are all kept a certain distance apart from one another, they all travel around at the same speed.
- → If you change gear, it is as if more resistance has been included in an electrical circuit. All the chain links (electrons) slow down. All the electrons in a circuit are affected by the amount of resistance present, not just the electrons before or after the resistors.

Parallel or branching circuits offer electrons more than one route on their journey from the negative to the positive terminal of the cell.

V V_1 V_2 V_3 $V = V_1 = V_2 = V_3$

As with series circuits, the bulbs in a parallel circuit do not get brighter as time goes on. So each time electrons go to pick up another packet of energy from the cell, they must have already given up all of their previous load. As each electron has only passed through one arm of the circuit (in this case, just one of the bulbs), it must have given up all its energy in that arm (to that bulb).

For **parallel circuits** $V = V_1 = V_2 = V_3 = \ldots$

Exam practice

answers: page 96

(a) Describe one model used to simplify an aspect of electrical theory.

(b) An ideal way to measure the number of people using the London Underground would involve not slowing the travellers down. Why is this? Describe, with a reason, the characteristics of an ideal ammeter. (20 min)

Action point

If you get the opportunity, use a multimeter to measure the electrical resistance in your body. Hold one probe between the thumb and forefinger of one hand and the other probe between the thumb and forefinger of your other hand.

Action point

Now use the multimeter as a voltmeter. Hold one probe against the temple of a friend's head; hold the other probe against his/her other temple. Now ask the volunteer a question and watch what happens to the voltmeter's display!

Checkpoint 3

What group of materials have very high electrical resistances?

The jargon

The phrase *constancy of current* refers to the fact that current is the same all the way around a series circuit. The electrons maintain a constant speed.

Checkpoint 4

Why are parallel circuits used in household wiring?

Grade booster

Marks are given for good use of English and spelling. Note the spellings of amps, amperes and ammeter.

Resistors and resistivity

Georg Simon Ohm found the relationship between p.d. and current in 1826. His father, Johann, had received no formal education yet he gave Georg an excellent introduction to science and mathematics that contrasted with the school-based teaching of that time. Georg's brilliance was recognized belatedly and the unit of resistance now bears his name.

Ohm's law

The resistance of a wire tells us how hard it is for electrons to flow along it. An ideal insulator would have infinite resistance whereas a superconductor has zero resistance. Ohm changed the p.d. across wires and then measured the current that flowed through them. He found that, provided the temperature remained constant,

$$\begin{pmatrix}\text{p.d. across the}\\\text{conductor}\end{pmatrix} = \begin{pmatrix}\text{current through}\\\text{the conductor}\end{pmatrix} \times \begin{pmatrix}\text{a constant}\end{pmatrix}$$

In other words, current I is proportional to p.d. V. This is **Ohm's law**. The constant in the equation above is resistance R, so:

$$V = IR$$

Links

See page 77 to remind yourself how resistance is defined by rearranging $V = IR$ into $R = V/I$.

Grade booster

The resistance of any component can be found as the ratio of V and I. The equation $V = IR$ is not in itself a statement of Ohm's law. To obey Ohm's law, the resistance must be constant.

I–V graphs

Fingerprints are used to identify criminals. I–V graphs are used to identify electrical components. Ohmic conductors are identified by I–V graphs that have a straight line going through the origin.

The jargon

Ohmic conductors are those that obey Ohm's law. A conductor that does not obey this law is called a *non-ohmic conductor*.

Watch out!

Some books plot graphs of V against I, rather than I against V as shown here.

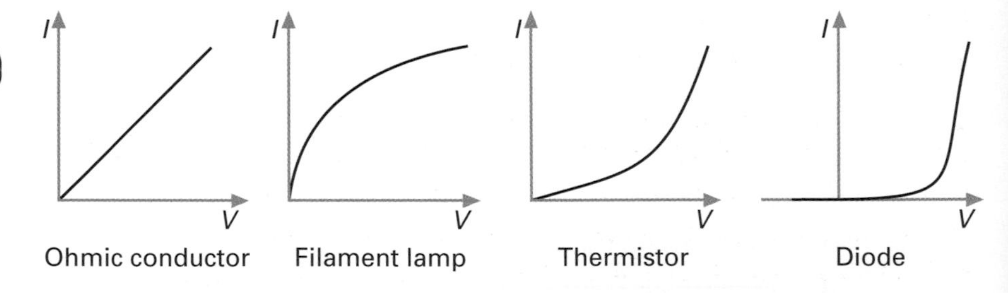

Resistivity

Ohm also investigated the factors that affect resistance. It was found that length l, cross-sectional area A and the material the conductor was made of affected its resistance. These results are summarized by:

$R = \rho l/A$ ρ is a property of the material: **resistivity**

- Properties ending in *-ance*, e.g. resist*ance*, vary between samples under investigation. A long, thin sample of Nichrome wire would have a larger resistance than a short, thick sample. Therefore you cannot look up the resistance of Nichrome in a book.
- Words that end in *-ity*, e.g. resistiv*ity*, refer to a material rather than an individual sample. Silver has a lower resistivity than iron. If we have two identically sized samples held at the same temperature, the silver sample would provide less resistance than the piece of iron.

Checkpoint 1

Rearrange $R = \rho l/A$ to make r the subject of the equation, i.e. ρ = ? Then use this new equation to find the units for ρ.

Checkpoint 2

Conductivity $\sigma = 1/\rho$. What are the units of conductivity? Explain whether you would expect to find a table of conductivities quoted in a table of physical constants.

The resistivity of materials varies with temperature. There are two effects:

- Increased temperature increases lattice vibrations making it harder for charges to move
- Increased temperature releases more charges from their bonds so they are 'free' to move.

With metals the first effect is larger, so their resistance increases with temperature, whereas with insulators and semi-conductors, the second effect is larger, so their resistance decreases with temperature.

Resistors in series and parallel

The total resistance R_T of three resistors R_1, R_2 and R_3 **in series** is:

$$R_T = R_1 + R_2 + R_3$$

The total resistance R_T if the resistors are **in parallel** is given by:

$$\frac{1}{R_T} = \frac{1}{R_1} + \frac{1}{R_2} + \frac{1}{R_3}$$

Using resistors

Resistors are used to reduce current. For example fixed resistors are connected in series with light emitting diodes (LEDs), used as standby indicators, to protect them from high currents. (LEDs are diodes that emit light when they conduct. They are smaller, more robust and take less current than filament bulbs.) Variable resistors are used in sound equipment to lower the volume by changing the voltage. Thermistors respond to changes in temperature so can be used to control temperature. Light dependent resistors (LDRs) are made from semiconductor materials. As more light falls on an LDR, more charges become 'free' so its resistance decreases. LDRs are used as light sensors.

Superconductors

- The resistance of most metals falls with temperature.
- If cooled sufficiently, some materials have no resistance at all.
- Some materials, based on oxides of bismuth, become superconductors (i.e. they have zero resistance) at –196 °C.
- This is their transition temperature.
- Liquid nitrogen has a boiling point of –196°C so can be used to keep superconductors superconducting.
- This technology is used in superconducting electromagnets.
- Their unusually stable magnetic fields are used in MRI scanners.

Exam practice answers: page 96

(a) Sketch *I–V* characteristic graphs of: a metallic conductor at constant temperature, the filament of a light bulb, a negative temperature coefficient thermistor and a diode. Explain the shape of each graph.

(b) Describe an experiment to find the resistivity of a metallic conductor.

(c) State three uses of superconductors. (25 min)

Checkpoint 3

The filament of the lamp is metal while the thermistor is a semiconductor. Explain how the graphs show how the resistance of the materials behaves with temperature.

The jargon

The thermistor described here is a *negative temperature coefficient* (NTC) thermistor – its resistance *decreases* with temperature. Positive temperature coefficient (PTC) thermistors can be designed which behave in the opposite way.

Checkpoint 4

Why is the total resistance of three 2 Ω resistors connected in series more than if they were connected in parallel?

Watch out!

$1/R$ is called conductance G and has the unit Ω^{-1} or *siemens* (S). The total conductance for a parallel combination is $G_1 + G_2 + G_3$.

Action point

If you have access to an LDR and a multimeter, perhaps at school, try this. Switch the multimeter to its ohmmeter setting and connect it to the LDR. Now cover the LDR with your thumb and watch its resistance increase!

Links

See *medical and health physics 2*, page 167 to find out more about Magnetic Resonance Imaging.

Examiner's secrets

When asked to sketch a graph, remember that scales are not required but labelled axes are!

Electrical energy and power

Electricity is a clean and convenient way of transferring energy around the country. At home we use electrical energy for doing a lot of our work, and for lighting, heating, and providing entertainment. If you have ever experienced a power cut you will know how much we have come to depend on it!

The jargon

Energy is a promise of work to be done in the future. It is the stored ability to do work. *Energy* is measured in joules (J).

Energy and power

Power tells us how quickly energy is changed from one form to another.

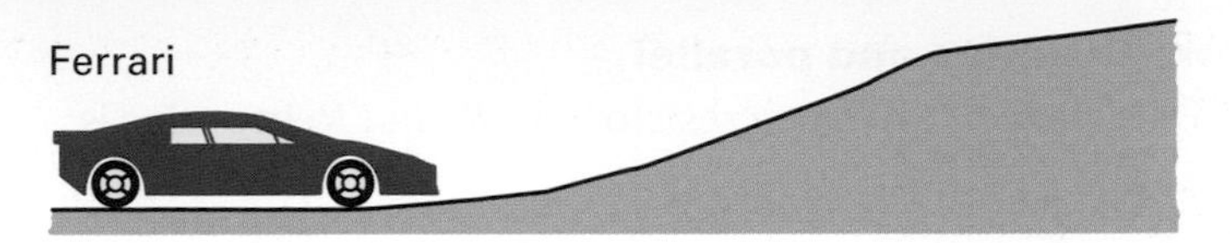

Links

To compare this work with the use of the terms energy and power in mechanics, see pages 26–7.

This Ferrari is very powerful – it can change chemical energy in its petrol into kinetic energy *very* quickly. So it goes uphill *very* quickly!

Energy and power in electrical circuits

The energy transferred to a component, e.g. a bulb, depends on:

- the p.d. across it (how much energy is dropped off at the component by the electrons)
- the current through it (how quickly the electrons are delivering their packets of energy)

Checkpoint 1

If all other factors were equal, why would you choose a 2 kW hairdryer rather than a 1500 W model?

So,

$$\text{power} = \text{potential difference} \times \text{current}$$
$$P = VI$$
$$\text{(watts)} = \text{(volts)} \times \text{(amps)}$$

Watch out!

In all these equations, *V* refers to the p.d. across the resistor and *I* refers to the current flowing through the resistor.

Power and energy equations

By using $V = IR$ and substituting for V, then I, this equation becomes $P = I^2R$ and $P = V^2/R$

There are also three forms of the equation for electrical energy. Power is how quickly energy can be converted, so:

$$P = E/t$$
$$\therefore E = Pt \qquad \text{where } t \text{ is time in seconds}$$

Using our expressions for power this becomes:

$$E = VIt = I^2Rt = V^2t/R$$

Grade booster

Remember to express power in W, p.d. in V, current in A, resistance in Ω, and time in s before using any of these expressions. Fewer mistakes are made this way.

Checkpoint 2

What is the power produced by a power station that feeds a current of 100 A into the National Grid at a potential of 440 kV? The resistance per unit length of a 20 km section of the power lines is 0.2 Ω km^{-1}. Calculate the power loss in this section.

Fuses

The word fuse means *to melt*. We have already seen that when a current flows through a wire, the temperature of the wire increases. If the temperature of the wire exceeds the wire's melting point then the wire will obviously melt. Fuses are short lengths of wire, often copper covered with tin, that melt when the current flowing through them exceeds a predetermined maximum. Fuses are used to protect circuits from dangerously high currents. They are deliberate weak links in a circuit, included for safety reasons. Fuses are labelled with the maximum current they can carry without melting. Only certain values are commercially available; e.g. 3 A, 5 A, 13 A.

The jargon

A *fuse rating* is the maximum current that can flow through the fuse without causing it to melt.

Checkpoint 3

A hairdryer, labelled 1 000 W, is connected to the 240 V mains. What fuse should be used to protect it?

Paying for electricity

We have already made the point that we pay for the energy delivered by electrons rather than the electrons themselves.

If you check an electricity bill, the number of joules (J) of energy that you have used will not be quoted. 1 J is a very small amount of energy. Just lifting an apple from the floor up to table height involves a transfer of about 5 J! A kilowatt-hour (kW h) is a much bigger unit of energy (3.6 MJ) and so it is more convenient.

energy in joules = power in watts × time in seconds
energy in kWh = power in kW × time in hours

Study this example.

A 3 kW heater is left on for 4 hours. Each kilowatt-hour costs 10 p. What was the cost of the energy used?

energy cost = (number of kW h) × (cost of 1 kW h)
= (number of kW) × (number of hours) × (10 p)
= 3 kW × 4 hours × 10 p
= 120 p

Checkpoint 4

What are the electrical energy costs of using a 5 kW oven for 2.5 hours if 1 kW h costs 10 p?

Exam practice answers: pages 96–97

(a) Calculate the energy costs of using a 100 W lamp for 1 day. (1 kW h = 10 p)

(b) How much heat is produced by a 20 Ω resistor carrying 5 A for 60 s?

(c) Calculate the heat produced in 10 min by a pair of 15 Ω resistors connected in parallel with a p.d. of 3 V across the combination. (15 min)

Kirchhoff's laws

Kirchhoff's laws for current and voltage built on the work of Georg Ohm. The two fundamental concepts explained in this spread allow voltage and current to be calculated at any point in a circuit. Kirchhoff's work was not confined to electricity. For example, his work on black-body radiation laid the foundations for the quantum theory.

The jargon

Current is a vector quantity. *Algebraic* sum means that you take the direction and therefore *sign* of the current into account.

Kirchhoff's first law

This law is based on the fact that current (or charge) cannot build up in a wire. A formal statement of this law is that:

- The (algebraic) sum of the currents into a point equals the sum of the currents out of that point.

This law can be expressed in a mathematical form as $\Sigma I = 0$.

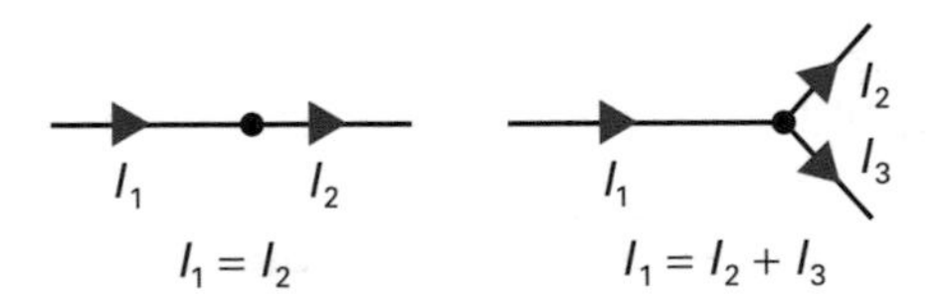

Checkpoint 1

Calculate the unknown current in the diagram below.

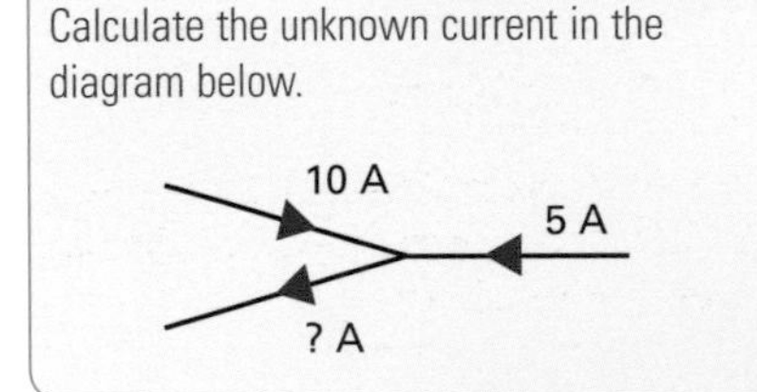

Kirchhoff's first law is a consequence of conservation of charge. For example, when electrons go into a point in a circuit they do not disappear, rather they come out of the other side. So the current that enters a point equals the current that leaves that point.

Don't forget

Conventional current flows from positive to negative.

Kirchhoff's second law

This law is based on the fact that all the energy that electrons pick up in the cell or battery is dropped off as they travel around the circuit. This means that:

- The sum of the EMFs in any closed loop in a circuit is equal to the sum of the p.d.s around that loop.

Mathematically this can be written as $\Sigma E = \Sigma IR$; i.e. in a closed loop, the sum of the EMFs equals the sum of the IR products.

The jargon

An EMF (in volts) is the energy *gained* by 1 C of charge as it passes through an energy source. A p.d. is taken to be the energy *lost* by 1 C of charge as it moves through a resistor. ($1 \text{ V} = 1 \text{ J C}^{-1}$)

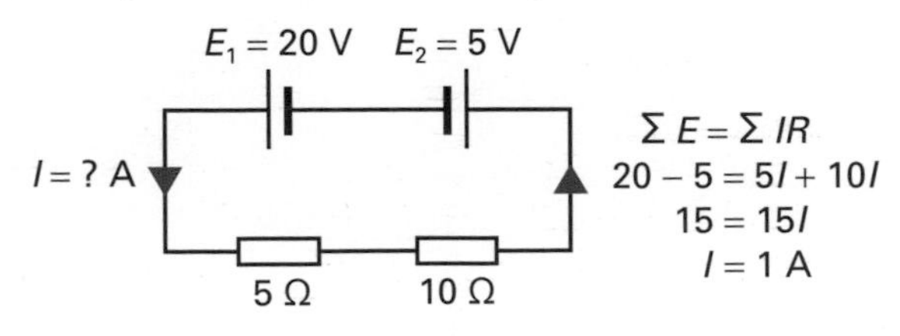

Checkpoint 2

In the example shown on the right, why do we assume that the direction of the unknown current is anticlockwise?

Kirchhoff's second law is a consequence of conservation of energy. Remember these important points about this law:

- energy gained by electrons moving through the cells equals the energy lost by the electrons as they move through the resistors
- in all the calculations involving Kirchhoff's second law, it is assumed that the amount of energy lost in the wires is so small that it can be ignored

Using Kirchhoff's laws

Work your way through this example to find I_1, I_2 and I_3.

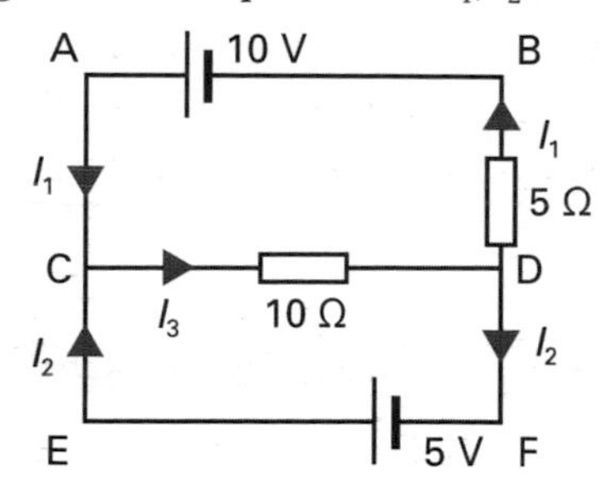

→ Indicate the direction of the currents with arrows (see above). (Make a best guess regarding the directions! You will know if you have made a mistake, negative numbers will appear as the answers.)

→ Apply Kirchhoff's first law:

at C $\quad I_1 + I_2 = I_3$ [1]

→ Select a closed loop and apply Kirchhoff's second law:

for ABCD $\quad 10 = 10I_3 + 5I_1$ [2]

→ Apply Kirchhoff's second law to another closed loop:

For CDEF $\quad 5 = 10I_3$

$I_3 = 5/10 = 0.5$ A

→ Substitute this value for I_3 into equation 2:

$10 = 5 + 5I_1$

$I_1 = 1$ A

→ Substitute these values for I_1 and I_3 into equation 1:

$1 + I_2 = 0.5$

$I_2 = -0.5$ A

→ This means that I_2 flows in the opposite direction to that shown in the diagram and has a magnitude of 0.5 A.

Exam practice

answers: page 97

(a) Use Kirchhoff's second law to find the value of R in the circuit below.

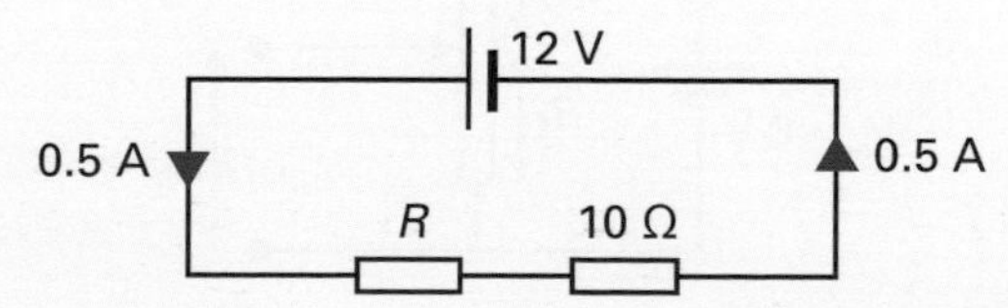

(b) Use Kirchhoff's laws to find the value of the unknown quantities in the circuit below.

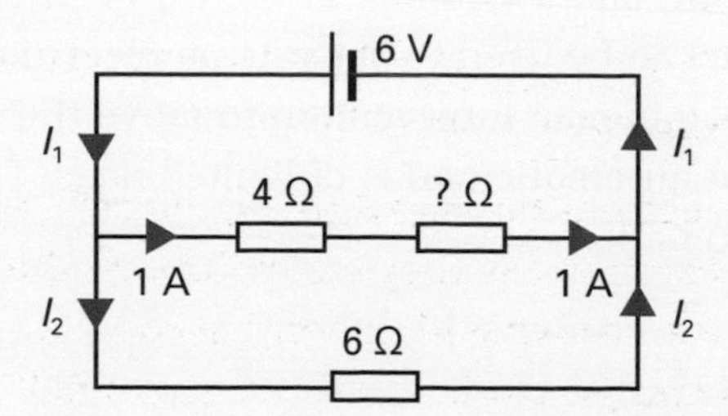

(10 min)

Examiner's secrets

Check your syllabus carefully. Not all expect you to be able to use simultaneous equations, as shown here.

The jargon

Simultaneous equations are a set of equations that are all satisfied by the same set of variables.

Grade booster

Always set out your work in a neat, logical order (see the example on the right).

Checkpoint 3

Kirchhoff's second law, for closed loops, is rather like someone going on a circular walk. Explain this analogy.

Examiner's secrets

It is unlikely that you will need to state Kirchhoff's laws, but you will certainly need to be able to use them.

Potential dividers and their uses

Potential dividers use resistors to divide a battery's voltage so that a portion of it can be used. Potential dividers are often used in automatic electronic circuits. These circuits can be used to control the temperature in a fish tank, operate lights that come on when it gets dark, maintain the temperature in a premature baby's incubator, and so on.

The jargon

It is better to use the term potential difference rather than *voltage*. But electronic engineers often use it for convenience and so we will use it in this spread.

Principle of potential dividers

This circuit shows how a small voltage can be produced from a larger one. The equation for the smaller (output) voltage V_{out} is important.

Checkpoint 1

Calculate the output voltage from this circuit if V_{in} = 12 V, R_1 = 200 Ω and R_2 = 100 Ω.

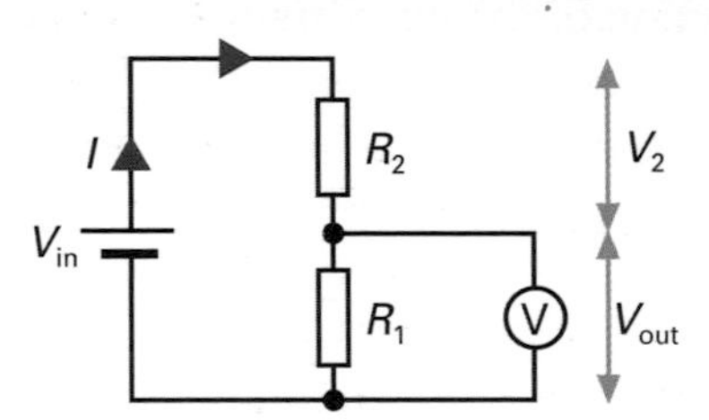

Links

For more on *Kirchhoff's laws* for current and voltage see pages 82–3.

The voltmeter has such a high resistance that very little current goes through it. Therefore, we can consider this circuit as a simple series circuit comprising a cell in series with two resistors R_1 and R_2.

$$\begin{aligned} V_{in} &= V_{out} + V_2 \\ &= IR_1 + IR_2 \qquad \text{(Kirchhoff's second law)} \\ &= I(R_1 + R_2) \end{aligned}$$

As $V_{out} = IR_1$:

$$\frac{V_{out}}{V_{in}} = \frac{IR_1}{I(R_1 + R_2)} = \frac{R_1}{R_1 + R_2}$$

$$V_{out} = \frac{V_{in}R_1}{R_1 + R_2}$$

Action point

Can you show $\frac{V_1}{V_2} = \frac{R_1}{R_2}$? This is a really useful relationship when solving problems involving potential dividers. Learn it!

A more useful circuit

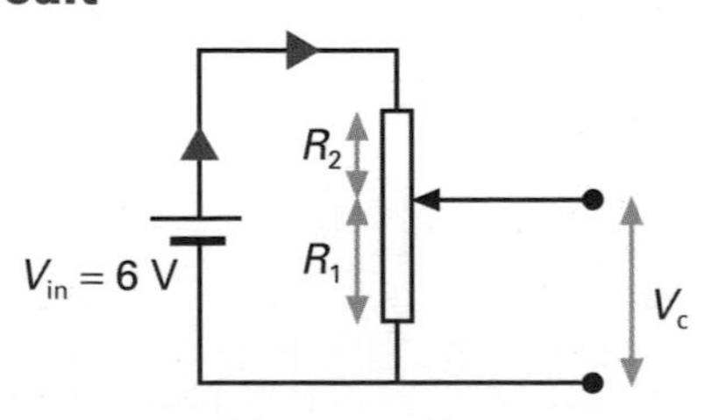

The jargon

Transistors are semiconductor devices that can switch (or amplify) electronic signals. Most are found in integrated circuits. Flash memory cards, used in digital cameras for example, can contain millions of transistors.

- By moving the sliding contact, any value for V_{out} between 0 V (slider at the bottom) and 6 V (slider at the top) can be obtained.
- The output voltage can be used to activate an electronic switch.
- This circuit relies on human intervention to move the sliding contact, so it is not automatic and is of limited use. (The next circuit is truly automatic.)

Real uses of potential dividers

The introduction to this spread mentioned some applications of potential-divider circuits. In each example, temperature or light intensity has to be monitored automatically. Thermistors and light-dependent resistors are the sensors used to carry out this job.

Circuits using thermistors

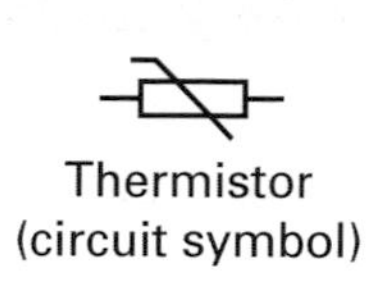

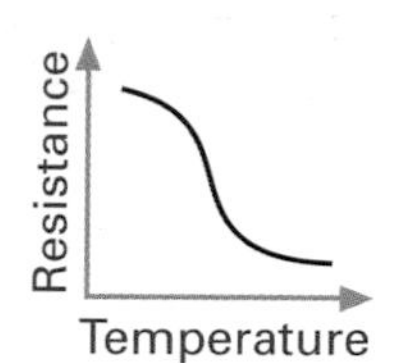

The resistance of thermistors varies with temperature. They are made from semiconductor materials. When hot, more free electrons are released inside a thermistor so its resistance falls. Therefore thermistors automatically sense temperature changes. If a thermistor is used, a potential divider produces a p.d. that depends on temperature.

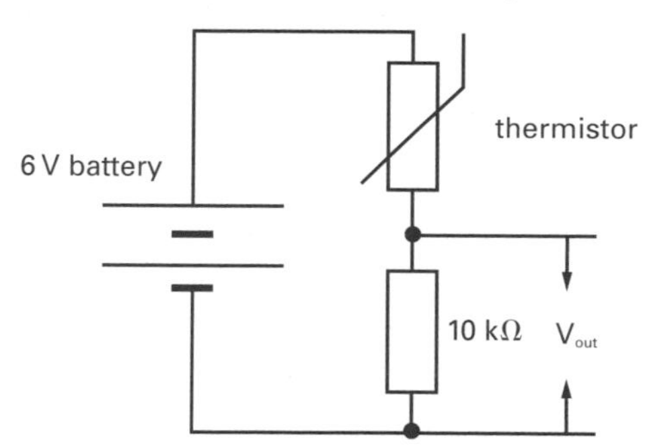

This circuit shows an automatic fire alarm. When hot, the resistance of the thermistor falls so that the p.d. across it falls and the p.d. V_{out} across the 10 kΩ resistor increases. When it is high enough it can operate an electronic switch that turns on the alarm.

Circuits involving light-dependent resistors (LDRs)

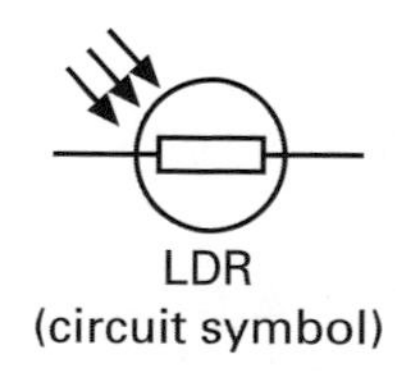

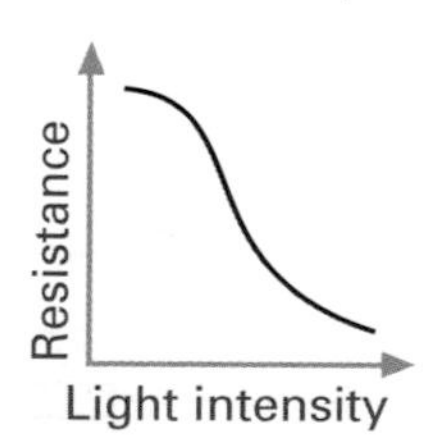

The resistance of an LDR depends on light intensity. More light energy falling on an LDR can release more electrons so that the resistance of this semiconductor-based component falls. If an LDR is used, a potential divider produces a p.d. that depends on light intensity.

Checkpoint 2

What other word does *thermistor* remind you of? What is the connection between these two words?

Checkpoint 3

What are the three essential components of an automatic decision-making circuit?

Checkpoint 4

State two advantages of this circuit.

Action point

Make lists of all the uses of potential-divider circuits mentioned in this spread that use:
(a) thermistors
(b) LDRs.
Do some research to add to these lists.

Grade booster

If you get the first part of a question wrong, you may still get marks in the second part. Examiners mark numerical questions using an *error carried forward*. So don't give up if you struggle with the early calculations in questions.

Exam practice answers: page 97

1 Draw, and explain, how a potential-divider circuit could be used to automatically control a light so that it turned on in dark conditions. (15 min)

2 Two 10 kΩ resistors are connected as a potential divider with a supply p.d. of 10 mV. What is the current flowing and the p.d. across each resistor? (5 min)

EMF and internal resistance

> *"Prediction is very difficult, especially about the future."*
>
> Niels Bohr

'If it moves it's biology, if it smells it's chemistry and if it doesn't work it's physics!' Why is it that if you select 6 V on a power pack and then check it with a voltmeter, there always seems to be a discrepancy? The power pack may not be very precise. But you will also discover that the p.d. across it depends on the circuit it is connected to.

Electromotive force (EMF) *E*

An EMF (in volts) is the electrical energy *gained* by 1 C of charge as it passes through an energy source. A p.d. (in volts) measures the electrical energy *lost* by 1 C of charges as it moves through a resistor. (In a resistor the electrical energy is transformed into heat or thermal energy.) Usually it is assumed that the connecting wires have no resistance so no energy is transferred to them.

Watch out!

EMF is a misnomer! This phrase is very misleading. It suggests that the cell pushes along the electrons. Instead, think of energy changes. Chemical energy is converted into electrical energy (i.e. the KE of the electrons).

Internal resistance r

You may have noticed that the bigger the current drawn from a cell, the lower the voltage across its terminals. This is because the cell has some electrical resistance in between its terminals due to the materials from which it is made. As electrons go through the cell, energy is transferred to them from chemicals in the cell. The chemicals also provide some resistance to the movement of the electrons. This is called **internal resistance**. All power supplies – photovoltaic or solar cells, dynamos, power packs – have internal resistance.

Internal resistance cannot be separated from the power supply but is normally shown in circuit diagrams as a separate resistor, r, in series with the *EMF*, E. Sometimes a dotted line or a circle is drawn around the symbols for E and r. From now on we will replace our old cell symbol with the more accurate one shown on the right.

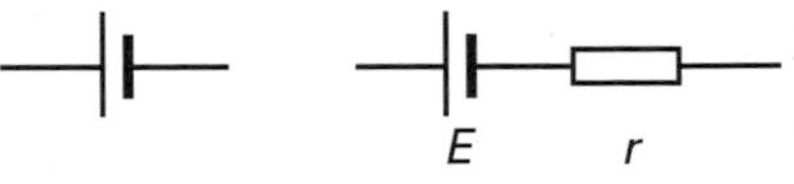

Connecting a voltmeter across a cell tells us about the energy gain per coulomb provided by the EMF, E minus the energy loss per coulomb due to the internal resistance, r.

Action point

Study this spread then read pages 80–1. Now produce a concept map for both spreads. Put the word *energy* in a box in the centre of a piece of paper. Write down as many important, relevant words as you can. Link up the words that are connected in some way and then jot down their connection on the line you have drawn to join them.

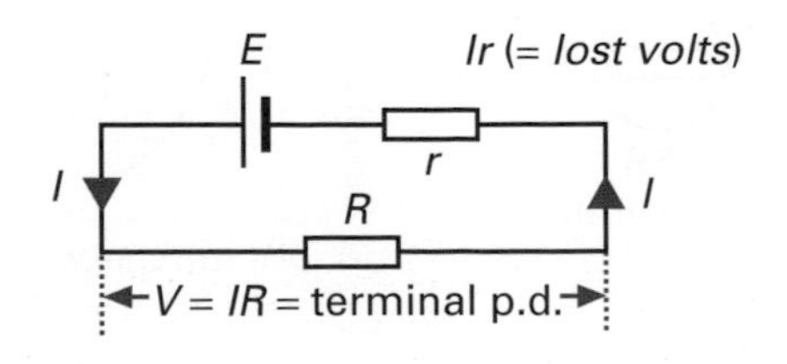

Checkpoint 1

Calculate the amount of energy transferred to 15 C of charge as it passes through a cell of 4 V.

Applying Kirchhoff's second law to the circuit above gives:

$$E = IR + Ir \qquad \text{or} \qquad E = I(R + r)$$

The terminal p.d. V (= IR) is also equal to:

$$V = E - Ir$$

hence the bigger the current, the bigger the lost volts and the smaller the terminal p.d.

If the current from a cell is zero, it is said to be on **open circuit** and the terminal p.d. will equal the EMF. So a very high resistance voltmeter connected across a cell will give a reading which is pretty close to its EMF.

There is no way to separate the internal resistance from the EMF, and so there is no way to measure r directly. To measure E and r indirectly, the circuit below can be used.

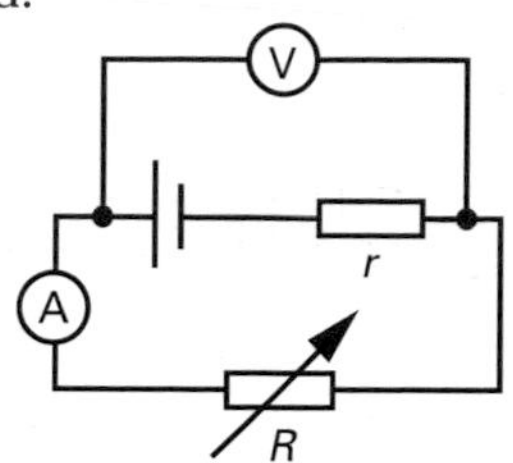

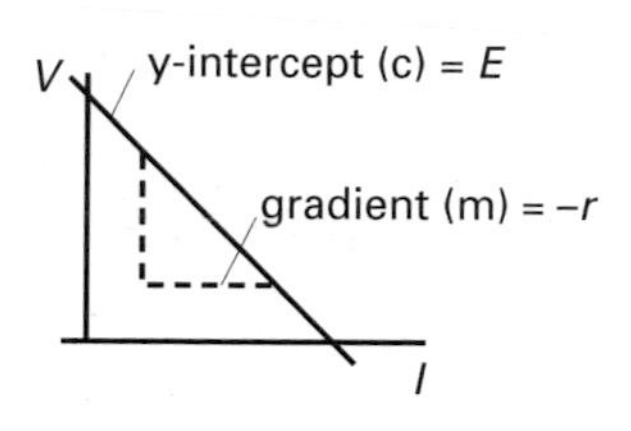

The resistance R is varied so that a series of values for terminal p.d. V and current I can be recorded using a voltmeter and ammeter as shown. If we compare our equation for this circuit ($V = -Ir + E$) with the equation for a straight line ($y = mx + c$), we can see that a graph of V against I will be a straight line, as shown above. .

- Lost volts are the reason why power supplies warm up when in use.
- A cell is short-circuited when $R = 0$. This means that $E = Ir$ and all the electrical energy becomes heat energy in the internal resistance. So the cell gets very hot!
- If a driver tries to start a car with the lights on, the starter motor will take a large current from the battery, the battery's terminal p.d. will fall and the car lights will dim.
- To get maximum power from a power supply, r = R but to transfer maximum energy to the external circuit, r<<R

Cells in series and parallel

For cells in series,
$E = E_1 + E_2 + etc$
and $r = r_1 + r_2 +$ etc

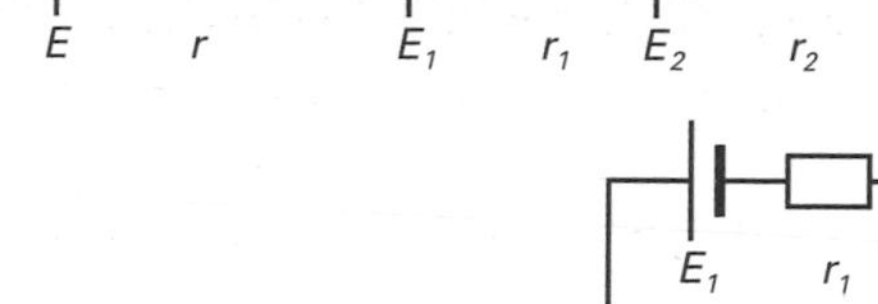

For similar cells in parallel, $E = E_1 = E_2 = etc$
and $1/r = 1/r_1 + 1/r_2 + etc$

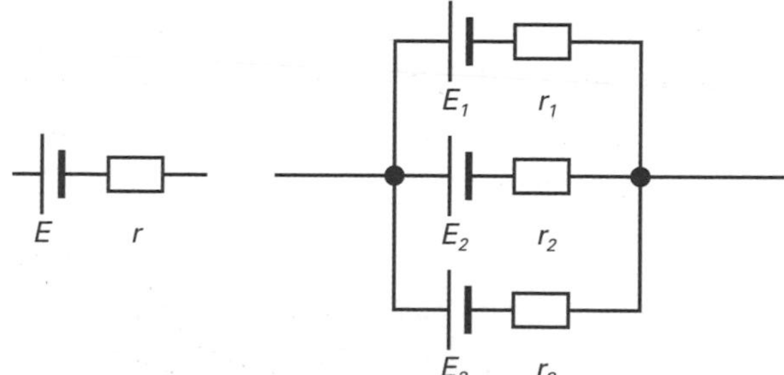

Exam practice

answers: page 97

1 A typical car battery has an EMF of 12 V and an internal resistance of 0.05 Ω. The battery has to produce a current of 80 A to the starter motor.
(a) Explain why internal resistance of the car battery must be very low.
(b) Find the p.d. across the internal resistance when the starter motor is running.
(c) Why is starting the car with the headlights on likely to affect their brightness?

2 A torch uses two 1.5 V cells. Its bulb is rated at 2.5 V, 0.5 A.
(a) What is the total EMF of the battery?
(b) Explain how the p.d. across the bulb is 2.5 V.
(c) Assuming each cell is identical, work out their internal resistance.

(20 min)

The jargon

The p.d. between one terminal of the cell and the other is called the *terminal* p.d.

The jargon

The p.d. caused by the internal resistance, *Ir*, is called the *lost volts.*

Links

See page 82 for Kirchhoff's second law.

Checkpoint 2

Why is there no way to separate the internal resistance from the EMF of a cell?

Grade booster

Check your syllabus to find out what are the mathematical requirements for your course.

Checkpoint 3

Why does the terminal p.d. of a cell depend on the circuit the cell is connected to?

Checkpoint 4

If you measure the EMF of an old non-rechargeable 1.5 V cell, it may still read close to 1.5 V. Try to use it, however, and you may find that it does not work. How can you explain this? (*Hint* what actually happens to a cell when it runs down?)

Capacitors

In the 18th century, Stephen Gray hung a workhouse boy from silk threads and connected him to a static electric charge. He wanted to show that the boy could lift pieces of paper from the ground beneath him using electrostatic attraction! The boy acted as a capacitor – anything that can store charge is a capacitor.

Watch out!

Charge is a quantity of electricity, carried by electrons, *not* a force.

Charging and discharging

Capacitors store charge and energy. This energy can be used as a backup for computers, as a power supply for camera flash bulbs, etc.

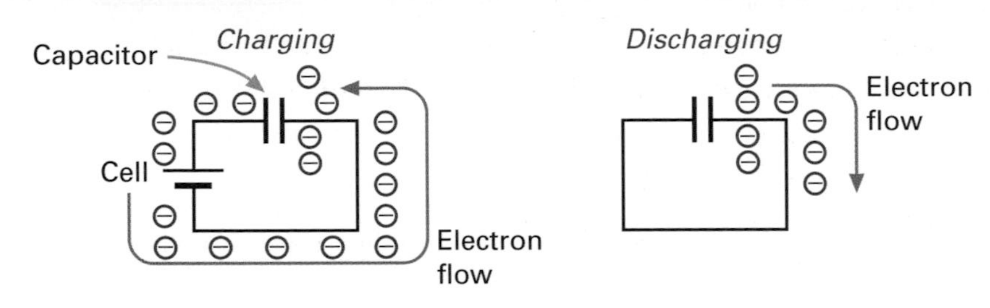

The jargon

A *dielectric* is an insulator, placed between the plates of a capacitor so that it can store more charge.

- ➜ Capacitors have two conducting plates separated by an insulator.
- ➜ The insulator, or dielectric, can be polythene, waxed paper, etc.
- ➜ To save space, capacitors are usually rolled up like a Swiss roll.
- ➜ When charging, the cell pulls electrons off one plate and pushes them on to the other (as shown above).
- ➜ Initially, electrons join the right-hand plate quickly (high current).
- ➜ As the plate fills, a growing repulsive force eventually stops any more electrons arriving. The capacitor is fully charged. No current.
- ➜ At this point, p.d. across the capacitor equals p.d. across the cell. (The clockwise force equals the anticlockwise force.)

The jargon

Permittivity of free space, $\varepsilon_0 = 8.85 \times 10^{-12}\ \mathrm{F\,m^{-1}}$.

The jargon

The *relative permittivity* ε_R of a dielectric tells us how much more charge can be stored with the dielectric inside a capacitor, compared with empty space between its plates. ε_R of mica is 7, so a mica-filled capacitor can store seven times as much charge than a similar vacuum-filled capacitor. Its capacitance increases by a factor of 7.

Capacitance

The maximum charge that can be stored Q is proportional to V. So,

$Q = \text{constant} \times V = CV$ where C = capacitance

$C = Q/V$ (in units, 1 farad (F) = 1 coulomb per volt ($\mathrm{C\,V^{-1}}$))

Capacitance is defined as charge per unit p.d. The factors that affect capacitance are shown in this equation:

$$C = \frac{\varepsilon_0 \varepsilon_R A}{d}$$

where A = plate area ($\mathrm{m^2}$), d = plate spacing (m),
ε_0 = a constant, the permittivity of free space ($\mathrm{F\,m^{-1}}$)
ε_R = relative permittivity (no units)

Checkpoint 1

Calculate the capacitance of a pair of parallel plates, each measuring 0.2 m × 0.1 m, when they are separated by an air gap of 3 mm. (ε_R of air is 1.0005.)

Two or more capacitors are often connected to give the required capacitance. The following equations are used to calculate the combined capacitance C_T of three capacitors (C_1, C_2 and C_3):

In series $$\frac{1}{C_T} = \frac{1}{C_1} + \frac{1}{C_2} + \frac{1}{C_3}$$ *In parallel* $$C_T = C_1 + C_2 + C_3$$

Checkpoint 2

Calculate the combined capacitance of three 50 μF capacitors connected in series.

Links

Compare with the equations for resistors in series and parallel on page 79.

The energy stored in a capacitor

Capacitors store energy because work is done to push electrons on to their plates. Look at this graph of charge Q against p.d. V:

We can substitute $Q = CV$ in $E = \frac{1}{2}QV$ to give $E = \frac{1}{2}CV \times V = \frac{1}{2}CV^2$.

We can substitute $V = Q/C$ in $E = \frac{1}{2}QV$ so $E = \frac{1}{2}Q \times Q/C = \frac{1}{2}Q^2/C$.

The mathematics of charging and discharging

When a capacitor and a resistor are connected in series, the time for

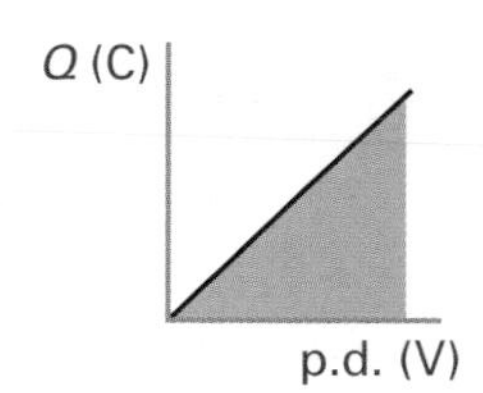

The energy stored by a capacitor equals the area under the graph. When it is charged to Q coulombs by a p.d. of V volts, then energy = $\frac{1}{2}QV$

the capacitor to charge or discharge increases as the current in the circuit decreases. The circuit below is used to study charging and discharging.

- With the switch as shown, the battery charges the capacitor.

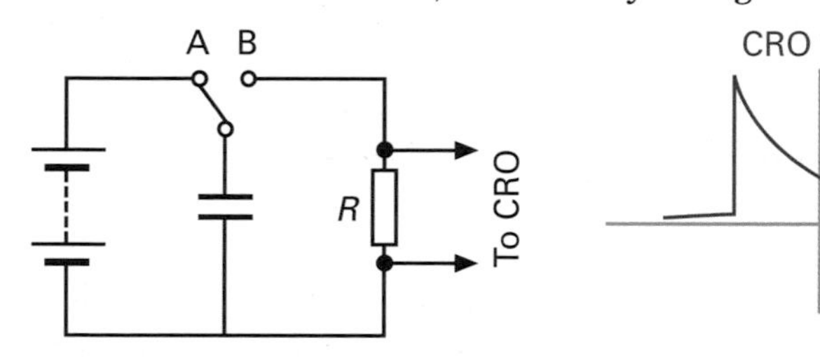

- If the switch is flicked to B, the capacitor discharges through R.
- A CRO trace is used to show how the p.d. across R changes.
- As $V \propto I$ (remember $V = IR$), the trace also shows how I changes.
- The shape of the curve is **exponential**.
- The time taken for the current to halve (e.g. to go from maximum current I_0 to $I_0/2$ or from $I_0/2$ to $I_0/4$) is always the same.
- The **time constant** RC (resistance × capacitance) is equal to the time taken for the charge to fall to 1/e (37%) of its initial value.
- As $Q = It$, $I = Q/t = Q/(RC)$.
- The charge Q on a capacitor at time $t = Q_0 e^{-t/(RC)}$, where Q_0 is the initial charge on the capacitor.

And for the non-mathematically minded . . .

The rate of charge leaving from, or arriving on, a capacitor depends on how much charge is already there. Pushing more electrons on to a partially charged capacitor is harder than putting them on an uncharged capacitor. When a capacitor is discharging, the strong repulsive force provided by lots of electrons means that the exodus of electrons (i.e. the current) is quickest when the capacitor is fully charged. Charging and discharging are exponential changes.

Checkpoint 3

Calculate the maximum energy that can be stored in a 10 000 μF capacitor when a p.d. of 20 V is applied across its plates.

The jargon

The *working voltage* of a capacitor is the maximum voltage that should be applied across it.

Checkpoint 4

What advantage does a CRO have, in comparison with a normal voltmeter, when used to study the charging characteristics of a capacitor?

Action point

Use pages 52–3 to draw up a table of similarities and differences between capacitor discharge and radioactive decay.

Checkpoint 5

Explain why the discharging of a capacitor is an exponential change.

Examiner's secrets

Capacitors deliver charge much the same as batteries and cells, but with one important difference: they cannot provide a constant current. You should learn this difference.

Exam practice answers: pages 97–98

A 5000 μF capacitor is charged to 6 V then discharged through a 100 Ω resistor.

(a) Show that the time constant is 0.5 s.

(b) Sketch a graph of current against time for this discharge. On your graph indicate the current at t = 0 and t = 0.5 s. (15 min)

Electromagnetism

It has long been known that certain rocks (lodestone, an iron-rich ore) can attract or repel each other. In 1821, Oersted discovered that current in a wire exerts a magnetic force.

Magnetic fields

Magnetic fields are produced in two ways.

- *Permanent magnets* The movement of individual electrons in their atoms causes weak magnetic fields. In ferrous materials, e.g. iron, all the weak fields combine to generate a strong field.
- *Electromagnets* This type of magnetism is temporary because the field is only produced when a current flows. *Soft* iron cores are used to increase the electromagnet's field strength.

The jargon

A *magnetic field* is the area around a magnet in which it can exert forces on magnetic materials.

The jargon

Soft magnetic materials are easy to magnetize and easy to demagnetize.

Checkpoint 1

Explain why a named *hard* magnetic material would not be used as the core of an electromagnet.

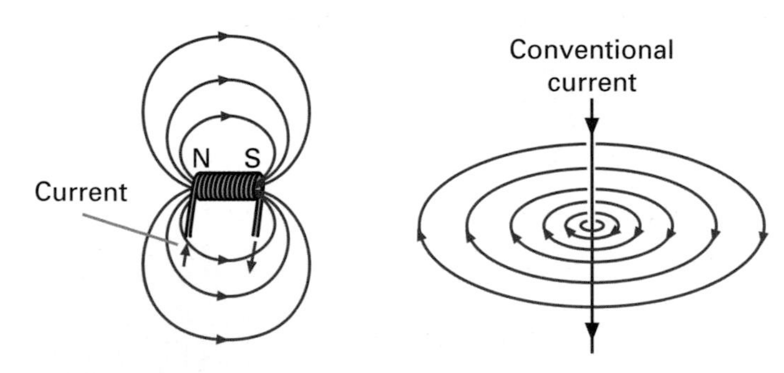

These diagrams show the magnetic fields around a current-carrying coil (a solenoid) and a current-carrying, straight wire. The field around the straight wire is weaker. The fields change direction if the current is reversed.

Magnetic force

Oersted discovered that when he switched on an electric current, a nearby compass needle moved. The temporary magnetic field around the current was interacting with the permanent magnet just as any two magnets would.

Checkpoint 2

Describe and explain the magnetic field surrounding a long, straight, current-carrying wire.

Fleming's left-hand rule

Grade booster

You will lose marks if you draw magnetic field lines crossing or touching. (Think what would happen to a compass needle if they did.)

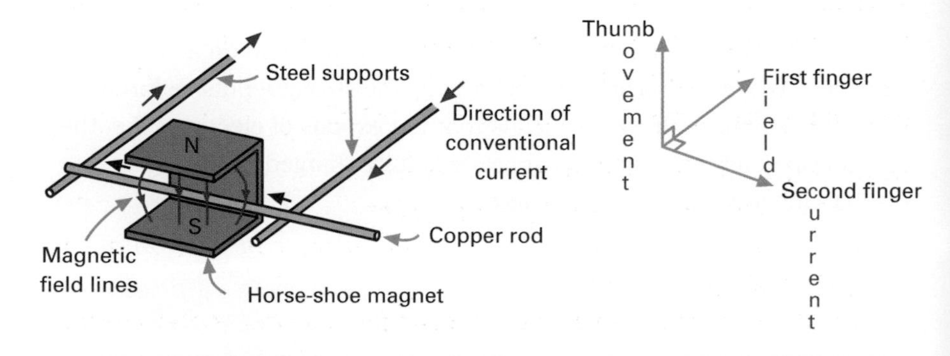

The direction of the force (or movement of the copper rod) can be predicted by keeping the thumb and first two fingers of your left hand at 90° to one another, see the previous diagram. Point your first finger in the direction of the permanent magnet's field (from North to South) and your second finger in the direction of conventional current; your thumb now shows the direction of the movement, or force, produced.

The jargon

The force produced by a permanent magnet interacting with an electromagnet, as in the last diagram, is called the *motor effect*.

Magnetic flux ϕ and magnetic flux density B

Magnetic flux is an imaginary fluid that flows from the North pole of a magnet to the South. It flows along the field or **flux** lines. B tells us how tightly packed the flux, or how strong the magnetic field, is.

Magnetic flux density = flux per unit area
$B = \phi/A$
$1\ T = 1\ Wb\ m^{-2}$
1 tesla =1 weber per square metre

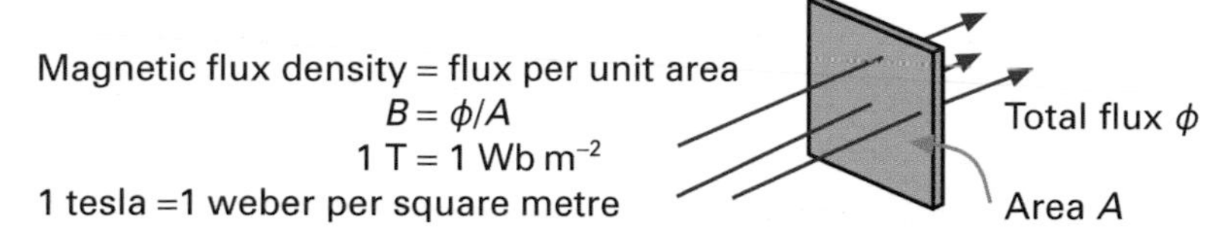

Magnetic force

- $F = BIl$ — force on a wire of length l, carrying a current I.
- $F = BIl \sin \theta$ — current-carrying wire at angle θ to field.
- $F = BQv$ — force on a charge Q moving with speed v.
- $F = BQv \sin \theta$ — charge Q moving at angle θ to field.

Definition of magnetic flux density B

$F = BIl$ so $B = F/Il$

A magnetic field has a strength of 1 tesla (1 T) if it exerts a 1 N force on a conductor 1 m long, carrying a current of 1 A at right angles to the field.

Measuring and calculating magnetic flux density B

Hall probes can measure B. They contain a semiconductor, across which a potential difference is established when the probe is placed in a magnetic field. A stronger magnetic field will produce a larger potential difference. Pre-calibrated probes convert this potential difference into a reading of B. Just hold the probe so that the magnetic field lines are passing at 90° through its flat face. Equations for calculating B include:

$B = \mu_0 nI$ — at the centre of a long solenoid, where n = number of turns per metre on the solenoid

$B = \mu_0 I/2\pi r$ — at a small distance r from a long straight wire

Force between current carrying conductors

Like currents attract, unlike currents repel.
To show that there is an attractive force between two wires carrying currents in the same direction:

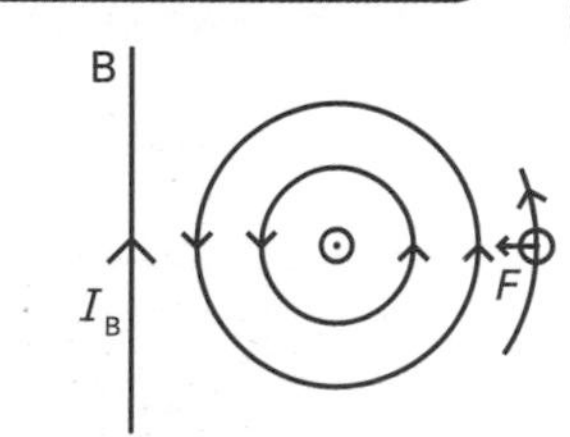

- Draw the wires from above and mark in the field around wire A (given by $B = \mu_0 I_A/2\pi r$)
- Use Fleming's left hand rule to work out the force on wire B (given by $F = BI_B l = \mu_0 I_A I_B l/2\pi r$)
- Wire A will have an equal and opposite force by Newton's third law.

Exam practice — answers: page 98

(a) Explain how the direction of magnetic flux lines can be determined.
(b) Describe an experiment to study force on a current-carrying conductor in a magnetic field.
(c) How can the direction of the magnetic force on a charge moving through a magnetic field be predicted? Why does the charge follow a circular path?
(35 min)

Checkpoint 3

Use Fleming's left-hand rule to predict the direction in which the copper rod in the diagram at the bottom of page 90 will roll.

Examiner's secrets

One of the strangest, and most frequent, sights in physics exams are row upon row of candidates grappling with Fleming's left-hand rule. Learn this rule! It crops up a lot!

The jargon

Flux means a flow or discharge. Think of an imaginary fluid that flows from North to South poles.

Checkpoint 4

Rearrange any of the equations for magnetic force to make B the subject of the equation. Is B a vector or a scalar quantity? Give a reason for your answer.

The jargon

μ_0 is a constant, the *permeability of free space*. $\mu_0 = 4\pi \times 10^{-7}\ H\ m^{-1}$.

Watch out!

Magnetic flux density and magnetic field strength are different names for the same quantity (B). Don't be fooled by this!

Action point

The force between current carrying conductors gives rise to the definition of the ampere as that current which when flowing in two infinitely long parallel wires 1 m apart in a vacuum produces on them a force of 2×10^{-7} N per metre of their length. Substitute into the equation for the force between current carrying conductors to show where this comes from.

Action point

Show that two wires carrying currents in opposite directions must repel.

Electromagnetic induction

Take note

Michael Faraday was appointed at the Royal Institution only after a fight in the main lecture theatre led to the dismissal of his predecessor!

Check the net

To find out more about Michael Faraday, go to www.rigb.org/heritage/faradaypage.isp

The jargon

Flux linkage is defined as $N\phi$, where N is the number of turns on a coil ($N = 1$ for a straight wire) and ϕ is the amount of magnetic flux. Think of flux linkage as the amount of overlap there is between the conductor and the flux lines. If this amount changes, the flux lines must be being cut and an EMF will be induced.

The jargon

When changing magnetic flux causes currents to flow within a solid conductor they are called *eddy currents.* These induced currents oppose the change producing them (Lenz's law) so any solid conductor moving through a magnetic field will experience a braking effect (like the magnet falling through the aluminium tube.)

Checkpoint 1

Rephrase Faraday's law of electromagnetic induction using the phrase *flux linkage.*

Checkpoint 2

Read the formal statement of Lenz's law again and then state what change would be observed if the magnet was pushed into and then pulled out of the coil, as shown above (right).

When Michael Faraday discovered electromagnetic induction he paved the way for electric transformers and generators, and for the whole electrical industry.

Generating electricity

If the following equipment is set up, we can show that:

- → if either the wire or magnet moves vertically, a current is induced
- → the current direction reverses if the wire moves up not down
- → doubling the speed of movement doubles the current induced

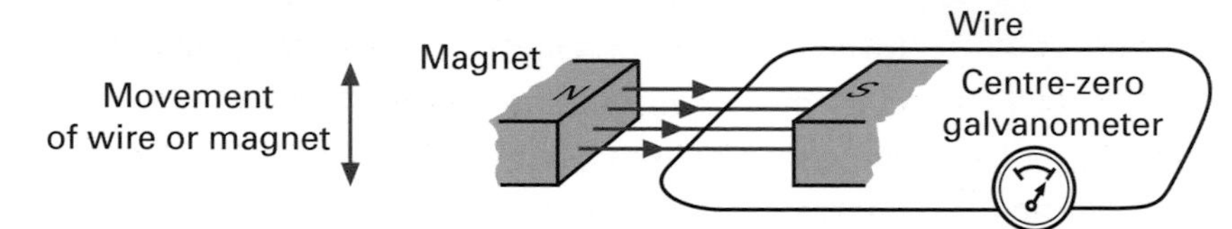

Short-hand ways to explain electromagnetic induction

- → An EMF is induced when a conductor cuts magnetic flux lines.
- → An EMF is induced when the amount of flux linkage changes.
- → When the conductor forms part of a complete circuit, the induced EMF can cause an induced current to flow around the circuit.
- → magnetism + movement → electricity

Laws of electromagnetic induction

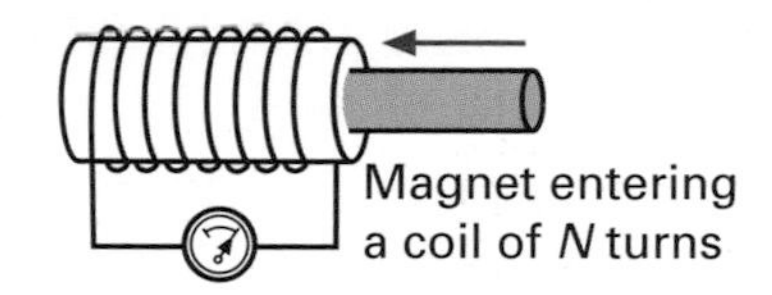

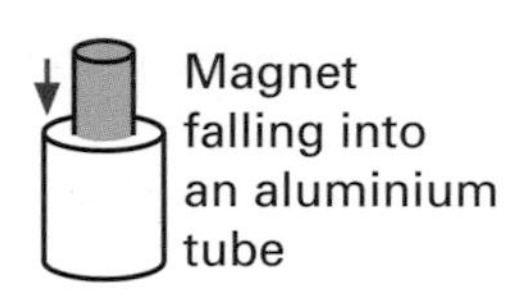

The experiment above should remind you of the first experiment on this page. An EMF is induced when there is relative movement between the coil and magnet. So, it does not matter whether the magnet or the coil moves. Faraday found that:

- → the magnitude of the EMF, E induced in a conductor is proportional to the rate at which magnetic flux is cut by the conductor, i.e.

$$E \propto d\phi/dt$$

The set-up shown above (right) is used to demonstrate Lenz's law. When the magnet is dropped it travels very slowly. Not as much of its potential energy is converted into kinetic energy as you might expect. Some is changed into electrical energy as an EMF is set up in the tube. Stated formally, **Lenz's law** is:

- → the direction of the induced EMF is such that it opposes the change that caused it (in this example, the EMF is directed up)

Both laws can be combined in one equation:

$$E = -d\phi/dt$$

Or, for a coil consisting of N turns of wire:

$E = -N\,d\phi/dt$ (minus signs represents Lenz's law)

Explaining electromagnetic induction

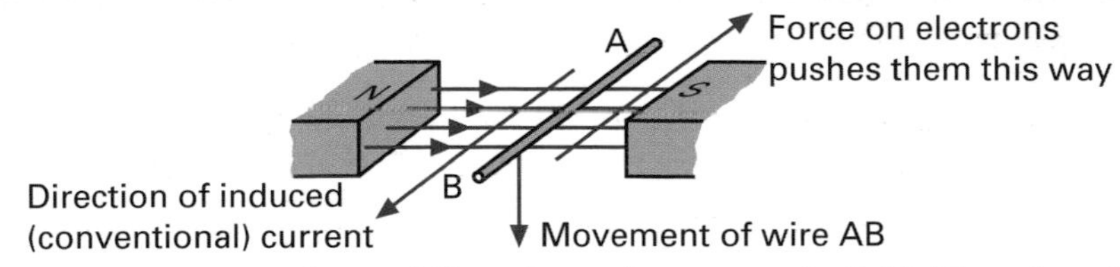

- → A straight wire is shown falling through a magnetic field. Therefore the free electrons in the wire are moving down too.
- → Free electrons move in the opposite direction to conventional current, and so the direction of conventional current must be up.
- → We know the direction of the magnetic field and the conventional current, and so we can use Fleming's left-hand rule to predict the direction of the resulting magnetic force. It pushes the electrons within the wire from B to A.
- → So, a conventional current has been induced (persuaded) to flow in the wire from A to B.

Transformers

Power stations generate electricity that is then stepped up to be distributed at hundreds of thousands of volts, as this wastes less energy. A network of electricity lines criss-cross the country rather like a spider's web. The voltages are then stepped down to different voltages to meet the needs of various users. Transformers step voltages up and down.

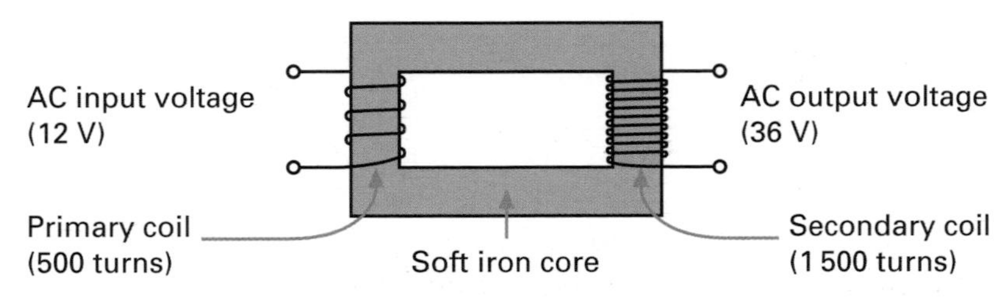

The alternating current in the primary coil produces a magnetic field that changes as the current changes. The soft iron core increases the magnetic field and links it with the secondary coil so there is a change in flux linkage in the secondary coil. This induces an alternating EMF in the secondary coil.

Since $V_S/V_P = N_S/N_P$, if there are more turns on the secondary than the primary, then the secondary voltage will be bigger and the transformer is a step-up transformer. Step-down transformers have more turns on the primary and step down the voltage. For an ideal transformer (that is one which has no power losses so is 100% efficient)

- → power out = power in
- → $V_S I_S = V_P I_P$ giving $V_S/V_P = I_P/I_S = N_S/N_P$ so for a transformer the current is stepped down (or up) by the same ratio that the voltage is stepped up (or down).

Exam practice

answers: page 98

(a) Why is Lenz's law an example of conservation of energy?

(b) Explain why eddy currents occur in the core of a transformer and how this alters the flux in the core. (10 min)

Checkpoint 3

An aircraft is flying at 925 km h^{-1} in a region where the vertical component of the Earth's magnetic field is 4.1×10^{-5} T. Its wing span is 29 m. Calculate the EMF induced between the wing tips.

The jargon

A *search coil* is a small flat coil with many turns. When it is placed in a varying magnetic field, an EMF is induced which is proportional to the flux density of the field. A search coil connected to a CRO can be calibrated in a known field so unknown fields can be measured.

Grade booster

Check whether your syllabus requires you to know about transformers.

Links

See pages 94–5 for more on alternating currents.

Action point

Use the transformer equation to check that the figures quoted in the diagram on the left are correct.

Action point

You should know why transformers do not work with DC.

Take note

The transformers used in the National Grid system can be up to 99% efficient. The reasons for their inefficiency are because:

- → not all of the primary flux links the secondary coil
- → resistance of the coils causes heating
- → eddy currents in the core cause heating
- → repeated magnetisation and demagnetisation of the core causes heating.

Alternating currents

Thomas Edison, the inventor of the electric light bulb, was a very wealthy man at the head of a DC-based empire when an employee, Nikola Tesla, suggested that AC would be more efficient. Edison invented the electric chair and electrocuted cats and dogs to show how dangerous AC was.

Check the net

To find out more about the fight for AC go to www.parascope.com/en/0996/tesla2.htm

AC generators

Checkpoint 1

Would it make more sense to move the magnet or the conductor in a power station?

Most mains electricity is generated in power stations by alternating current generators. The simplest generator is a coil of wire which is made to rotate in a magnetic field. The ends of the coil are connected to the rest of the circuit by slip rings that rub against carbon brushes as the coil rotates.

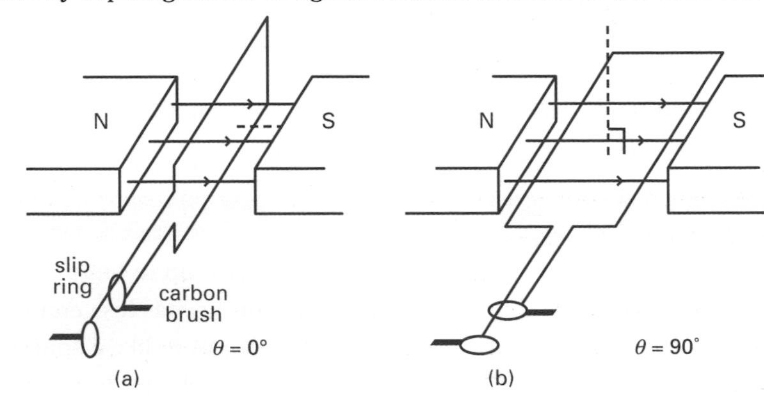

Checkpoint 2

Use $T = 1/f$ to find the time period T of an alternating supply of frequency $f = 50$ Hz.

- As the coil rotates, it cuts magnetic flux causing an induced EMF across the ends
- There is maximum flux linking the coil when it is at right angles to the field (see (a))
- There is zero flux linking the coil when it is parallel to the field (see (b))
- Flux linkage is given by $N\phi = NBA \cos\theta$ where θ is the angle between the normal to the plane of the coil and the field.
- EMF induced, $E = -N\mathrm{d}\phi/\mathrm{d}t = NBA\omega \sin \omega t$

The jargon

The *root-mean-square* (RMS) value of an alternating current, I_{RMS}, is the direct current that delivers the same average power as the alternating current. The RMS value is also known as the *effective* value.

Action point

Why $\sqrt{2}$?
Peak power is IV. Power varies between IV and zero, so average power = $IV/2$.
Can you see that $IV/2$ is the same as $\frac{I_{RMS}}{\sqrt{2}} \times \frac{V_{RMS}}{\sqrt{2}}$?

Alternating current (AC) and voltage

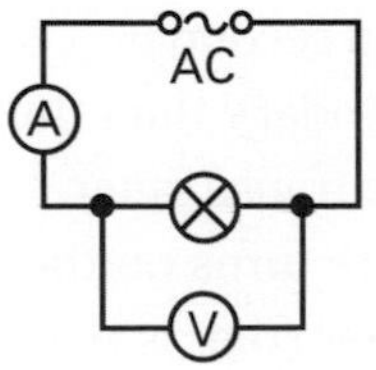

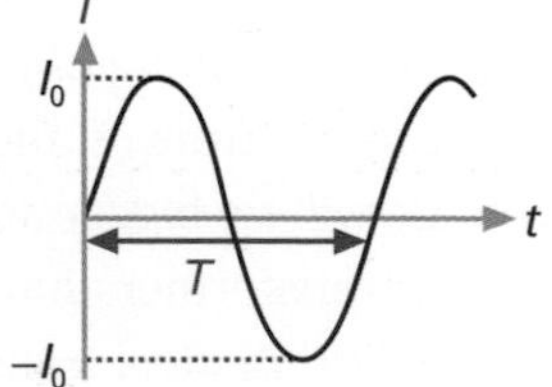

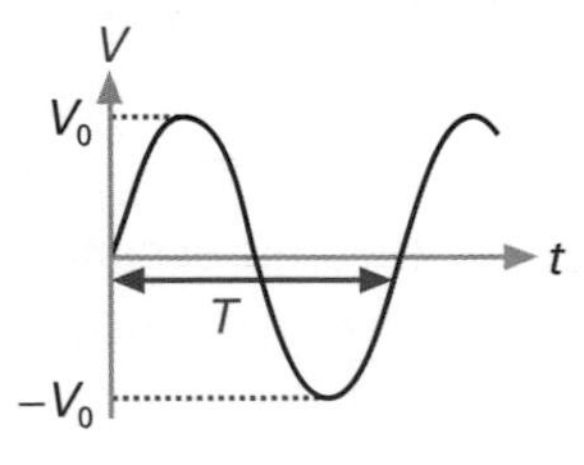

- AC is constantly changing direction.
- Frequency of mains AC in the UK is 50 Hz (changes direction 50 times/s).
- AC varies sinusoidally, $I = I_0 \sin 2\pi ft$, where f = frequency.
- In simple circuits, I and V vary in step or in phase.
- You cannot see a mains-operated bulb flickering as the energy to it varies, because 50 Hz is quite fast. The filament does not have time to cool before the next delivery of energy arrives.

Checkpoint 3

100 kW is delivered to a town along power lines that have a total resistance of 0.4 Ω. Calculate the power loss if the electricity is distributed at a voltage of (a) 240 V and (b) 24 kV.

Checkpoint 4

Explain, in a sentence, the benefits that come from using AC and transformers in the distribution of electricity.

For alternating current:

- Peak value is the maximum value, I_0
- Peak-to-peak value is $2 \times I_0$

- Average value is 0
- Root mean square value, $I_{RMS} = I_0/\sqrt{2} = 0.707\ I_0$.
- The same applies to alternating voltage.

Power

From the graph below you can see that the energy delivered varies.
Average AC power = $I_{RMS} \times V_{RMS}$.

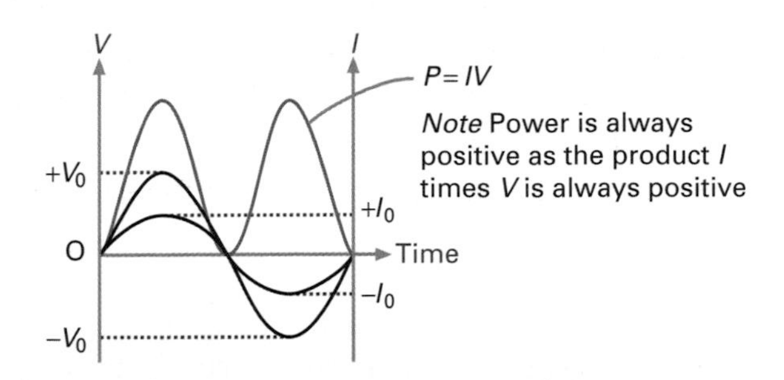

Cathode-ray oscilloscopes

A CRO is a high-resistance voltmeter.

- The voltage can be calculated from the height of the trace if the number of volts per scale division (voltage sensitivity) is known.
- The current can be calculated using $I = V/R$ if the voltage is measured across a known resistor.
- The time for a cycle of a periodic voltage can be measured from the time-base setting (the number of seconds per division), and its frequency calculated using $f = 1/T$.

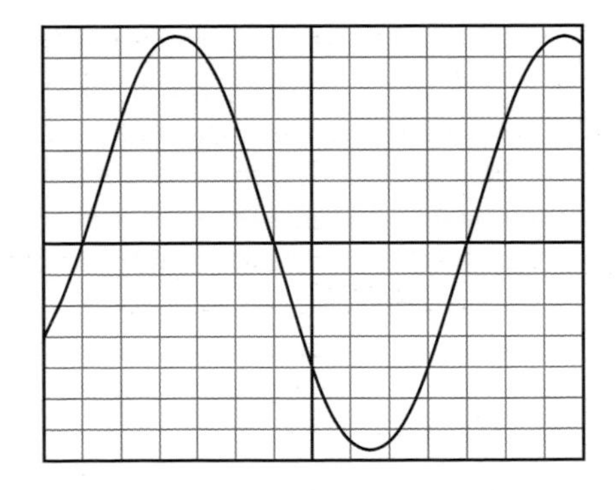

Checkpoint 5

What would be the benefit of using a cathode-ray oscilloscope (CRO) rather than the conventional voltmeter shown in the circuit on the right?

Checkpoint 6

When does the country's demand for electricity surge rapidly? How is this sudden demand satisfied?

Checkpoint 7

Why does mains electricity use AC?

Exam practice

answers: page 84

An AC voltage of 5 V_{rms} and mains frequency 50 Hz is displayed on a CRO.

(a) Calculate the peak-to-peak voltage.

(b) The screen of the CRO is shown.
Calculate
(i) the voltage sensitivity of the CRO
(ii) the time base setting of the CRO. (20 min)

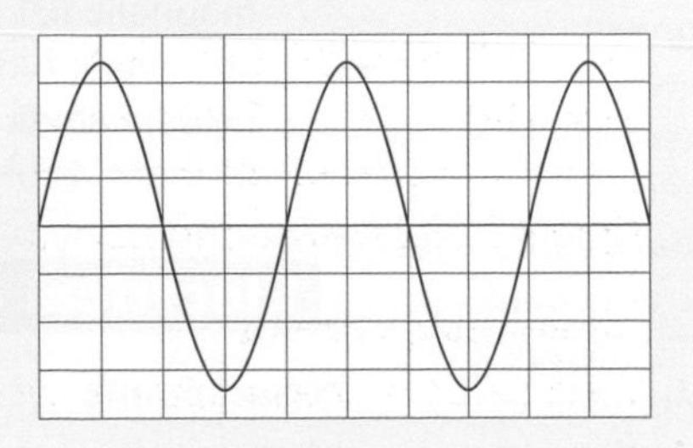

Answers

Electricity and electromagnetism

Current as a flow of charge

Checkpoints

1 Electrons are moved, then remain *stat*ionary in *static* electricity. *Electr*ons flow like a river in current *electr*icity.
2 $Q = It = 12 \times 3 \times 60 \times 60 = 129\,600$ C
3 Resistance of an ideal ammeter is 0 Ω to avoid changing the current in question.

Exam practice

(a) $\Delta t = \Delta Q/I = 4.68 \div 26 \times 10^{-3} = 180$ s
(b) Normally a conductor would contain many more free electrons than a semiconductor. If energy is transferred to the semiconductor (e.g. by heating a thermistor) many more free electrons can be generated.
(c) $v = I/nAe = 7 \div (10^{29} \times \pi r^2 \times 1.6 \times 10^{-19})$
$= 7 \div (10^{29} \times \pi r^2 \times 1.6 \times 10^{-19}) = 5.6 \times 10^{-4}$ m s^{-1}

Current, p.d. and resistance

Checkpoints

1 (i) The wires have no resistance so the graph is flat in sections that relate to the wires as no energy is being transferred.
(ii) The electrons do not experience any resistance when flowing through the chemicals in the cell. The EMF of the cell is quoted as 6 V but the electrons would lose some of the energy they pick up just because of the cell's internal resistance.
(iii) Current can flow, even when the circuit has been disconnected.
(iv) Electrons give up all their energy before returning to the cell, otherwise the bulbs would get increasingly bright.
2 An ideal voltmeter has infinite resistance.
3 Insulators have very high resistances.
4 If one component in a parallel circuit blows only the arm containing that component will be affected.

Exam practice

(a) In the water-pump model the cell is represented by a water pump, wires = water pipes, electrons = water, ammeter = water meter, voltmeter = pressure gauge, resistor = pipe narrows. One disadvantage of this model is that it suggests that only cross-sectional area affects the resistance of a wire, ignoring the effects of resistivity (i.e. the material), length, temperature and impurities.
(b) If you tried to count the number of passengers entering the Underground, slowing them down in any way could cause a queue that might put other travellers off thereby reducing the number of passengers using the service. An ideal ammeter would have zero resistance to achieve a 'God's-eye view'. Current could be checked without changing it, just as God is believed to be able to watch over the Earth without intervening in the actions of humans.

Resistors and resistivity

Checkpoints

1 $\rho = RA/l = \Omega$ m^2/m $= \Omega$ m
2 Ω^{-1} m^{-1}. Conductivities could be given in a table as conductivity is a property of materials not individual samples.
3 The gradient of the I–V graph is the reciprocal of resistance. For the filament lamp, the gradient decreases with increasing current. But its temperature will increase with increasing current. So the resistance of a metal decreases as its temperature increases. The reverse is true for a thermistor which is made from semiconducting material.
4 If the resistors are connected in series, electrons pass through all three, so the total resistance is 6 Ω. If the resistors were joined in parallel, each electron would only pass through one resistor so the total resistance would be less.

Exam practice

(a) The graphs are on page 78. A metallic conductor, at constant temperature, is characterized by a straight line through the origin so current is proportional to p.d. This is a statement of Ohm's law. A metallic conductor, at constant temperature, is known as an ohmic conductor. The filament lamp graph is non-linear; a lamp is not an ohmic conductor. As current rises, the filament gets hotter, its resistance increases and the graph gets flatter. A thermistor gets hotter when more current flows. This causes it to release more free electrons; its resistance decreases. Diodes only allow a significant current to flow in one, 'forward' direction. Diodes have high resistances in the 'reverse' direction.
(b) Resistivity $\rho = RA/l$, so we must measure each of these values. The cross-sectional area A of a wire can be deduced after measuring its diameter with a micrometer screw gauge. NB $A = \pi r^2$ and r = diameter/2. The length of the wire can be measured with a metre rule and a multimeter, on its ohms range, can measure resistance.
(c) Superconductors are used to make superconducting magnets that produce exceptionally stable magnetic fields used in MRI scanners. Frictionless bearings are made possible by lowering a superconductor onto a magnet. The magnet induces a current in the superconductor, large enough (because the superconductor has zero resistance) to produce a magnetic field that is repelled from the original magnet so levitation (floating) is achieved. Superconducting power cables could transfer energy without power losses as there would be no heating effects in the cables.

Electrical energy and power

Checkpoints

1 The more powerful hairdryer can change electrical energy into other forms more quickly, and so your hair would dry faster.

2 $P = VI = 440 \times 10^3 \times 100 = 44$ MW. We are given the potential produced by the power station, not the p.d. across the section of wire involved. So we must use $P = I^2R$ not $P = VI$. $P = I^2R = 100^2 \times (20 \times 0.2) = 40$ kW.
3 $I = P/V = 4.2$ A, so use a 5 A fuse.
4 Electrical energy cost = kW h × cost of 1 kW h = 5 × 2.5 × 10 p = £1.25

Exam practice

(a) 0.1 kW × 24 h × 10 p = 24 p
(b) $E = I^2Rt = 25 \times 20 \times 60 = 30$ kJ
(c) $E = V^2/R \times t = 2 \times (9/15) \times 10 \times 60 = 720$ J

Kirchhoff's laws

Checkpoints

1 15 A
2 Because the EMF of the 20 V cell is greater than the EMF of the 5 V cell.
3 On the walk, the walker's height above sea level might change. However, his/her final height above sea level (and so his/her PE) remains unchanged. As an electron goes round a circuit its potential also changes. As with the walker, when the electron returns to its starting point in the cell, its potential will once again be the same.

Exam practice

(a) Σ EMF = ΣIR so $12 = 0.5R + (10 \times 0.5)$ and $R = 14\ \Omega$
(b) Label the circuit with A,B,C,D,E and F as in the example on page 83. Applying Kirchhoff's first law at junction C gives $I_1 = 1 + I_2$ (eqn 1) so $I_3 = I_1$. Applying Kirchhoff's second law to loop ABEF gives $6 = 6I_2$ so $I_2 =$ 1 A. Substituting this value in eqn 1 gives $I_1 = 1 + 1 = 2$ A. Applying the second law to ABCD gives $6 = 4 + (? \times 1)$ so ? = 2 Ω.

Potential dividers and their uses

Checkpoints

1 $V_{out} = V_{in} \times (R_1/R_1 + R_2) = 8$ V
2 Thermistor is similar to thermometer; they are both associated with temperature.
3 A sensor (e.g. a thermistor), automatic switch (e.g. a transistor) and an actuator (e.g. a heater) are required.
4 The circuit is practical and cost effective as it senses and alerts automatically.

Exam practice

1

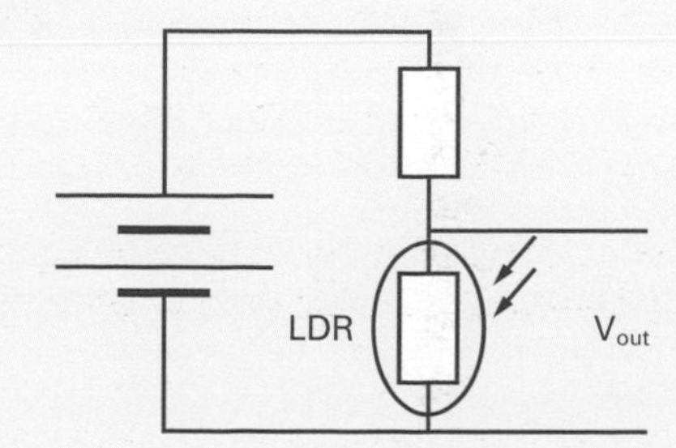

In daylight, the LDR has a low resistance and therefore takes a low share of the battery's 'voltage'. As it gets dark, the LDR's resistance and share of the battery's 'voltage' increase. This 'voltage' makes an electronic switch operate which turns on a light bulb.

2 Current = p.d./total resistance
$= 10 \times 10^{-3}/20 \times 10^3$
$= 0.5 \times 10^{-6}$ A
p.d. $= I \times R$
$= 0.5 \times 10^{-6} \times 10 \times 10^3$
= 5 mV

EMF and internal resistance

Checkpoints

1 60 J
2 An EMF is produced as electrons travel through chemicals in a cell but the chemicals simultaneously provide resistance to the motion of the electrons.
3 The smaller the resistance of the external circuit, R, the larger the current, $I = E/(R + r)$. This means the lost volts, Ir, will be large so the terminal p.d., $V = E - Ir$, will be less.
4 As a dry cell 'runs down' the internal resistance increases. Measuring the open circuit terminal p.d. will not indicate the extent of this increase as no current flows. As soon as a current in drawn from the cell, these will be 'lost volts' across the internal resistance which reduces the output p.d.

Exam practice

1 **(a)** The internal resistance must be low so the battery can produce a large current from an EMF of only 12 V.
(b) lost volts, $V = Ir = 4$V
(c) The terminal p.d. will be 8 V which will dim lights rated at 12 V.

2 **(a)** 1.5 × 2 = 3 V
(b) When drawing a current there will be a voltage drop across the internal resistance so the terminal p.d. will be less than 3 V.
(c) Using $V = E - Ir$, $2.5 = 3 - 0.5r$, $r = 1$ so the internal resistance of each cell is 0.5 Ω.

Capacitors

Checkpoints

1 5.90×10^{-11} F (59 pF)
2 16.7 μF
3 2 J
4 A CRO provides a visual record that makes comparison of alternating currents easy. It shows how changes occur against time.
5 The rate of capacitor discharge at any particular time depends on how much charge remains at that time, and so it is an exponential change.

Exam practice

(a) $CR = 5000 \times 10^{-6} \times 100 = 0.5$ s

(b) At $t = 0$, initial current $I_0 = V/R = 6/100 = 0.06$ A (60 mA). At 0.5 s, $I = I_0/e = 22.1$ mA
The graph is an exponential decay curve.

Electromagnetism

Checkpoints

1 Hard magnetic materials, e.g. steel, do not lose their magnetism when the current is switched off.

2 The magnetic field lines are in concentric circles. The spacing between circles increases further from the wire indicating that the magnetic field is getting weaker.

3 The rod rolls forward out of the magnet.

4 e.g. $B = F/qv$. F and v are vectors, so B is too.

Exam practice

(a) Magnetic flux lines flow from north to south; i.e. the way a small, imaginary north pole would move.

(b) A magnet is placed on top of a top-pan balance. Copper wire, connected to a low-voltage power pack, is lowered into the magnet's field. When no current flows, the balance shows only the mass of the magnet. When current flows, the force between the wire and the magnet increases or decreases the meter reading, depending on current direction.

(c) Use Fleming's left-hand rule (page 90). As the force is always perpendicular to the velocity of the charge, the particle will follow a circular path

Electromagnetic induction

Checkpoints

1 The magnitude of an induced EMF is proportional to the rate of change of flux linkage.

2 An induced current flows in the opposite direction if a magnet is pulled out of, rather than pushed into, a coil.

3 Speed = 925 000 / 3600 = 257 m s^{-1}
EMF = rate of change of magnetic flux
$= BA / t = 4.1 \times 10^{-5} \times 29 \times 257 = 0.31$ V

Exam practice

(a) When a magnet falls into an aluminium tube, some PE is changed into KE, some into electricity. As energy is conserved there is less KE than had the magnet fallen to Earth without the involvement of the aluminium tube.

(b) There is a change of flux in the core, which induces an EMF in the core, causing currents to flow, which oppose the original change of flux by Lenz's law so reducing the flux in the core.

Alternating currents

Checkpoints

1 Most practical large AC generators have static coils and a rotating electromagnet. The advantage is that the slip rings and brushes only have to carry a small current needed to magnetise the rotating electromagnet instead of a large induced current.

2 0.02 s

3 (a) $I = P/V = 417$ A, $P = I^2R = 69.4$ kW (b) $I = 4.2$ A, $P = 6.9$ W so much less power is lost!

4 Transformers can be used with AC to distribute electricity at high voltages so less energy is wasted.

5 A CRO provides a visual, analogue display that allows easy analysis of AC.

6 During intervals in or at the ends of popular TV programmes, eg World Cup Finals, millions of people switch kettles on for a cup of tea for example. Electricity cannot be stored so has to be made to suit the demand, which is why electricity distributors keep a note of TV listings as well as the weather forecast. Hydroelectric power stations can satisfy a surge in demand.

7 Power stations are usually not near towns or cities so the power has to be distributed by the power lines of the National Grid. Power loss from current-carrying wires = I^2R; i.e. $\propto I^2$. Transformers are used to step the voltage, which means the current is stepped down and less power is lost in transmission. At towns or cities, transformers are used to step down the voltage. Electricity is transmitted at up to 440kV. This can be stepped down to 33kV for heavy industry, 11kV for light industry or 240V for homes, using transformers that work on AC.

Exam practice

(a) peak voltage = $V_{RMS} \times \sqrt{2} = 7.07$ so peak-to-peak voltage = 14 V

(b) (i) 7 divisions = 14 V so voltage sensitivity = 2 V div^{-1}
(ii) 4 divisions = 1 cycle = 0.02 s
so time base = 5 ms div^{-1}

Kinetic theory

The kinetic theory of matter should be very familiar to you. This theory pictures solids, liquids and gases as being made from particles that are constantly moving. Previously you will have used this theory to explain properties such as changes of state, gas pressure and diffusion. The kinetic theory model is used in both physics and chemistry. This section concentrates largely on gases. A more quantitative approach, as required at A-level, is developed.

Exam themes

- *Applying equations* to calculate physical quantities associated with gases.
- *Evidence* in support of the kinetic theory, such as Brownian motion.
- *Macroscopic properties* Using kinetic theory to explain properties like gas pressure.
- *Quantitative expressions* Applying the theory to get expressions for pressure and temperature.
- *Developing links* between kinetic theory and energy.

Topic checklist

	Edexcel		AQA/A		AQA/B		OCR/A		OCR/B		WJEC		CCEA	
	AS	A2	AS	A2	AS	A2	AS	A2	AS	A2	AS	A2	AS	A2
Behaviour of gases – Experiment		●		●		●		●		●	○			●
Behaviour of gases – Theory		●		●		●		●		●	○			●
Behaviour of gases – in bulk		●		●		●		●		●	○			●
Internal energy		●		●		●		●		●	○			●
Specific and latent heat capacities		●		●		●		●		●	○			●

Behaviour of gases - Experiment

"It doesn't matter how beautiful your theory is, it doesn't matter how smart you are. If it doesn't agree with experiment, it's wrong."

Richard P. Feynman

This section takes you back to the seventeenth and eighteenth centuries, when the relationships between pressure, volume and temperature of a fixed mass of gas were investigated. The laws that emerged are still in common use and gave rise also to a different way of considering that most basic of properties - temperature.

The gas laws

With three separate variables (assuming the mass of gas does not change), one must be kept constant in each investigation.

Boyle's law

Temperature is kept constant, and the relationship between pressure and volume is considered. The results gave:

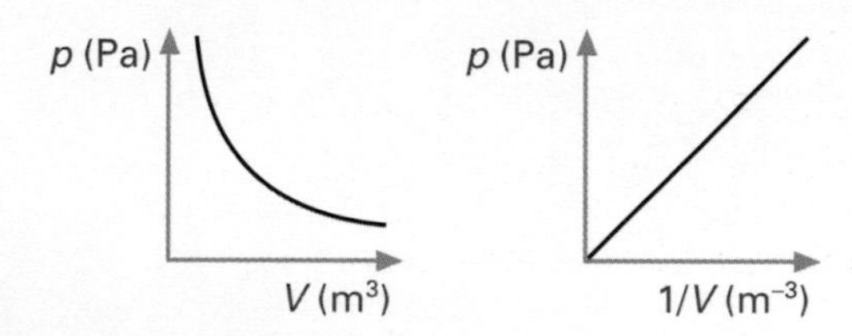

Checkpoint 1

How can Boyle's law be used to explain the movement of gases into, and out of, the lungs?

The outcome is summarized as **Boyle's law**:
p is proportional to $1/V$ or pV = constant or $p_1V_1 = p_2V_2 = p_3V_3 =$ etc.

Charles' law

Grade booster

Check your specification and if necessary, learn the details of each of these experiments.

Pressure is kept constant and the relationship between volume and temperature is investigated.
The results gave:

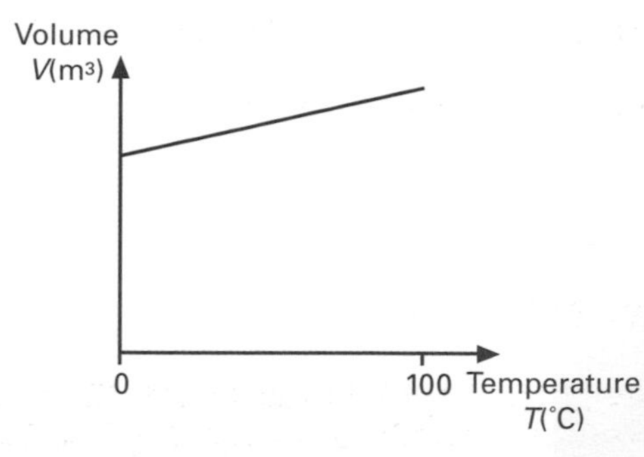

This shows that the volume increases as the temperature increases, in a linear fashion, though not in direct proportion since the line does not go through the origin. However, Kelvin had ideas about this (see below).

Pressure law

Pressure is kept constant and the relationship between volume and temperature is investigated. The results gave:

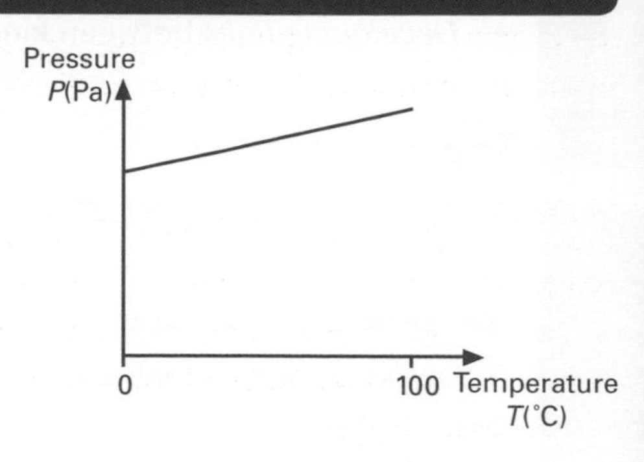

This shows that the pressure increases as the temperature increases, in a linear fashion, though not in direct proportion since the line does not go through the origin. However, Kelvin also had ideas about this (see below).

The Kelvin scale of temperature

In the middle of the nineteenth century, scientists (especially William Thomson, Lord Kelvin) recognized that extending the two graphs above to lower and lower temperatures would produce an interesting result. At a certain temperature it seems that the volume and pressure would fall to zero and if the graphs are re-plotted, we get:

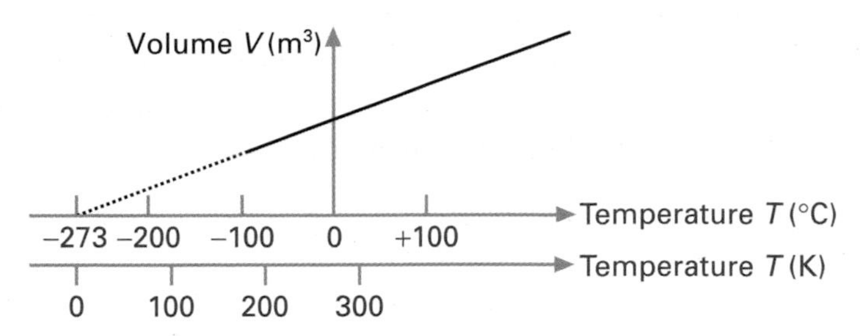

Here the lower scale on the temperature axis is the **Kelvin** or **absolute** temperature, and:

- temperature (°C) = temperature (K) – 273.15
- temperature (K) = temperature (°C) + 273.15

As long as T is in K, then Charles' Law becomes:

V is proportional to T, or $V = kT$, or V/T = constant, or $V_1/T_1 = V_2/T_2 = V_3/T_3$ = etc

and the Pressure Law is:

P is proportional to T, or $P = kT$, or P/T = constant, or $P_1/T_1 = P_2/T_2 = P_3/T_3 =$ etc

These can be combined into one equation, the **Ideal Gas Equation**:

pV/T = constant (R, the molar gas constant, for one mole of gas), so that

$pV = RT$ for one mole
$pV = nRT$ for n moles

$R = 8.31\ \text{Jmol}^{-1}\text{K}^{-1}$.

Ideal gases

An **ideal gas** obeys all three gas laws. The laws work well for gases like air, helium and nitrogen when studied at normal atmospheric pressure and around room temperature. But real gases can be found at lower temperatures and higher pressures. Real gases do not obey these laws.

Checkpoint 2

What problems would arise in extending the V-T and p-T graphs to lower and lower temperatures?

Examiner's secrets

Many candidates make careless mistakes in calculations. One of the most common is ignoring the fact that in this topic temperature is always measured in kelvin (K).

The jargon

This equation is sometimes called the Equation of State for an Ideal Gas

Watch out!

Even though ideal gases play an important role in physics, none exists! You might wonder why we bother with them. Well, at high temperatures many real gases behave like they are ideal.

Exam practice

answers: page 110

(a) 5×10^{-2} m³ of a gas are held at a temperature of 300 K and a pressure of 1.5×10^6 Pa. If the pressure is increased to 2.0×10^6 Pa, while the temperature remains constant, calculate the new volume occupied by the gas.

(b) A cylinder of volume 3×10^{-3} m³ holds gas at 1×10^6 Pa and 327 K. Calculate the number of moles and then the number of molecules present.

(10 mins)

Behaviour of gases – Theory

"Perfect clarity would profit the intellect but damage the will."

Pascal

What would happen if you had the ability to 'see' air? What would you notice? How could you convince others to believe in your vision? Perhaps you could make use of the Invisible Man story. If an invisible man walked through a crowded street, he might be invisible, but his existence could be proved. How? It is his effect on others that would give him away. So, too, with gases.

Brownian motion

If your audience was receptive, you might convince them that air existed by measuring the mass of a plastic bag, then re-measuring it again after blowing air into it, the difference being due to the air. Robert Brown's audience of fellow scientists was not easily persuaded.

In 1827, Brown watched pollen grains moving randomly in apparently still water. He concluded that the pollen grains were being bombarded from all sides by tiny, invisible water molecules.

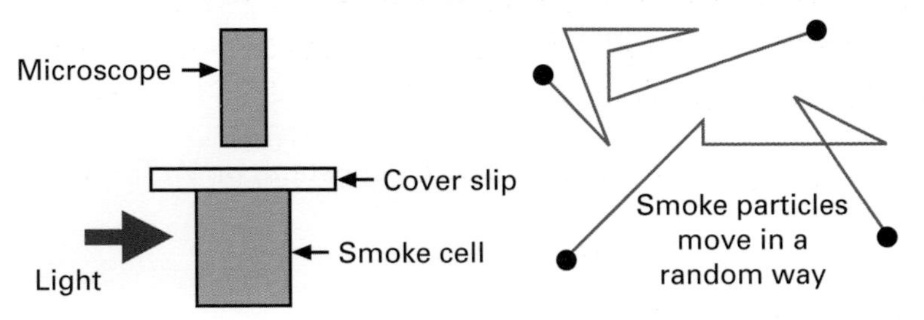

In the updated version of Brown's work, shown above, smoke molecules (people in a busy street) are being bombarded by tiny, invisible air molecules (invisible man). The random nature of the smoke particles' motion, as shown above, provides indirect proof of the existence of air. (People falling over for no apparent reason betray the existence of the invisible man.) It was not until 1906, when Einstein fully explained it, that many scientists finally believed in Brownian motion.

Checkpoint 1

Explain whether the observed movement of the smoke particles can be explained by:
(a) convection currents
(b) gravitational attraction between particles, or between particles and the Earth

Checkpoint 2

What would happen to the motion of the smoke particles if the air was heated?

Kinetic model (as applied to gases)

This theoretical model attempts to explain the behaviour of gases in terms of the motion of the particles that make it up. As with all models, it starts with certain assumptions:

- the volume of the molecules is small compared to the volume of the gas (this avoids having to consider different-sized molecules!)
- there are no intermolecular attractions (which would decrease the pressure produced on the walls of the gas container)
- all collisions are elastic (otherwise pressure would fall)
- the impact time is much less than the time between collisions (this allows us to consider force as rate of change of momentum)

Checkpoint 3

A consequence of one of these assumptions is that an ideal gas would never be allowed to change into a liquid. Which assumption causes this consequence?

Macroscopic and microscopic

Large-scale properties of materials are called **macroscopic** because they can be felt or measured by instruments. One of the great strengths of the kinetic theory is that it is a **microscopic** (small-scale) model that can be used to explain many macroscopic properties, like pressure and density.

Kinetic Theory equation

The outcome of applying the basic laws of mechanics to the behaviour of gases is called the kinetic theory equation, which gives an expression of the pressure of a gas in terms of its density and the speed of the molecules. It is:

$$p = {}^1\!/_3\rho\langle c^2\rangle$$

where $\langle c^2\rangle$ is the average of the squares of the speeds of the molecules (mean square speed).
Alternatively, this can be written:

$$p = {}^1\!/_3Nm\langle c^2\rangle/V \qquad \text{(or } pV = {}^1\!/_3Nm\langle c^2\rangle\text{)}$$

for N molecules, each of mass m, occupying a volume V. This is because Nm/V is another way of writing the density.

Watch out!

Even though ideal gases play an important role in physics, none exists! You might wonder why we bother with them. Well, at high temperatures many real gases behave like they are ideal.

Examiner's secrets

Doubling the temperature (in K) increases the average speed of the molecules by about 1.4. This is a common *show that* question. ($\sqrt{2} = 1.414$)

What is needed for theory to predict experiment?

Any theory will only be accepted if it accurately predicts the outcome of experiments. For the kinetic theory, this means that the equation above must agree with the ideal gas equation, which was derived from experiments.
For this to be so, both expressions for pV must be equal:

$${}^1\!/_3Nm\langle c^2\rangle = nRT$$

This can be manipulated to give:

$${}^2\!/_3N \times {}^1\!/_2m\langle c^2\rangle = nRT$$

Here ${}^1\!/_2m\langle c^2\rangle$ is the average translational kinetic energy of a molecule, and N, n and R are constants. Hence, all that is required for theory to predict experiment is that the average translational kinetic energy of the molecules is proportional to the Kelvin temperature. This does not seem unreasonable and, as a consequence, the theory was accepted.
The equation is often written as:

average translational KE = $3/2\ kT$

where k is the Boltzmann constant (= molar gas constant/the Avogadro constant (R/N_A)). Its value is 1.38×10^{-23} JK^{-1}.

The jargon

Translational kinetic energy refers to the energy a molecule has because it is moving along. (Molecules may also spin or tumble – rotational kinetic energy.)

Exam practice

answers: page 110

(a) What qualitative deduction can be made about the speed of air particles (mass = 1×10^{-25} kg) that collide with smoke particles (mass = 1×10^{-15} kg) in a Brownian motion demonstration? How did you arrive at this answer?

(b) There are six molecules in a container. They have speeds of 1, 2, 3, 4, 5 and 6 units respectively.
- (i) What is the mean speed?
- (ii) What is the mean square speed?
- (iii) The square root of this is called the root mean square (r. m. s) speed. What is its value?

(c) In general, solids are denser than liquids. Suggest an exception to this trend. Why does this example behave in such an unusual way? (20 min)

Behaviour of gases – in bulk

"The steam engine has given more to science than science has given to the steam engine."

William Thompson, Lord Kelvin

We have already considered the experiments that were carried out on gases and the theory that was built up in order to explain the observations. These two pages look at bulk gases in more detail – starting with quantities familiar from GCSE, but then extending into heat engines, which are devices that are essential to life in the 21st century in taking hot gases and using them to do useful work; car engines and power stations are very common examples.

Watch out!

Make sure that you use the correct units in any pressure and density calculations

Gas pressure

We think of atoms in gases as fast moving, hard spheres. We know that they create gas pressure when they collide with each other, or with the walls of their container. Pressure is measured in pascals (Pa).

$$\text{pressure} = \frac{\text{force}}{\text{area}} \text{ or } P = \frac{F}{A} \quad (\textit{Note } 1\text{ Pa} = 1\text{ Nm}^{-2})$$

Density

Density tells us how tightly packed atoms in a material are found to be. The density of a substance is the mass of 1 m^3 of that substance.

$$\text{density} = \frac{\text{mass}}{\text{area}} \text{ or } \rho = \frac{m}{V} \quad (\text{Units kgm}^{-3})$$

Checkpoint 1

A sample of helium-4 has a mass of 4.0 kg. How many atoms are contained within this sample?

The mole and Avogadro's constant

A **mole** is just a quantity of atoms or molecules. By definition the number of particles (atoms or molecules) in a mole is always 6.02×10^{23}. This number is called **Avogadro's constant** (or the Avogadro constant).

It can be useful to have a benchmark for comparison. Athletes often compare each performance against their personal best. Scientists compare the mass of atoms or molecules against the mass of 1 mole (that is 6.02×10^{23} atoms) of carbon-12. Carbon was chosen as the benchmark because it is quite common and easy to work with. One mole of carbon-12 atoms has a mass of exactly 12 g or 0.012 kg.

Work done by, or on, a gas

When a gas expands, say in a steam engine, it does work. If an external pressure squeezes a gas, work is done on the gas. In either case, we say:

$$\text{work done} = \text{pressure} \times \text{change in volume}$$
$$W = p\Delta V$$

The work done by a gas is equal to the area under a p–V graph.

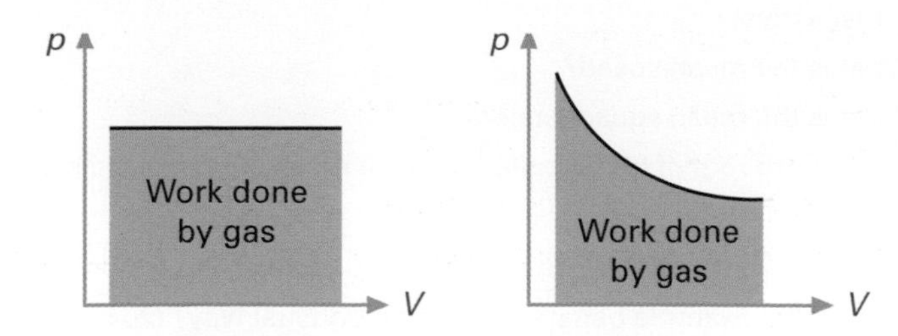

Jargon

Thermodynamics is the study of energy changes that involve heating or cooling.

Heat engines

These are devices that take thermal energy from a material at a high temperature ("hot source") and pass it on to a material at a lower temperature ("cold sink"). On the way, some of the energy is converted to useful work. This diagram summarizes the process:

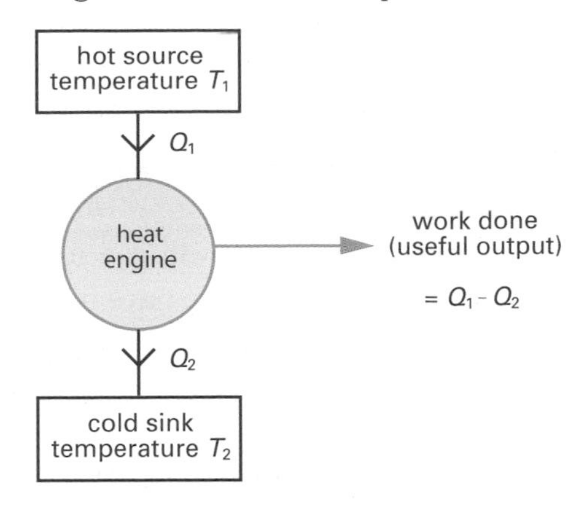

Examiner's secrets

Remember the shape of this diagram. It often comes up in exams.

The efficiency of a heat engine is given by:

$$\text{Efficiency} = \text{useful energy output/energy input} = W/Q_1 = (Q_1 - Q_2)/Q_1 = 1 - Q_2/Q_1$$

Temperatures (in kelvin!) can be used instead of thermal energy giving:

$$\text{Efficiency} = 1 - T_2 / T_1$$

So, to make a heat engine as efficient as possible, make T_2 as small as you can and T_1 as big as you can. This is the reason why car designers try to run their engines (the hot source) at as high a temperature as possible, and why cooling towers are usually seen around power stations – to reduce the temperature of the cold sink.

Checkpoint 2

What are the hot source and cold sink for a petrol/diesel engine?

Test yourself

A refrigerator is a type of heat pump - a heat engine working in reverse. Draw a diagram like the one above, but for a heat pump. Think about the directions of the arrows.

Exam practice

answers: page 110

The efficiency of a heat engine is given by:

Efficiency = $1 - T_2 / T_1$

(a) What is represented by T_1?
(b) What is represented by T_2?

A power station works at an efficiency of 56% and releases steam into the atmosphere at a temperature of 100°C.

(c) What is the initial temperature of the steam in °C?
(d) Give two ways in which the efficiency of this power station could be improved.

Grade booster

Many physical processes can be described on a *microscopic* scale.
It is often useful to relate these to a *macroscopic* scale, that is on a scale that we can measure.

Internal energy

This is a concept that it is important to try to understand for A-level. It is not complicated, but will require you to adapt your vocabulary, because internal energy is a quantity that you have previously perhaps known simply as "heat". It is sensible to try to use that word only to describe the process of *transferring* energy – use internal energy to describe the change that takes place within the object being "heated".

The jargon

Heat is a process, not a form of energy! Heating or cooling an object changes its internal energy.

What is internal energy?

At any temperature above 0 K, the particles within a substance vibrate randomly. They have kinetic energy. They also have potential energy because this motion squeezes and stretches the bonds holding them together. This diagram shows that the particles in a solid are tightly packed.

Here are some other important points about internal energy:

The jargon

Internal energy is the sum of the kinetic energy of all the particles and the potential energy stored in all the bonds.

→ solids, liquids and gases all have internal energy
→ the energy within a substance is continually changing between potential and kinetic; e.g. if the particles move further apart, their bonds oppose the motion, and so kinetic energy decreases and potential energy increases
→ the total internal energy of a substance is constant at a particular temperature (as long as it doesn't change state)
→ at absolute zero (0 K) all objects have minimum internal energy (internal energy is the sum of all potential and kinetic energies)

Action point

Rephrase the points in this list into an essay-style paragraph.

Two ways to increase the internal energy of a gas

→ *Heat it* the walls of the container get hot, making the particles inside move more quickly so their kinetic energy increases.
→ *Do work on it* the walls of the container move in so that the gas particles bounce off more quickly, increasing their kinetic energy.

The first law of thermodynamics

Can you remember the conservation of energy principle? (Energy can never be made or destroyed, but only changed from one form to another.) The first law of thermodynamics just restates the conservation of energy!

$$\Delta U = \Delta Q + \Delta W$$

where ΔU is the change in internal energy of the system
ΔQ is the energy transferred thermally (by "heating")
ΔW is the work done on the system

The jargon

Thermodynamics is the study of energy changes that involve heating or cooling.

Examiner's secrets

This is another favourite topic. Make sure that you are clear about the signs.

Note that ΔQ is *positive* if thermal energy is transferred *to* the system and *negative* if thermal energy is transferred *from* the system. Similarly, ΔW is *positive* if work is done *on* the system and *negative* if work is done *by* the system.

As long as the state does not change, ΔU is *positive* if the temperature goes *up* and *negative* if the temperature goes *down*.

What is temperature?

This sounds a very simple question, and the answer, too, is simple, but you will have to think about it. Temperature is simply a number on a scale that tells you the direction of the transfer of thermal energy – it will *always* flow from hot objects to colder objects. The scale is just an agreed system form describing temperature consistently; *any* system would do. We have adopted the Celsius scale, but the USA stick with Fahrenheit.

It's no more complicated than that, and is the easy way of deciding whether an energy change is "heating" or "working". Just ask yourself "is a temperature difference necessary to make the energy move?" If the answer is "yes", then it is **heating**; if it is "no", it is **working**.

Checkpoint 1

Predict what will happen to the temperature of a rubber band, previously stretched, if it is subsequently released.

Checkpoint 2

It has been rumoured that James Joule wanted to measure the water temperature at the top and bottom of a waterfall while on honeymoon in Switzerland. What scientific link was he trying to prove?

Exam practice answers: page 110

A 40 W filament bulb is operating at its normal working temperature of 1600°C. Apply the equation $\Delta U = \Delta Q + \Delta W$ to the filament for a period of 5 seconds, stating and explaining the size and sign of ΔU, ΔQ and ΔW. Would the *signs* be different if you were considering the first millisecond after the bulb was switched on? (10 min)

Specific and latent heat capacities

How could you speed up making a cup of tea? If the kettle had to supply less energy that would help! You could heat less water or not wait for the water to reach 100 °C. (Or you could use alcohol rather than water. The same mass of alcohol requires less energy than water to reach 100 °C – strange tea though!)

The jargon

Specific refers to 1 kg of a substance and means per unit mass.

Watch out!

Specific heat capacity is now often known as *specific thermal capacity* to get away from the old idea that *heat* is the energy inside matter.

Grade booster

Marks are available for good spelling, punctuation and grammar. For example, when a metre is a unit of measurement it ends in *–re*. When a meter is a measuring instrument it ends in *–er*. The joulemeter can be replaced by an ammeter, voltmeter and stop-clock, of course.

Action point

Rearrange $c = \Delta Q/(m\Delta\theta)$ to give an expression for the amount of energy ΔQ required to raise the temperature of a substance.

Watch out!

Several solids, such as ice, contract on melting. The important point is that the particles in a solid stay in fixed relative positions.

Specific heat capacity *c*

If you supply the same amount of energy to 1 kg of alcohol, it will get hotter than 1 kg of water treated the same way.

- The **specific heat capacity** of a substance is defined as the amount of energy required to raise the temperature of 1 kg of the substance by 1 °C.

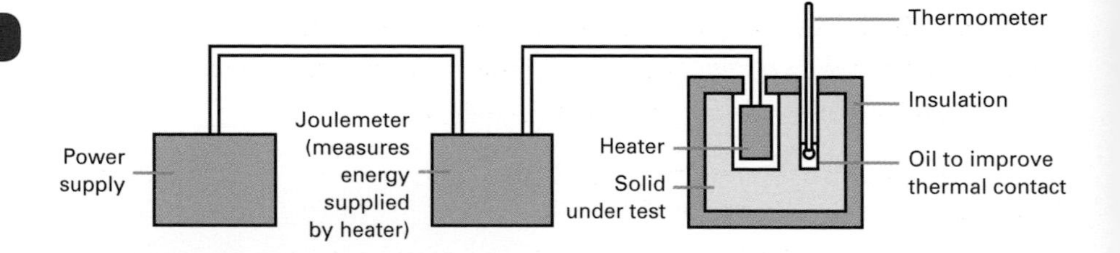

Determining a solid's specific heat capacity *c*

- Find the mass m of the solid.
- Record its initial temperature.
- Switch on the power and the joulemeter.
- Record the change in temperature $\Delta\theta$ in a given time.
- Record how much energy ΔQ has been supplied in this time.
- Calculate specific heat capacity c using $c = \Delta Q/m\Delta\theta$.

Changes of state

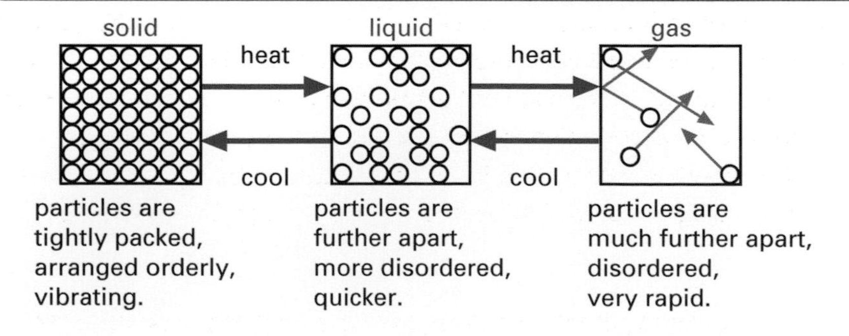

As a solid is heated, the network of bonds that holds its particles together breaks down. In liquids, the particles can wander around. This allows liquids to flow into any shape. Gas particles have the greatest kinetic energy, because they move so quickly. They also have the greatest potential energy because they are the furthest spread apart.

If we could supply energy to ice at a steady rate we would get idealized results and a graph like this:

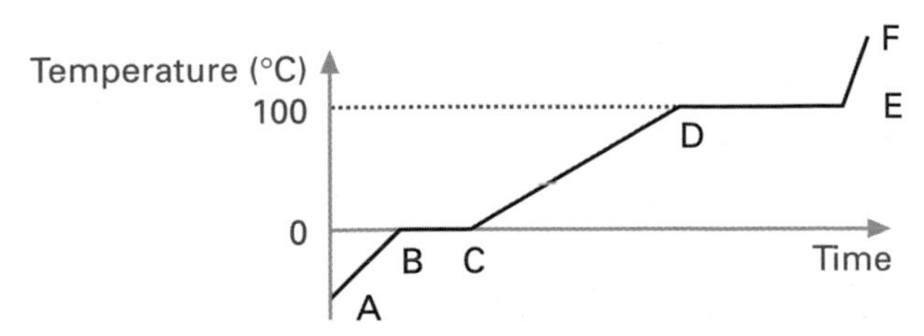

You can see from the graph that there is no change in temperature during changes of state (between B–C and D–E). The energy transferred to the water, during these periods, pulled particles apart by breaking the bonds between neighbours. It increased the potential energy of the particles in the process.

Look at the other sections of the graph (A–B, C–D and E–F). You can see that between changes of state, the energy supplied to the water increased its temperature. This allowed the particles to move faster so that their kinetic energy increased. All the time that energy was transferred into the system, its internal energy was increasing.

Checkpoint 1

How can you tell from this graph that boiling water requires more energy than melting ice?

Specific latent heat *L*

→ The amount of energy required to melt or boil 1 kg of a substance, without making it warmer, is called the **specific latent heat of fusion** (solid to liquid) or **vaporization** (liquid to gas). The equation that you have to remember is:

$$L = \Delta Q/\Delta m \quad \text{or} \quad \Delta Q = L\Delta m$$

The jargon

Latent means hidden. When you melt or boil a substance, its temperature doesn't rise – the energy supplied seems to have disappeared!

Evaporation

Evaporation is not the same as boiling! It is a surface effect during which the liquid does not have to be heated to its boiling point. The gas formed by *evaporation* is a *vapour* – a gas below its boiling point. When you perspire, the sweat evaporates from the surface of your skin making use of evaporation's cooling effect. There are two ways to explain this effect. You can say that the sweat takes latent heat from your body so that it can start evaporating, thus lowering your temperature. Or, you can explain that it is the most energetic particles that leave the surface of the liquid, reducing the average energy of those left behind so that the liquid is now colder.

Checkpoint 2

Why does an aerosol can cool down when you use it?

Exam practice answers: page 110

(a) A 20 kg hot water tank, made of copper ($c = 390$ J kg^{-1} °C^{-1}), holds 70 kg of water. Calculate how much energy is required to raise the temperature of the tank by 1 °C, first when it is empty and then again when it is full of water ($c = 4\,200$ J kg^{-1} °C^{-1}).

(b) Why does melting ice require much less energy than boiling the same mass of water?

(c) Draw diagrams to show the differences between evaporation and boiling. (20 min)

Grade booster

You should always be prepared to answer questions based on your practical work. There are a number of common class practicals involving heat capacity and latent heat. Look back over your lab book!

Answers
Kinetic theory

Behaviour of gases - Experiment

Checkpoints

1 During inhalation, the diaphragm (a wall of muscle below the lungs) moves down, increasing the lungs' volume so their internal pressure falls (Boyle's law). As external pressure is now greater than the pressure inside the lungs, air is forced in. The reverse is true for exhalation.
2 The gas would turn to liquid at some stage, and hence the extended graphs are shown with a dashed line.

Exam practice

(a) 0.0375 m^3
(b) 1.10 moles, 6.62×10^{23} molecules

Behaviour of gases – Theory

Checkpoints

1 Convection and gravity would both cause the smoke particles to move along predictable routes; neither can explain the haphazard motion that is observed.
2 More violent collisions would force the smoke to be even more agitated.
3 No intermolecular forces of attraction.

Exam practice

(a) The air molecules are much lighter than the smoke particles, yet they still push them around! The air molecules must be moving a lot faster than the smoke particles.
(b) (i) Mean speed $= \frac{1+2+3+4+5+6}{6} = \frac{21}{6} = 3.5$ units

(ii) Mean square speed $= \frac{1^2+2^2+3^2+4^2+5^2+6^2}{6}$

$= \frac{91}{6} = 15.71$ units

(iii) r.m.s speed $= \sqrt{\frac{91}{6}} = 3.9$ units (i.e. bigger than mean)

Behaviour of gases – in bulk

Checkpoints

1 6.02×10^{26}.
2 Hot source is burning fuel/air mixture; cold sink is atmosphere.

Exam practice

(a) Temperature of the hot source.
(b) Temperature of the cold sink.
(c) Eff = 1 - T_2 / T_1
hence $T_2 / T_1 = 1 - E = 1 - 0.56 = 0.44$
$T_1 = T_2/0.44$
$T_1 = 373(K)/0.44 = 848$ K = 575 °C
(d) Increase temperature of hot source and decrease temperature of cold sink.

Internal energy

Checkpoints

1 It becomes cooler.
2 A link between energy and temperature.

Exam practice

ΔU is zero (temperature is constant); ΔW is +200J (40W x 5 s), since electrical work is being done *on* the filament; ΔQ = –200J (from First Law of Thermodynamics), since the filament is "heating" its surroundings.
For the first millisecond: ΔU is positive (temperature of filament is rising); ΔW is positive (electrical work being done *on* the filament; ΔQ is still negative (filament hotter than surroundings, so energy transferred *from* it).

Specific and latent heat capacities

Checkpoints

1 Section DE is longer than BC. So it takes longer to boil the water than melt the ice. As energy is supplied at a constant rate, more is supplied to boil the water.
2 In evaporation, energetic particles escape first, forming a gas and reducing the average energy of the liquid particles. The liquid, and its container, both cool.

Exam practice

(a) Empty = 7 800 J, full = 301 800 J.
(b) When a solid melts, only one or two bonds are broken per molecule. Each molecule is still bonded to most of its neighbours. When a liquid boils, each molecule has to break free from all its neighbours; eight or nine bonds may have to be broken so more energy is required to boil rather than melt.
(c)

Evaporation is a surface effect

Boiling is a bulk effect, it happens throughout the liquid

Heat

Waves and oscillations

We live in a wavy world! If you look around, listen to a radio or speak on a telephone your life involves waves. Whether you use a microwave oven, dip your toe in a bath or play a musical instrument, it is hard to avoid waves. Waves are very important in physics (and science generally). Health physics, astrophysics, particle physics, electricity, light, sound . . . the list goes on and on. What do they have in common? Waves!

Exam themes

- *Mathematical description* of waves, e.g. the wave–speed equation.
- *Wave properties* Description and understanding of wave properties, e.g. diffraction, interference.
- *Natural world* Examination of aspects that involve waves, such as light and sound.
- *Applications in modern physics* of wave ideas, e.g. wave–particle duality.

Topic checklist

	Edexcel		AQA/A		AQA/B		OCR/A		OCR/B		WJEC		CCEA	
	AS	A2	AS	A2	AS	A2	AS	A2	AS	A2	AS	A2	AS	A2
Types of waves and their properties	○		○		○		○		○		○		○	
Electromagnetic spectrum	○		○		○		○		○		○		○	
Reflection and refraction	○		○		○		○		○		○		○	
Applications of reflection and refraction	○		○		○		○		○		○		○	
Diffraction	○		○		○		○		○		○		○	
Superposition	○		○		○		○		○		○		○	
Interference	○		○		○		○		○		○		○	
Standing waves	○		○		○		○		○		○		○	
Photoelectric effect	○			●		●	○		○		○		○	
Atomic line spectra		●	○	●		●	○		○		○		○	
De Broglie's equation and atomic models		●	○	●		●	○	●			○		○	

Types of waves and their properties

"Almost everything that distinguishes the modern world from earlier centuries is attributable to science."

Bertrand Russell

It has been said that if the population of China jumped up and down in harmony, they would set up a tidal wave that could sweep the Earth. Waves carry energy from place to place.

Transverse waves

Transverse waves are caused by vibrations moving at right angles to the direction of the wave motion. All electromagnetic waves, such as light waves, are transverse. When a stone is dropped into water, the waves that spread out are transverse.

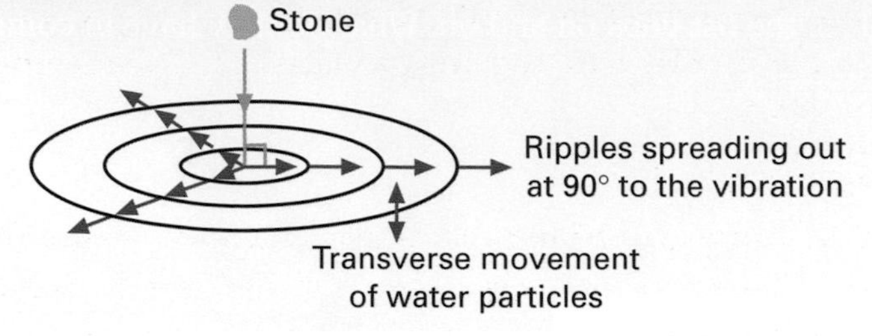

Jargon

A *slinky* is a large diameter, flexible spring.

You can also use a slinky to demonstrate transverse waves.

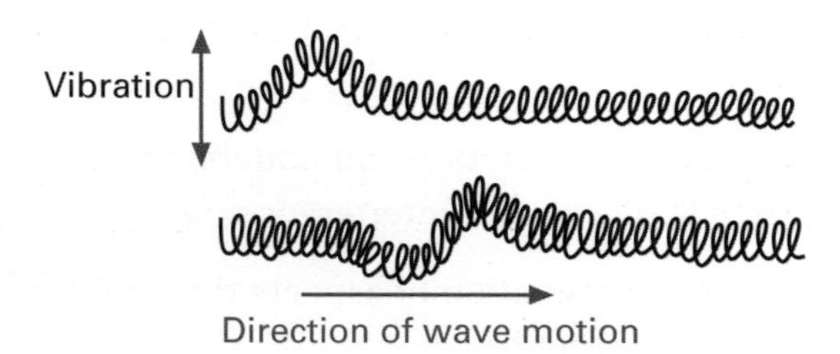

Checkpoint 1

Earthquakes also produce transverse waves. What would happen to both types of seismic waves if they reached a region of denser rock within the Earth?

Checkpoint 2

When waves are produced, do they transfer energy or matter?

Longitudinal waves

Vibrations moving in the same direction as the wave motion cause **longitudinal waves**. When your vocal cords vibrate back and forth, the resulting sound waves are longitudinal. The fastest (primary) waves produced by earthquakes are longitudinal. A slinky can also be used to demonstrate longitudinal waves:

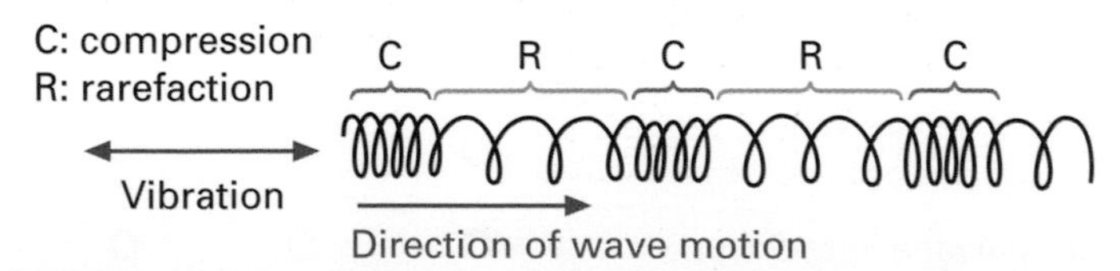

Describing waves

You need to learn the following to fully describe waves:

- **Displacement** x the distance between a point on the wave and the line of zero disturbance. It is measured in metres.
- **Amplitude** a maximum displacement. It is also in metres.
- **Period** T the time taken for one complete vibration to be made. It is measured in seconds.
- **Frequency** f the number of vibrations made per second. It is measured in hertz (Hz), 1 Hz = 1 vibration per second. ($f = 1/T$)
- **Wavelength** λ the distance between two similar points on a wave (e.g. crest to crest or trough to trough). It is usually in metres.
- **Speed** c for a wave, speed equals distance/time.

Examiner's secrets

The examiner expects you to include an appropriate unit when you define a physical quantity, e.g. speed in m s^{-1}

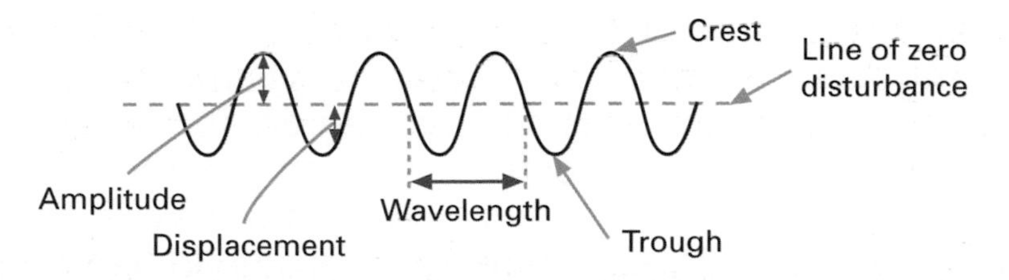

Classifying waves

We have already met one way to distinguish between waves: they are transverse or longitudinal. Waves can also be described as:

- **Mechanical** requiring a material to travel through, e.g. sound.
- **Electromagnetic** (EM) not requiring a material to travel through, e.g. light waves.

Waves can also be thought of as:

- **Progressive** spread out energy from the source vibration into the surrounding space, e.g. when a ripple is made on the surface of a puddle. As the energy spreads out, so its intensity decreases.
- **Stationary** or **standing** waves. The positions of their peaks and troughs do not move. Think of a guitar string. Some parts of the string vibrate while other parts (e.g. the ends) do not.

Wave properties

All waves can be **reflected** (bounced back), **refracted** (bent) and **diffracted** (spread out). As waves can spread out, they can also **interfere** with one another. All these properties are dealt with later in this section. Transverse waves, but not longitudinal, can be **polarized**. This means that they can be forced to oscillate in one fixed direction only. In this example, the transverse wave has been polarized in the vertical plane:

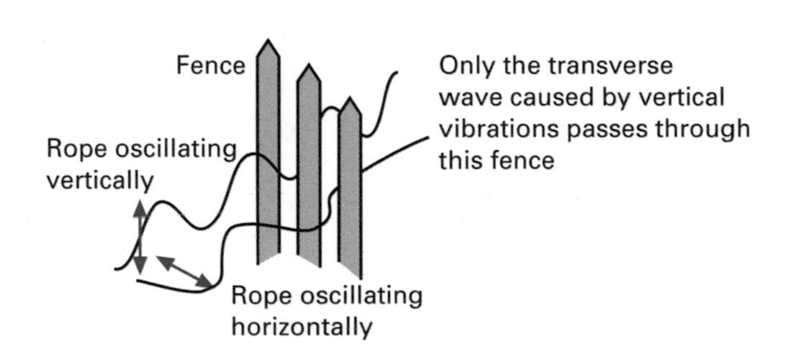

The wave equation

The speed, frequency and wavelength of any wave are related by:

speed = frequency × wavelength or $c = f\lambda$

Exam practice answers: page 134

(a) Describe how transverse and longitudinal waves differ.

(b) Explain why it is impossible to polarize one of these wave types.

(c) Give an example of both transverse and longitudinal waves.

(d) Give one example of a standing transverse wave, and then name a standing longitudinal wave. (20 min)

Test yourself

Study the terms and diagram used to describe waves. Then cover this page with a blank sheet of paper and see how much you can remember.

Action point

Divide a page into four squares. Label the rows *transverse* and *longitudinal*. Label the columns *electromagnetic* and *mechanical*. Now place all the waves you know in the appropriate squares, but remember, the waves must match the labels. Where would you place a *Mexican* wave?!

Links

For more on EM waves see pages 114–5.

Watch out!

Don't get confused by the term *stationary wave*. It does not mean that nothing is moving.

Checkpoint 3

What are the units of each of the physical quantities in the wave equation?

Electromagnetic spectrum

"First we had the luminiferous ether. Then we had the electromagnetic ether. And now we haven't e(i)ther"

The Strange Story of the Quantum

It was Newton who first discovered that white light is composed of seven colours, the visible spectrum. Now we know that this visible region forms part of the electromagnetic spectrum, which contains other similar types of wave. Electromagnetic waves are all around us, and have many properties in common. These pages list those properties, then consider more detailed characteristics and uses of the different types of electromagnetic waves.

Properties of all EM waves

All EM waves share common features:

- they transfer energy
- shorter wavelength waves are more energetic and dangerous
- they can be reflected, refracted or diffracted
- they are transverse waves
- they can all travel through a vacuum at 3.0×10^8 m s^{-1}
- wave speed = frequency × wavelength

Electromagnetic (EM) waves

EM waves are produced when charged particles (e.g. electrons) vibrate or lose some of their energy. You need to know about the main radiations listed below.

Radiation	*Wavelength (m)*	*Produced by*
Radio	$>10^6$ to 10^{-1}	electrons oscillating in a transmitting aerial
Microwave	10^{-1} to 10^{-3}	Klystron (electron tube) oscillators
Infrared (IR)	10^{-3} to 7×10^{-7}	hot solids emit IR; in fact all objects emit some IR because of the motion of their particles
Visible	7×10^{-7} (red) to 4×10^{-7} (violet)	the Sun and vibrating atoms in other light sources, e.g. bulbs
Ultraviolet (UV)	4×10^{-7} to 10^{-8}	high-temperature matter emits some of its energy as UV beyond visible violet, hence ultraviolet
X-rays	10^{-8} to 10^{-13}	bombarding metal targets with fast moving electrons
Gamma rays	10^{-10} to 10^{-16}	nuclear processes such as radioactive decay

Features and uses of EM waves

The features, and therefore uses, of EM waves gradually change as their wavelengths, frequencies and energies change. You are expected to know about the main features of each region of the EM spectrum.

Test yourself

Try to memorize the wavelength range for all the principal radiations shown here.

Links

For more on X-rays see the *medical and health physics 2*, pages 166–7. For gamma rays see *properties of ionizing radiation*, pages 50–1.

Don't forget

$10^0 = 1$ and you must key 10^7 (for example) into your calculator as 1 EXP 7.

Checkpoint 1

After studying this double-page spread, try to establish what the only real difference is between an X-ray and a gamma ray of the same wavelength.

Grade booster

The usual way to represent the electromagnetic spectrum is on a *logarithmic* scale. The numbers do not go up in even steps, but in equal ratios. A linear scale for this would stretch as far as Pluto! You should be able to explain what a *logarithmic* scale is and why it is useful.

Radio waves

Stars produce radio waves that can cause hissing noises in televisions or radios that have not been tuned to a particular station.

- *Ultra high frequency* (UHF) waves transmit TV signals.
- *Very high frequency* (VHF) waves are used by local radio, police and ambulance communications.
- *Medium and long wave* radio waves send messages over longer distances. As they have long wavelengths, they can diffract around the curve of the Earth and any hills that might be in their path.

Action point

Use $c = f\lambda$ to check that as velocity is constant for all EM waves, a shorter wavelength means a higher frequency.

Microwaves

Microwaves are sometimes considered to be a sub-group of radio waves. They are usually associated with cooking but do not forget about:

- *TV and communications satellites* Microwaves are the most energetic radio waves and are therefore ideal for this application.
- *Radar* originally used in World War Two, radar is now used for air-traffic control around the world.

Action point

Use $E = hf$, where E is energy, h is a constant and f is frequency, to check that higher frequencies are more energetic.

Watch out!

Photon energy is also given by $E = hc/\lambda$, which you can get by combining $E = hf$ and $c = f\lambda$.

Infrared (IR) radiation

IR cameras are used by fire fighters to detect IR radiation coming from warm objects, like humans! IR radiation is also used in:

- *Optical fibres* IR radiation can be used in telephone networks to code and send information, such as speech, along these fibres.

Watch out!

Short wavelength IR is often called IR *light* even though it isn't visible!

Light

This is the only form of radiation that is visible to the human eye. Therefore it is very important in communicating images or ideas.

Ultraviolet (UV) radiation

UV rays can cause tanning, skin cancer, eye damage and fluorescence.

Grade booster

Try to be specific when giving information. Don't say that UV radiation causes cancer; make sure you mention skin cancer.

X-rays and gamma rays

X-rays can cause cancer, but in concentrated beams they can be used to kill cancer cells. Gamma rays have similar effects and uses.

The nature of electromagnetic radiation

An EM wave is a disturbance in electric and magnetic fields in space. A change in an electric field can cause the associated magnetic field to change too. The resulting continuous cycle can keep the EM wave going. But the wave aspects of EM radiation are only half the story; you can start to find out about the particle aspects on page 132.

Exam practice answers: page 134

The development of an understanding of our physical environment has taken place over many centuries. With reference to one area of physics, describe how the earlier work of scientists led the way for those who followed them. (25 min)

Reflection and refraction

> *"God created everything by number, weight and measure."*
>
> Isaac Newton

Physics is about explaining the world around us. Why is it that when an explosion damages property in a narrow street, we often find that only every other window, on each side of the street, has been broken? How are rainbows formed after heavy rain? The phenomena used to explain these examples are reflection and refraction.

Checkpoint 1

Why can we only normally see the Moon by night?

The laws of reflection

Light travels in straight lines called **rays**, with many rays forming a **beam**. This can be seen when we examine laser light or see sunlight streaming through trees. Luminous objects, which are normally hot (e.g. the Sun), give out their own light. But most objects are non-luminous (e.g. the Moon); we can only see them if light bounces off them (reflects) and enters our eyes.

Action point

This diagram shows *regular* reflection, with light bouncing off a smooth surface. Draw a similar diagram to show *diffuse* reflection from a crumpled surface, e.g. cooking foil.

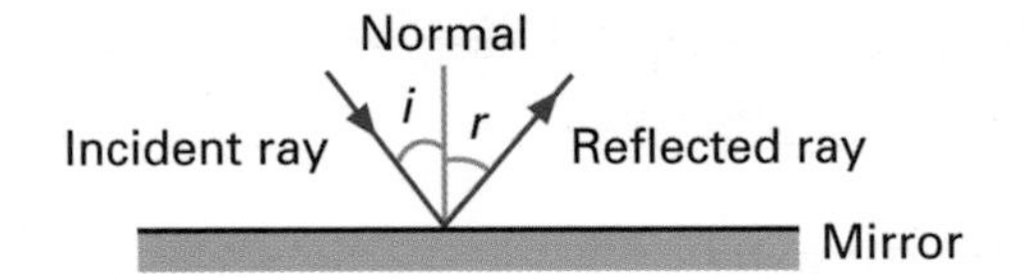

The jargon

If three lines all lie in the same plane, they are said to be *co-planar*. They can all be drawn on one flat sheet of paper.

The **laws of reflection** are that:

- The angle of incidence i is equal to the angle of reflection r.
- The normal, incident and reflected rays all lie in the same plane.

Refraction

The jargon

A *slanting angle* is taken to be an angle of incidence other than 0°.

Why does light bend when it leaves one *medium* (e.g. water) to enter another (e.g. air) at a *slanting angle*? You may have seen this when a wooden rod (such as a pencil) is held under the surface of a beaker of water and appears to be bent. This effect is due to the fact that the speed of light changes and is called **refraction**.

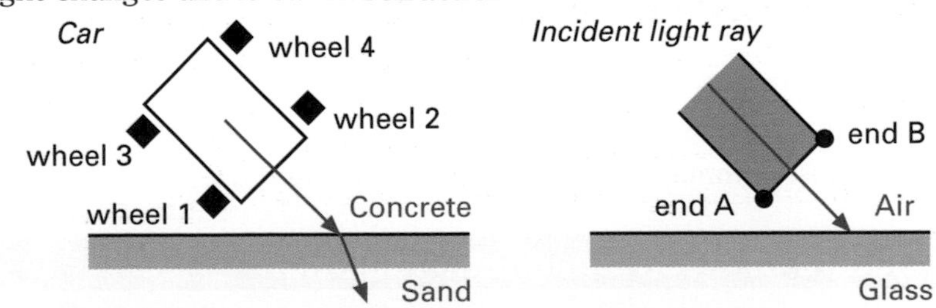

Checkpoint 2

What would happen to the car, or ray, if the angle of incidence was 0°?

Imagine a car leaving a road to enter a beach, as shown above. As each wheel enters the sand, it will slow down. The diagram above shows a bird's-eye view of the scene. We can see that wheel 1 will reach the sand first and we can imagine that the car will change direction as shown. Consider a light ray entering glass:

- Which end of the ray enters the glass first (A or B)?
- What will happen to the speed of that end, on entry to the glass?
- What will happen to the path of the light ray?

Examiner's secrets

By understanding how and why phenomena like refraction happen, you will find it easier to remember the equations associated with them.

Think again of a car driving into sand; you should see that if the speed of the wheels changed more dramatically on going from one medium to another, then the car would swerve even more. The ratio of the two speeds, called the **refractive index** (n), tells us about the degree of bending.

The laws of refraction

Try to remember that when a ray enters an *optically* denser medium, e.g. travelling from air into glass, it bends towards the normal.

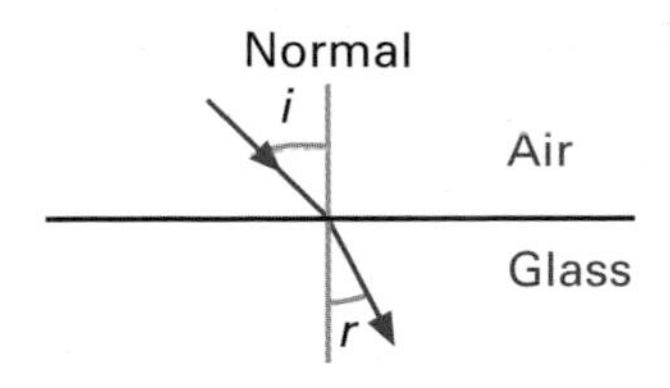

The **laws of refraction** are:

➜ $n_1 = \dfrac{\text{speed of light in a vacuum } c}{\text{speed of light in medium 1 } c_1}$

Where n_1 is called the **absolute refractive index**.

➜ refractive index $_1n_2 = \dfrac{\text{speed in one medium } c_1}{\text{speed in second medium } c_2}$

Where $_1n_2$ is the refractive index for light, or the car, travelling from medium 1 into medium 2.

➜ $_1n_2 = \dfrac{\sin i}{\sin r}$ (Snell's law) and $_1n_2 = \dfrac{\text{real depth}}{\text{apparent depth}}$

➜ $_1n_2 = \dfrac{n_2}{n_1}$ and $_2n_1 = \dfrac{1}{_1n_2}$

Example

An ultrasound wave travelling through muscle approaches bone at an angle of 15°. As $_{\text{muscle}}n_{\text{bone}} = 0.377$, calculate the angle of refraction.

$$_1n_2 = \sin i \div \sin r$$

$$_{\text{muscle}}n_{\text{bone}} = \sin 15° \div \sin r = 0.377$$

$$\sin r = \sin 15° \div 0.377 \quad \text{so} \quad r = 43.4°$$

Waves: reflecting and refracting

Ripple tanks can be used to show reflection and refraction of waves. The thin lines, in the diagrams below, represent wave crests. Notice that when the waves refract in the second diagram, there is a change in wavelength due to a change in speed in the shallow water.

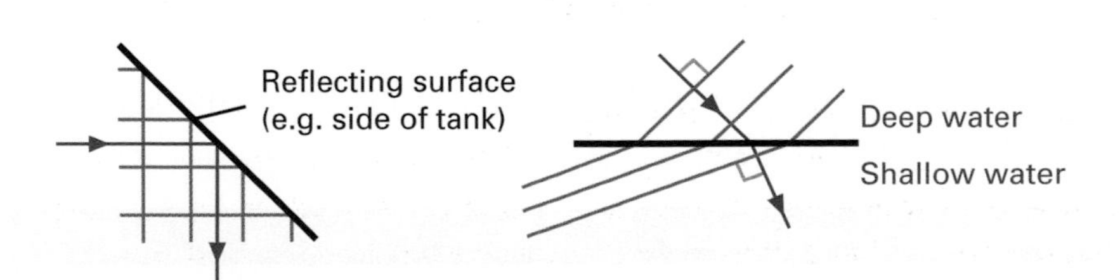

Exam practice

answers: page 126

(a) The refractive index of glycerol is 1.47. Calculate the angle of refraction if light enters glycerol at an angle of incidence of (i) 5°, (ii) 15°.

(b) Violet light has a wavelength of 400 nm in a vacuum. When it enters an unknown material, its speed decreases to 2.5×10^8 m s^{-1}. Calculate the frequency of violet in both media and its wavelength in the unknown material. (15 min)

Checkpoint 3

In what direction would a ray bend if it entered an optically less dense medium at a slanting angle?

Watch out!

Don't get confused! If you look up the value of a refractive index in a book, you will find the absolute refractive index quoted.

"What is it that breathes fire into the equations and makes a universe for them to describe?"

Stephen Hawking

Examiner's secrets

The work of many students can be let down by howlers. This candidate was describing a refraction experiment: 'We *filled* the ripple tank with *shallow* water.'

Grade booster

Every year examiners report that many students fail to gain marks because of the poor quality of their diagrams. Make sure you are not one of these students by practising your drawings until they are clear and accurate.

Applications of reflection and refraction

"I think there's a world market for about five computers."

Thomas Watson, Founder of IBM

There are many applications of the effects introduced on the previous page and two are considered here. Firstly, a property that is very important in 21st century communication – total internal reflection and its use in fibre optics. The function of lenses in forming images is another essential feature of modern life – from glasses to correct defective sight to the more complex imaging systems in telescopes.

Total internal reflection (TIR) and the critical angle

When light travels from one medium (e.g. glass or plastic) into a less dense medium (e.g. air), it bends away from the normal. If the angle of incidence i is greater than a critical angle C, the light will be totally internally reflected at the boundary of the two media.

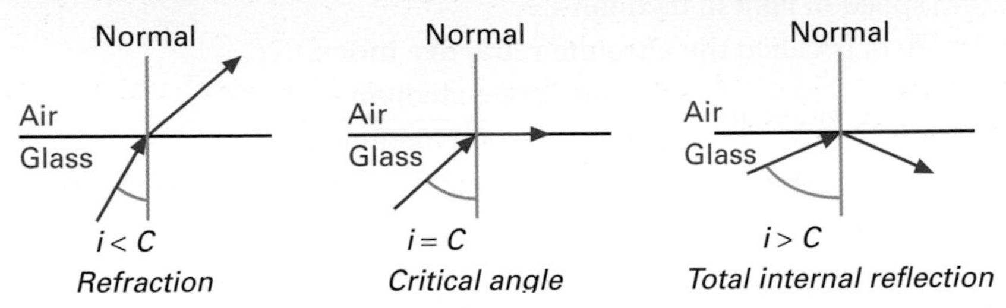

Consider the diagram above. When the angle of incidence equals the critical angle, the ray is refracted along the boundary of the two media:

$${}_{\text{glass}}n_{\text{air}} = \sin i \div \sin r = \sin C \div \sin 90^{\circ} = \sin C$$

We already know that ${}_1n_2 = 1/{}_2n_1$ so:

$${}_{\text{air}}n_{\text{glass}} = 1/\sin C$$

Checkpoint 1

Calculate the critical angle for the sample of glass shown in the central figure opposite, where ${}_{\text{air}}n_{\text{glass}} = 1.50$.

Grade booster

It is easy to get the ratios the wrong way up. If you get a refractive index less than 1, you must have made a mistake. Check again if this happens to you!

Fibre optics

One of the most important uses of total internal reflection is in the field of telecommunications. Information is sent, as pulses of light, along optical fibres. These optical fibres have some important advantages:

- each one can potentially carry more than 30 000 calls at a time
- they are lighter, smaller and cheaper than copper cables
- the signals carried are more secure, reliable and of a better quality than was previously the case
- the signals can travel tens of kilometres before they require additional amplification

Fibre optics – up close

Optical fibres are often grouped together in bundles. Each fibre is protected by a polyurethane cover, as shown below. Light, whose angle of incidence is greater than the critical angle, is 'trapped' and sent down each optical fibre.

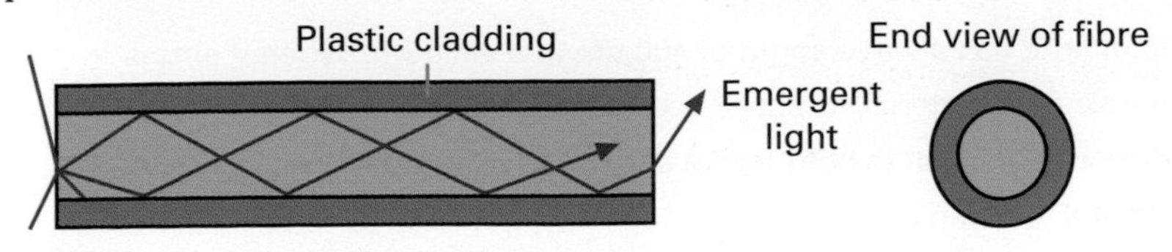

Optical fibres must let light pass through as easily as possible. Very pure sodium borosilicate glass is used, as a 20 km length is as transparent as a window pane. However, fibre optics do have practical problems.

- They must use monochromatic (single-wavelength) light to stop some pulses travelling faster than others. (Lasers are used to send monochromatic infrared pulses. They can be switched on and off at a rate of up to 2×10^8 Hz, meaning that millions of pulses can be sent every second.)
- The fibres must be very thin (no thicker than a human hair) to reduce the number of possible paths the light pulses can take. (This also helps to prevent some pulses arriving before others.)

Checkpoint 2

A material used to produce optical fibres absorbs 1% of the available light for each metre of its length. Calculate what percentage of the original light would reach a regenerator placed 50 m from the start of the fibre.

Checkpoint 3

State three changes that light undergoes when it travels from air into an optical fibre, at a slanting angle.

Lenses and the lens equation

Lenses crop up in a wide range of applications, and the way that they form images is shown below:

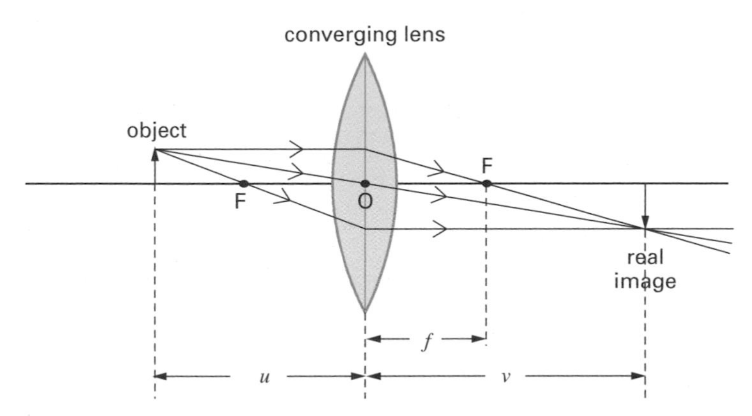

Checkpoint 4

If the refractive index of a glass fibre is 1.5, the speed of light in it will be 2×10^8 m s^{-1}. By how much does the wavelength in the fibre change?

There are three rules that you need to follow to draw ray diagrams to locate images formed by lenses:

- a ray parallel to the principal axis is refracted through F
- a ray through the optical centre (O) continues in a straight line
- a ray through F is refracted parallel to the principal axis

The convention that is used is called "real is positive, virtual is negative", and the equation that you will need is:

$$1/u + 1/v = 1/f$$

where u is the object distance, v is the image distance and f is the focal length. This works for converging and diverging lenses.
The magnification, m, is given by:

$$m = v/u$$

Exam practice answers: page 134

(a) Write down six points which could form the basis of an essay on 'The social, economic and technological changes which have arisen as a result of the development of fibre-optic transmission of information'.

(b) Name four other uses of fibre optics.

(c) Draw diagrams to illustrate total internal reflection and the critical angle. (20 min)

Diffraction

Why can we hear around doorways and yet we cannot see around them? It has been known for centuries that waves spread out when they pass barriers, as long as the gap is about the right size. This effect is called diffraction and the fact that it could not be observed in light was a big issue when the wave theory of light was first proposed. We now know that if we look closely, diffraction of light can be seen. Diffraction is a property that is a fundamental characteristic of a wave.

Watch out!

When using a ripple tank, don't worry if your results are slightly different from those shown in the diagram opposite. Idealized results are shown here.

Checkpoint 1

If you were trying to eavesdrop on two people chatting around a corner from you, would their sound waves reflecting off walls help or hinder you?

Grade booster

You should be able to describe what happens to waves passing through a gap as the gap is progressively narrowed.

Jargon

A *wavefront* is a surface at 90° to the wave's motion.

Diffraction of water waves

Ripple tanks can be used to show diffraction of water waves. Plane (straight) ripples are sent towards a gap or an edge of an obstacle. Waves spread out as they pass through the gap, or bend around the side of the obstacle. This effect is called **diffraction**.

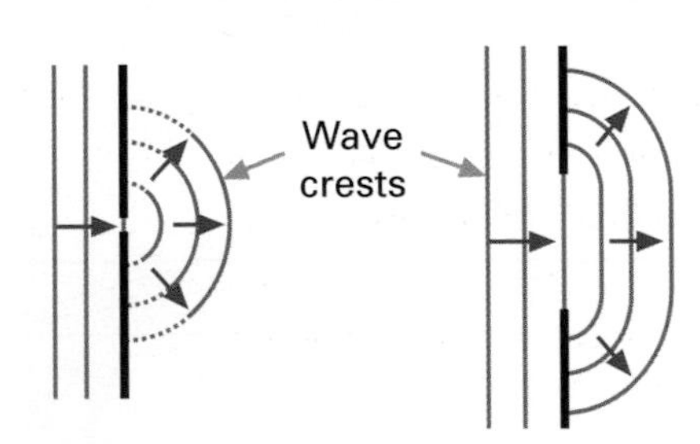

This diagram shows the relationship between gap width and wave length. *Maximum diffraction* occurs when the *gap width is equal to the wavelength*. The wavelength of sound waves is typically the same as the width of a doorway and so sound waves are noticeably diffracted, allowing you to hear around corners!

When a plane wave reaches a narrow gap, it makes the water in the gap go up and down. As far as the water in the gap is concerned, this is no different to a point source (like a stone dropping in the water), and so a circular wave is sent out.

Diffraction of light waves

The wavelength of visible light lies between 4×10^{-7} m and 7×10^{-7} m.

➜ A very narrow gap is required to diffract light successfully.

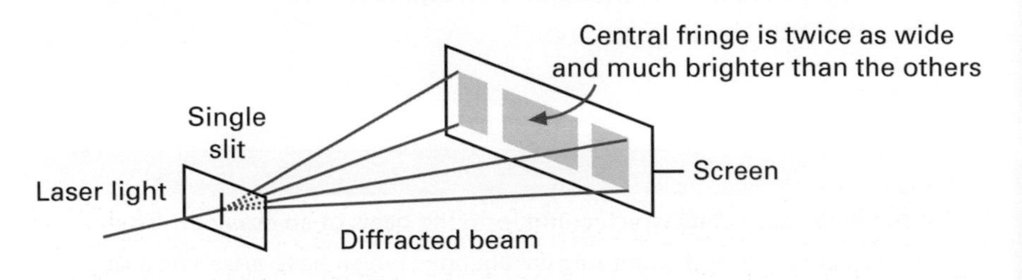

By adjusting the slit width and wavelength of light (by putting colour filters in front of the pin hole), in the experiment shown below, we find that diffraction increases:

- as slit width decreases (at constant wavelength)
- as wavelength increases (if slit width remains constant)

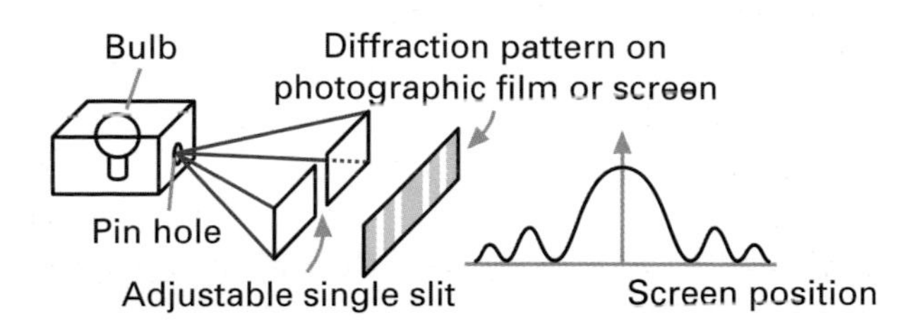

Links

To understand how diffraction produces bright and dark fringes see pages 122–5.

Action point

To observe diffraction of light waves, look at a distant street lamp through a pinhole in a piece of cooking foil. You should see rings of light around the pinhole.

Applications and uses of diffraction

Diffraction gratings

See *Interference* section, page 125.

Examiner's secrets

Drawing fringes in diffraction patterns can be confusing. You must label the light and dark bands so the examiner is clear which is which. Otherwise the examiner must assume a shaded band is a dark fringe. This may be the exact opposite to what you intended!

X-ray diffraction

X-rays are diffracted by the parallel planes of atoms found in crystals. The layers of atoms form the equivalent of a diffraction grating. This technique suggests that the layers are separated by approximately 0.1 nm. X-ray diffraction has been used to study complex molecules like DNA, providing fundamental information for the field of genetics.

Action point

Learn the definitions of *diffraction, refraction* and *reflection*. They sound similar, but have different meanings.

Microwave ovens

The metal mesh on the door uses a diffraction effect. The size of the mesh is chosen to be too small to let microwaves pass through, but will let light (much smaller wavelength) through so that you can see the food that is cooking.

Loudspeakers

Here, the design is such that diffraction is maximised, so that the sound spreads out in all directions.

Satellite transmitters

This is just the opposite to loudspeakers, because in communication satellites, you want the information to be restricted to a narrow beam so that it does not miss the target, which would be very inefficient.

Exam practice answers: page 135

(a) Why do harbours (natural and artificial) normally have circular shorelines? (Use a diagram in your answer.)

(b) A microwave oven uses microwaves whose wavelength is 0.12 m. The door of the oven contains a metal grid with grid gaps of 0.005 m. How does this design protect us and let us see inside the oven? (15 min)

Superposition

Have you ever been in a crowded room when your attention has suddenly been seized by a conversation taking place at the other side of the room? Almost amazingly we can shut out all other noises and listen in to this private chat! There are lots of sound waves travelling in all directions and yet we can listen in. Superposition can help to explain why.

Links

Superposition applies to all waves. It is used to explain diffraction (pages 120–1), interference (pages 124–5) and standing waves (pages 126–7).

What happens when two waves cross?

If two single waves (**pulses**) are sent down a slinky, simultaneously and from opposite ends of the spring, the *pulses pass through each other* and then continue to carry on their way as if the crossover had never happened! In just the same way, sound waves can cross through each other, allowing a private chat to pass through a sea of other sound waves before arriving at your ear!

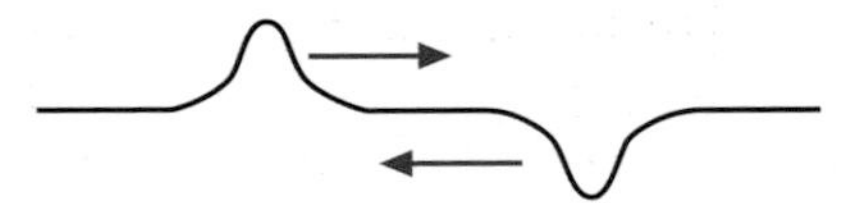

Links

To remind yourself of what moves when waves travel, see pages 112–3.

Combining waves

If the light beams from two torches are mixed they do not bounce off each other. Instead, just like sound waves, they pass straight through one another. Remember that energy moves when waves travel, not matter, and so it is not like two cars crashing into one another!

The principle of superposition

Checkpoint 1

If two up pulses (crests) are sent from opposite ends of a slinky, what would happen at the instant that they cross?

At the instant that the two wave pulses passed, in the diagram above, their displacements combined. For a fleeting moment a crest met an equal-sized trough and it was as if there were no waves travelling down the slinky – and then the two waves carried on. Consider an alternative:

When two troughs meet they produce . . .

. . . a 'supertrough'

Watch out!

Displacement is a vector quantity so remember to consider whether each displacement is positive or negative when using the principle of superposition.

A formal statement of the **principle of superposition** is that:

➜ the total displacement at a point equals the sum of the individual displacements at that point

In less formal language:

➜ crest + crest = supercrest
➜ trough + trough = supertrough
➜ crest + trough = zero

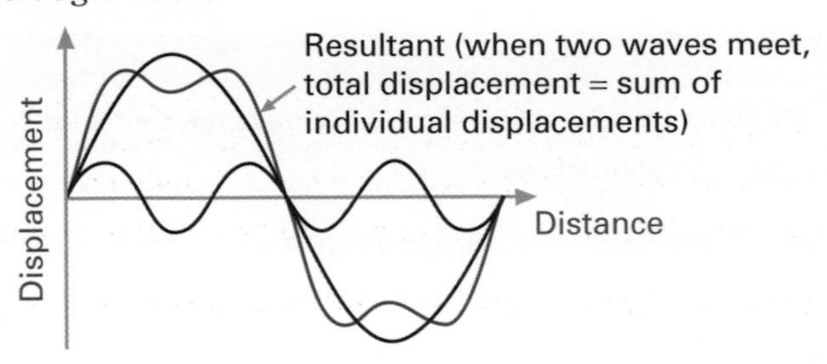

Explaining single-slit diffraction

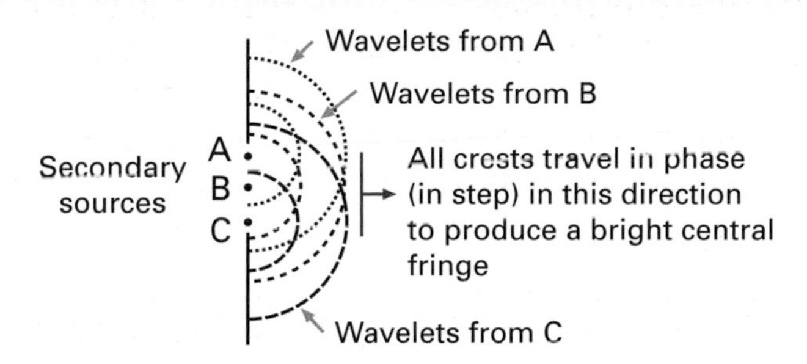

In this diagram, three secondary sources are sending out wavelets. We can use the principle of superposition to find the combined effect of all these wavelets in any direction. In some directions they add together to produce **bright fringes** (where crests arrive with other crests to form supercrests) and in other directions they cancel out to form dark fringes.

In the straight-ahead direction all the crests are travelling **in phase**. Each crest leaves A at exactly the same time as crests leave B, C and every other secondary source along the gap width. They all travel at the same speed and they all have the same distance to cover to reach the screen (**path difference** equals zero). Therefore they all reach the screen together to produce a supercrest, resulting in a bright fringe.

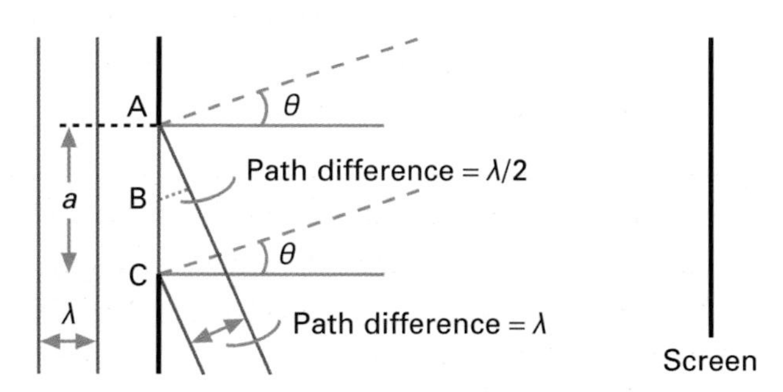

This diagram shows that at angle θ, wavelets from C have to travel further than wavelets from A in order to reach the screen. They have to travel an extra wavelength λ. Point B is halfway between A and C. The wavelets from A and B are in **antiphase**. So when a crest arrives at the screen from A, it is cancelled by a trough arriving from B. For every secondary source between A and B, there is another between B and C producing wavelets in antiphase causing a dark fringe. **Dark fringes** are produced at all angles θ when $\sin\theta = n\lambda/a$, where $n = 1, 2, 3 \ldots$. At points between these angles not all the light is cancelled and so light bands occur, the intensity of which decrease as θ increases.

The jargon

Travelling in phase means travelling in step. A *wavelet* is simply a small wave.

The jargon

Path difference means how much further one wave has to travel compared with a second wave.

Grade booster

If the path difference is a *whole number of whole wavelengths* ($n\lambda$), there will be a bright fringe. Dark fringes result from a path difference of an *odd number of half-wavelengths* $(2n + 1)\lambda/2$. (Take $n = 0, 1, 2 \ldots$ etc.)

Checkpoint 2

What is the path difference between wavelets sent from A and C, and from A and B?

Jargon

In antiphase means exactly out of step.

Exam practice answers: page 135

(a) Superposition of waves is used to combat noise pollution by making antisound. Describe how a sound wave could be cancelled in this way.

(b) Describe how hand movements can explain superposition of wave pulses.

(15 min)

Grade booster

Complete cancellation requires two waves to be in *antiphase*. Sometimes waves that superpose do not have exactly the same amplitude and so complete cancellation does not occur.

Interference

"The yeoman work in any science is done by the experimentalist, who must keep the theoreticians honest."

Michio Kaku

The wave theory of light was not readily accepted by others. Newton's proposal that light consists of tiny particles, or corpuscles, was preferred. It was not until 1801 that Thomas Young found evidence in support of the wave theory. Young's experiment could only be explained by accepting that light is a form of wave motion.

Checkpoint 1

Both loudspeakers are connected to the same signal generator operating at 1 000 Hz. The speed of sound in air is 330 m s^{-1}. Calculate the wavelength of the sound waves coming from both loudspeakers.

Interference of sound waves

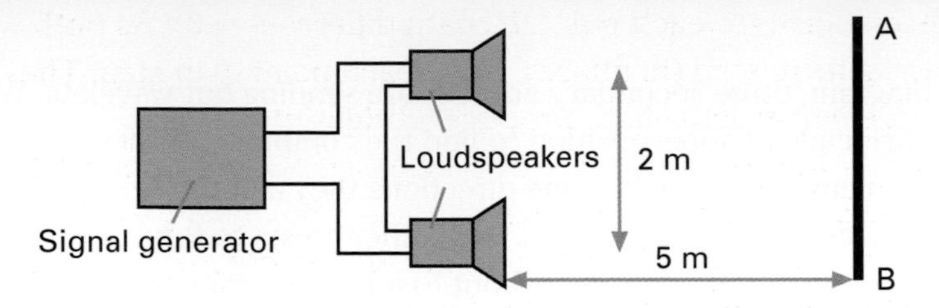

To demonstrate interference of sound waves:

- connect the equipment as shown, to produce waves in phase
- set the signal generator to 1 000 Hz
- ask observers to walk along the line AB

The results and conclusions are that loud and quiet regions exist.

- At *loud points*, two crests have arrived together (in phase) forming a supercrest. This is called **constructive interference**.
- At *quiet points*, a crest has arrived with a trough (antiphase). They cancel each other out. This is called **destructive interference**.

The jargon

Phase difference means how much one wave leads, or lags behind, another. If the *phase difference* equals zero, neither wave leads. They are said to be *in phase*.

Grade booster

Of course, lasers hadn't been invented in Young's day! If you don't have a *coherent* source in your set-up, you must have a single slit before the double slits.

A modern version of Young's experiment

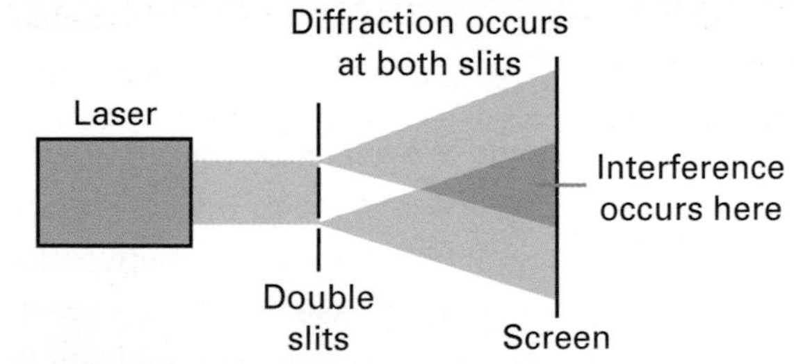

In this modern version of Young's experiment, laser light spreads out at each of the two slits. The light waves from each slit overlap and interfere with one another. The principle of superposition can be used to predict or explain the resulting interference pattern.

Using a laser allows us to get an interference pattern because:

- As only one laser is used, the waves from both slits are **coherent** (meaning that the phase difference between them remains constant).
- Lasers produce **monochromatic** (single wavelength) light. This ensures that both sets of waves have the same wavelength.

Examiner's secrets

This experiment frequently appears in examinations because of its importance in support of the wave theory of light.

Explaining the experiment

The interference pattern is caused by waves from either slit travelling different distances to reach each point on the screen. The following diagram can be used to help explain double-slit interference.

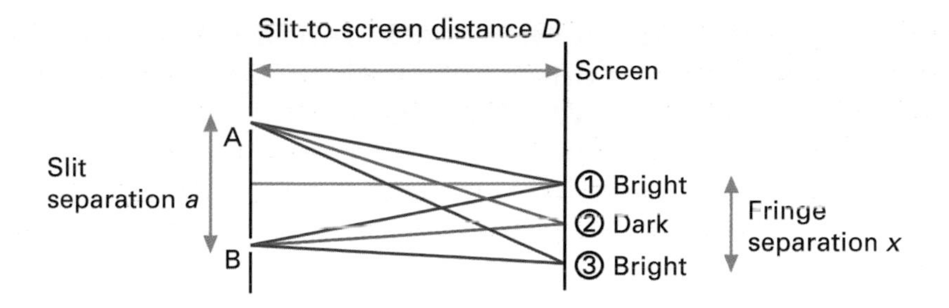

- Point ① is equidistant from both slits. Both sets of waves travel the same distance to reach point ① (path difference = 0). As both waves left the slits in step (in phase), they reach point ① in step. They will interfere constructively to produce a bright fringe here.
- Point ② is the centre of the first dark fringe. Light from slit A has had to travel an extra half wavelength to reach point ② (path difference = $\lambda/2$). As this light is $\lambda/2$ behind, it interferes destructively with light from slit B, causing a dark fringe here.
- Point ③ is at the centre of the next bright fringe. Light from A has had to travel a complete extra wavelength to reach this point. So, the two sets of waves arrive in step here to produce a bright fringe.

Other important points

- The double-slit experiment can be used to find the wavelength of light using the formula $\lambda = ax/D$ (see the diagram above).
- This experiment supports the wave theory of light because it can only be explained using wave properties: diffraction and interference.
- Similar experiments use a ripple tank or microwaves.

Diffraction gratings

Passing monochromatic light through more than two slits can also cause interference. Diffraction gratings have up to 10 000 slits per cm.

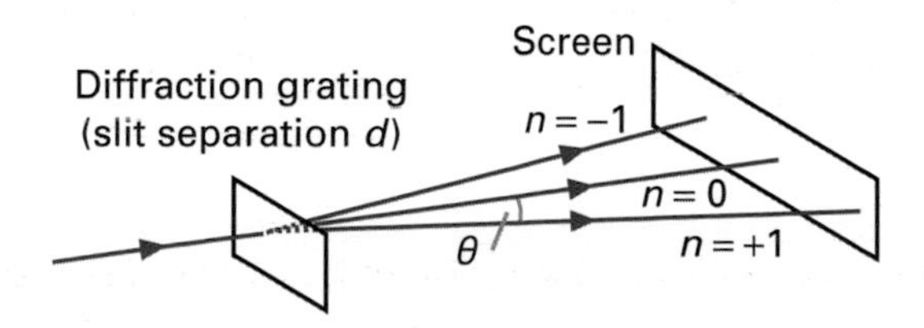

For diffraction gratings, wavelength λ and θ are related by:

$d \sin\theta = n\lambda$ $\quad$ d = slit separation, n = order of the maximum

Exam practice answers: page 127

(a) By explaining the term *coherence*, explain why we do not observe many more clear interference effects in our everyday lives.

(b) Give one example of both coherent and incoherent light sources. (15 min)

(c) The slit separation d of a diffraction grating is 0.4 mm. The wavelength of light is 600 nm. Calculate the angle made by the second order fringe. (5 min)

Checkpoint 2

The slits used in an experiment similar to that shown opposite were 1 mm apart. The distance between slits and screen was 3 m and the width of 10 fringes was 3 m. Calculate the wavelength of light used.

Test yourself

When you have finished working through this section, close the book and see how much you can remember. In a week's time, refresh your memory and then test yourself again!

Watch out!

If asked how many maxima will be produced by a certain diffraction grating, remember to include the zeroth-order maximum and all those for which n is negative.

Examiner's secrets

Each syllabus is slightly different. Does your syllabus require a derivation of $d \sin\theta = n\lambda$?

Action point

If a grating has a slit spacing of 0.25 mm, how many lines (slits) per m does it have? (*Hint* $N = 1/d$)

Standing waves

> *"There is no higher or lower knowledge, but only one, flowing out of experimentation."*
>
> Leonardo da Vinci

Links

The terms standing or stationary waves and transverse waves were first introduced on page 113.

Standing, or stationary, waves can be both useful and troublesome. Many musical instruments work because of standing waves. Standing waves can help explain why atoms have definite energy levels. But standing waves in aircraft wings and loudspeaker cabinets can cause problems for aeronautical and acoustic engineers.

Standing wave patterns

If you tie a slinky to a table leg, it is easy to see standing waves. Flick the other end of the slinky from side to side to send transverse waves along. They reflect off the fixed end and by varying the frequency you can get different patterns, as shown below.

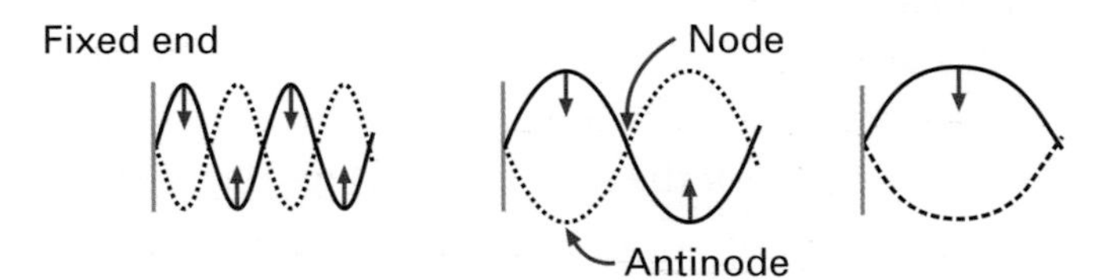

The jargon

The simplest way a guitar string can vibrate is with its *fundamental* frequency, sometimes called the *first harmonic*. The next frequency to produce a standing-wave pattern is called either the *first overtone* or the *second harmonic*.

Nodes and antinodes

The simplest way a guitar string can vibrate is shown below. As with all standing waves, a node is a point where the amplitude is zero. A point where it is a maximum is always called an antinode.

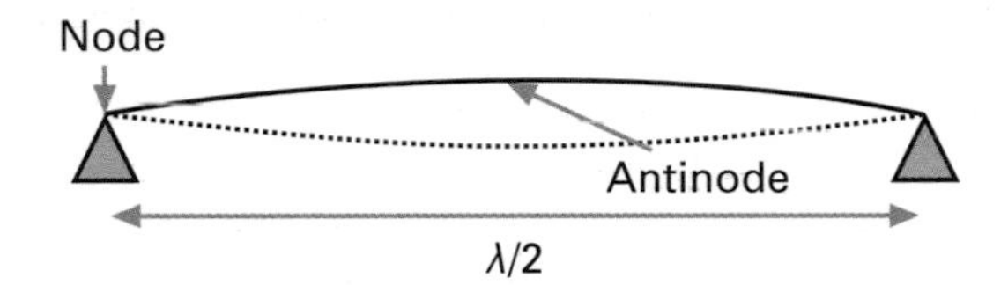

Checkpoint 1

Sketch a diagram showing a string vibrating at its second harmonic.

Other important points about nodes and antinodes include:

- the string appears as a series of loops separated by nodes (see the upper diagrams on this page too)
- adjacent sections of the string move in opposite directions (see arrows in the upper diagrams on this page)
- neighbouring nodes (or antinodes) are separated by $\lambda/2$

Grade booster

To get full marks when describing how to produce standing waves you must have three basic ingredients:

(i) Something to start the initial vibration.
(ii) A method of changing the frequency of vibration.
(iii) Knowledge that there could be different wave patterns.

Make sure you learn all of these.

Standing wave experiments

Stretched strings

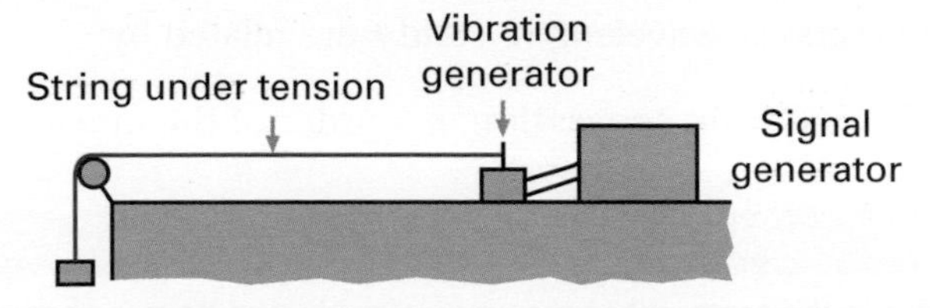

This diagram shows Melde's experiment. Weights attached to one side of the string, via a pulley, maintain the tension. As the frequency of the vibration generator is increased, all the standing-wave patterns discussed above (and more) can be seen. The wavelength of the waves can be calculated by measuring the distance between nodes. String length, tension and other factors can be studied.

Microwaves

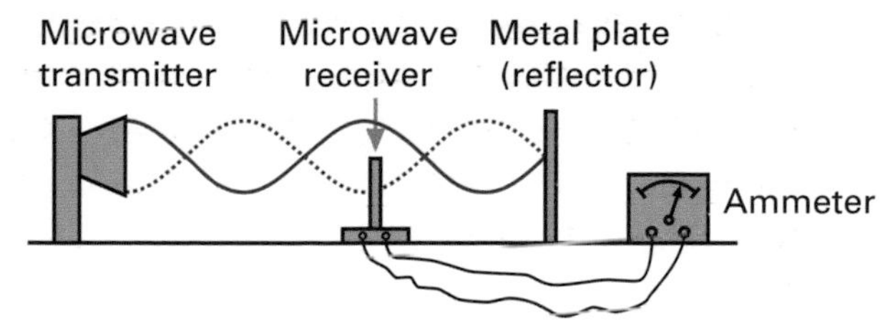

- This diagram shows how you could investigate standing waves produced by microwaves. Position the transmitter 50 cm from the reflector.
- Move the receiver around in the region between transmitter and receiver to observe alternating points of high and low intensity.

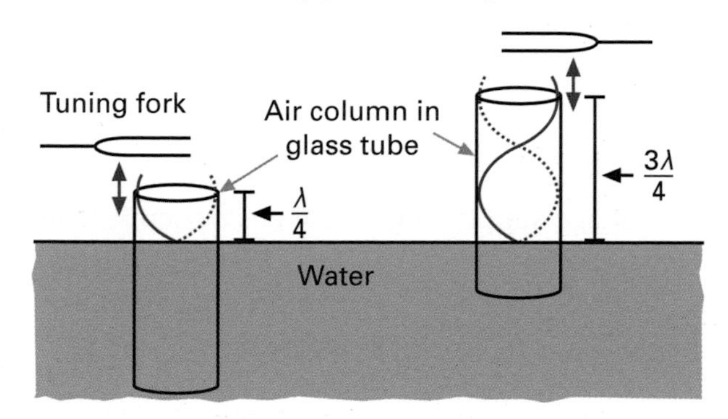

Sound waves

This apparatus can be used to deduce the wavelength of sound waves:

- hold a vibrating tuning fork over the glass tube
- move the tube up and down to vary the air column length
- the sound made by the tuning fork will get louder and quieter
- the shortest air column that produces a loud sound = $\lambda/4$

Explaining standing waves

Standing waves get their name because they do not *appear* to be travelling in either direction (forwards or backwards). In fact they are caused by two identical waves travelling in opposite directions: the original wave and a reflected wave. In the example above, the tuning fork sent out the original wave, which was reflected at the water surface.

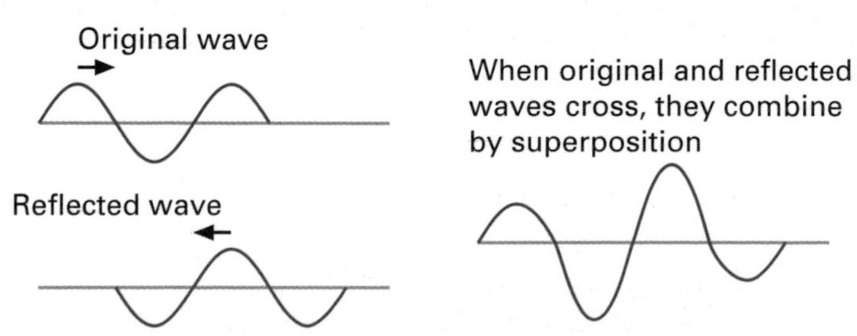

Nodes (and antinodes) are caused by destructive (and constructive) interference.

Exam practice answers: page 136

Write a series of revision notes showing how the wavelength and speed of sound can be found using standing waves. Mention the errors that are inherent in this experiment and describe an alternative experiment. (20 min)

Checkpoint 2

In an experiment like the one opposite, the shortest air column to produce a loud (resonant) sound was 16 cm long. What was the wavelength of the note produced? The tuning fork was labelled 512 Hz; what was the velocity of the sound? By examining the diagram carefully, explain why your calculations are really only estimates.

Action point

Kundt's tube can be used to demonstrate standing waves in pipes. Sound waves are sent down a horizontal tube, closed at one end. A layer of fine powder on the bottom of the tube moves into piles. Would the piles be at the nodes or the antinodes? Explain your answer.

Examiner's secrets

The equipment for practical exams normally has to be easily available to schools. Experiments that require standard equipment, like the one above, are often used.

Grade booster

Make sure you can convince your fellow students that you know the difference between *standing* waves and *progressive* waves.

Photoelectric effect

"Some say that the only thing quantum theory has going for it is that it is unquestionably correct."

Michio Kaku

It has been said that new theories are never accepted, it is just that old men die! Newton believed that light was composed of tiny particles. Young's experiment could only be explained using the wave theory. But when the photoelectric effect was discovered, the wave theory failed. An unknown, Albert Einstein, had the answer.

Check the net

To find out more about Einstein go to physics.gla.ac.uk/introPhy/Famous/einstein/einstein.html

The photoelectric effect

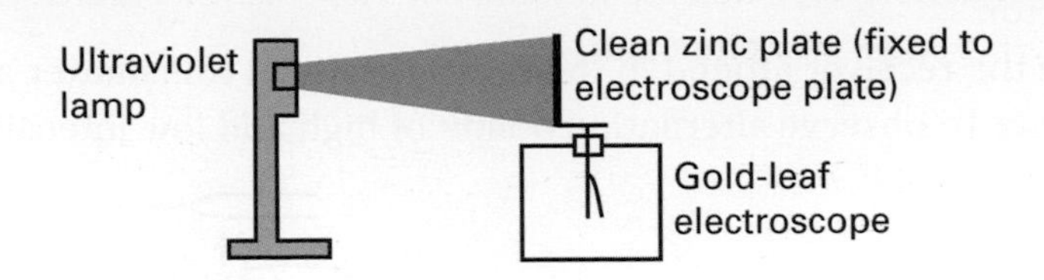

To demonstrate the photoelectric effect:

- use a high-voltage supply to give the electroscope a negative charge, the leaf will deflect
- shine an ultraviolet lamp on the zinc and the leaf will slowly fall
- mysteriously, the light helps the electrons supplied to escape, and so the leaf loses its charge and falls back down – the escaping electrons are called **photoelectrons**

Grade booster

If you are asked to describe the experiment shown opposite, make sure you state that the gold-leaf electroscope is first charged. By the way, shining light onto it does not charge it!

Observations that the wave theory could not explain

- Electrons were not emitted unless the frequency of light used was above a minimum value (which depended on the metal used). The wave theory suggested that if weak light waves were used for a long enough time they should be able to release an electron eventually.
- It did not seem to matter how bright the light was. Some electrons left the metal with more kinetic energy (KE) than others. The maximum kinetic energy (KE_{max}) was dependent on the frequency of the light used, but not its intensity.
- Even a very weak beam of high-frequency light released electrons almost immediately. How could such a weak wave, spread evenly over a large area, pick out just a few electrons for release?

Checkpoint 1

After reading about Einstein's explanation of the photoelectric effect, explain why the minimum frequency of light required depended upon the metal used.

Setting the scene

Planck had suggested that oscillating charges could only vibrate with certain set frequencies so that energy too would come in small packets. Einstein then said that radiation is quantized in small packets as well.

Action point

Construct a table with two columns. Make a note of each of these puzzling observations in one column and Einstein's explanations in the other.

Einstein's explanation

Einstein proposed that the electromagnetic radiation (ultraviolet in our demonstration) reached the metal in packets of energy – now called **photons**. He used ideas that had been suggested by Max Planck in 1900, from his work on black body radiation - namely that the energy carried by the radiation E was proportional to the frequency of the light f. Einstein's theory was that all a photon's energy is given to just one electron. Thus $E = hf$, where h is Planck's constant ($= 6.63 \times 10^{-34}$ Js)

It is quite significant that the photoelectric effect takes place from metals since they have so-called **free electrons** that are not bound to any one atom. If an electron is sufficiently loosely bound to its metal, some of an incoming photon's energy can be used to break it free. If there is any energy left over, the escaping electron can flee the surface of the metal as the remaining energy reappears as its KE. The minimum amount of energy required to let the electron reach the surface of the metal is called the **work function** *W*. A metal that holds onto its electrons more strongly would have a larger work function.

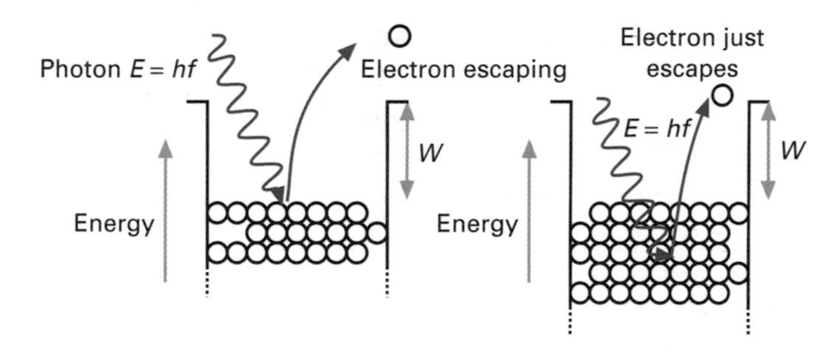

Electrons escape with a variety of kinetic energies

$$\begin{pmatrix}\text{energy of incoming}\\ \text{photon } E\end{pmatrix} = \begin{pmatrix}\text{work done to remove}\\ \text{electron from metal}\end{pmatrix} + \begin{pmatrix}\text{KE of escaping}\\ \text{electron}\end{pmatrix}$$

Electrons from the metal surface require least energy (the work function *W*) to break them free. Hence they have most kinetic energy, KE_{max}.

$$KE_{max} = E - W = hf - W \qquad \textbf{(Einstein's equation)}$$

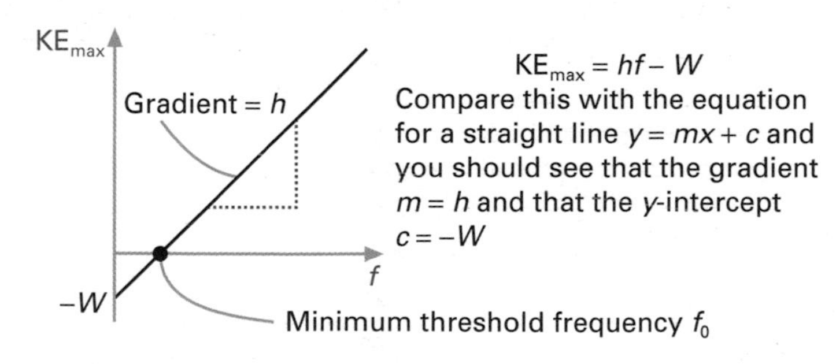

The graph above illustrates Einstein's photoelectric equation.

> *"Science can purify religion from error and superstition. Religion can purify science from idolatry and false absolute."*
>
> Pope John Paul II

Checkpoint 2

What other scientific phrase can be used to replace the term *free electrons*?

Watch out!

Work function is often quoted in electron-volts (eV). The symbol ϕ is used in some specifications.

Grade booster

The graph of KE against frequency for a different metal would have the same gradient but the intercept would be different.

Don't forget

To get the gradient of a graph you have to wise up! Gradient = *y*/*x*.

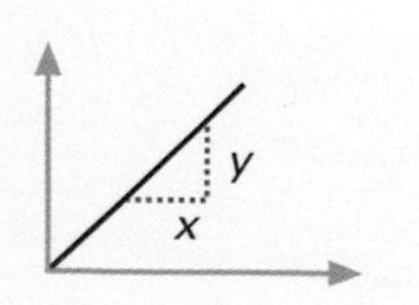

Exam practice answers: page 136

(a) Use the term *work function* to explain why photons are sometimes unable to remove certain electrons from metals, can get some electrons to the metal's surface and can release others with a variety of kinetic energies.

(b) Why is there such a short delay between shining appropriate light on a metal and the escape of the first electrons?

(c) Why does intensity not affect KE_{max} of photoelectrons? (20 min)

Atomic line spectra

"An expert is a man who has made all the mistakes which can be made, in a very narrow field."

Niels Bohr

Checkpoint 1

This may not be the first time that you have come across a technique that uses colour to identify chemicals. What are *flame tests*?

We know that white light consists of seven colours, the visible spectrum. By the mid-1800s astronomers were using the fact that hot gases emit only certain characteristic colours as fingerprints to identify the elements present in stars. But it was not until 1913 that Niels Bohr was able to use quantum theory to explain these characteristic spectra.

Line spectra

A narrow source of light containing a hot gas, such as sodium or neon, can produce a recognizable line spectrum containing some wavelengths of light but not others. Spectrometers are used to view line spectra.

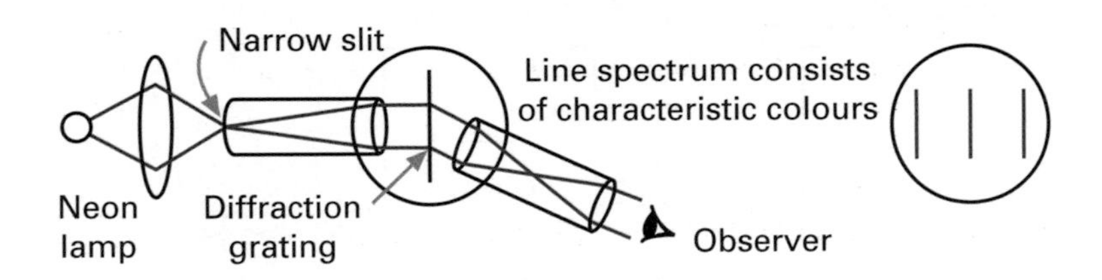

There are two types of line spectra.

- **Emission spectra** show the light emitted by hot gases, as previously described.
- **Absorption spectra** are obtained when white light passes through a cool gas. The gas absorbs certain wavelengths from within the white light. The absorbed wavelengths are characteristic of the gas and can therefore be used to identify it.

Action point

Produce a flow diagram to show how theories about the nature of light changed over time. Improve this chart by adding really important extras – breakthrough experiments, puzzles that new theories could explain, etc.

What causes line spectra?

It was obvious that atoms of a particular element could only emit or absorb certain wavelengths of light. The question was why?

Einstein had suggested that electromagnetic (including light) radiation is carried in small packets or **photons**. When a photon strikes an atom, the target atom's electrons absorb its energy. When electrons lose energy, they emit photons.

The fact that electrons in particular atoms can only absorb, or emit, very specific wavelengths of light means that they can only absorb, or emit, very specific amounts of energy carried in very specific photons.

Grade booster

It is a common error for students to get absorption and emission the wrong way round. *Down and out* might help you remember it!

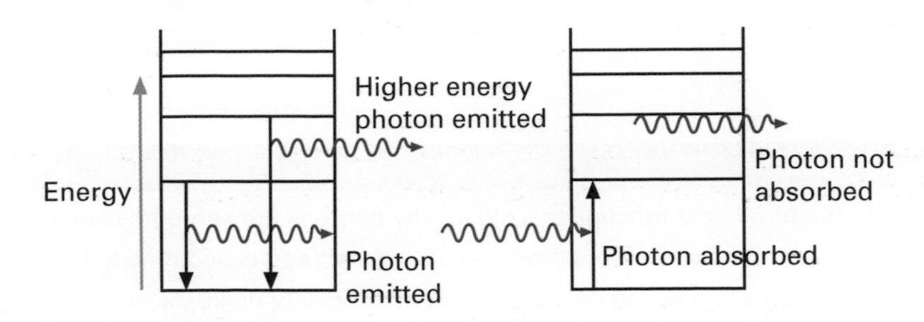

These **energy-level diagrams** (or *ladders*) show that electrons can only emit or absorb specific photons. Each element has its own unique energy-level diagram.

Action point

After reading this spread, draw a diagram, like the one opposite, to show how this imaginary element could absorb a more energetic photon.

Try to remember that:

- it is as if each element has its own ladder
- every ladder has rungs in different places so an atom's electrons can only have certain amounts of energy
- each element is characterized by the energy levels available to its electrons

When an atom emits light, one of its electrons falls from a higher to a lower energy level. A single photon is emitted whose energy equals the energy difference between the two levels. This means that a bigger fall for the electron will produce a more energetic photon.

$E_1 - E_0 = 1.63 \times 10^{-18}$ J
$= 10.2$ eV

E_1

E_0

10.2 eV photon emitted

This diagram shows an electron falling from E_1 to E_0 in a hydrogen atom. To "pump" the electron back up to E_1 (its **first excited state**) requires a photon of exactly 10.2 eV to be absorbed. The atom would then be unstable and the electron would very quickly fall back to E_0 (its **ground state**), emitting a 10.2 eV photon.

This idea of elements having energy ladders with characteristically fixed rungs, allowing only certain jumps or falls, explains why hot gases emit *signature* photons. As these photons have specific energies, the wavelengths of light that can be observed are limited.

Similar ideas apply to absorption spectra:

- if an incoming photon is to be absorbed, it must deliver exactly the right amount of energy to lift an electron from one energy level, or rung, to another
- if it does not deliver this amount, it will not be absorbed
- energy-level spacings are characteristic of elements, so each element can be identified by the photons it absorbs

Line spectra of solids and liquids

So far, we have only mentioned gases. The line spectra of solids and liquids are much more complicated because their atoms are more tightly packed. This allows electrons of neighbouring atoms to affect one another producing spectra with many, many wavelengths.

Ionization

If an atom absorbs a photon of sufficient energy, it can be possible for an outer electron to leave the atom altogether. This is called **ionization**.

The jargon

An *electron volt* (eV) is the amount of energy an electron gains when it is accelerated across a potential difference of 1 volt. 1 eV = 1.60×10^{-19} J

Checkpoint 2

Use $E = hf$ to calculate the frequency of the photon emitted in the diagram opposite. Now calculate its wavelength.

The jargon

The *ground state*, or *level*, is the lowest energy state available to an electron.

Examiner's secrets

Very often the highest energy level is not shown. This is because it has the value *zero*. An atom is ionized when an electron has enough energy to jump to this level from the ground state.

The jargon

The *first excited state* is the first energy level, above the ground state, that is available to an electron.

Speed learning

Try to understand how line spectra are produced; the theory will then be easier to remember.

Exam practice answers: pages 136–7

(a) Stars produce both emission and absorption spectra. With reference to energy-level diagrams, explain how both types of spectra are produced.

(b) Why can spectral analysis identify individual elements? (20 min)

De Broglie's equation and atomic models

"If one does not feel dizzy when discussing the implications of Planck's constant h it means that one does not know what one is talking about."

Niels Bohr

In 1913, Bohr's work had been a major breakthrough. For the next ten years or so, others tried to build on his results, without success. Then Prince Louis de Broglie took a huge step forward. He suggested that all particles, including electrons, had split personalities. They could exist not only as particles but also as waves!

Wave–particle duality of light

It is necessary to think of light as a wave to explain diffraction, interference and polarization. But a particle model (photons) must be used when considering the photoelectric effect and line spectra. Neither model is satisfactory on its own as light sometimes behaves as waves and at other times behaves as particles. This compromise is called the **wave–particle duality of light**.

You should use *two rules to decide how to think of light* (or any other form of electromagnetic radiation) in a given situation:

- use the particle model when light interacts with matter, e.g. when it strikes the surface of a metal in the photoelectric effect
- use the wave model when light goes through a gap of similar width to its own wavelength, e.g. single-slit diffraction

Checkpoint 1

Why can wave–particle duality be thought of as a complementary principle?

Action point

Work out your own de Broglie wavelength. Find the product of your mass and the fastest speed you can run. Divide this into Planck's constant. Hence explain why *you* can never diffract! (*Hint* diffraction is greatest when the gap is the same size as the wavelength.)

Check the net

To find out more about the work of Prince Louis de Broglie go to www.chembio.uoguelph.ca/educmat/chm386/rudiment/tourquan/broglie.htm

Wave–particle duality of all particles

De Broglie's great advance was to suggest that *all particles, not just light, might have a dual nature*. He related the wavelength λ of a particle to its momentum p in his equation:

$\lambda = h/p$ Where h is Planck's constant

This equation shows the dual nature of particles, as wavelength is a wave property whereas momentum is associated with particles. As

momentum = mass × velocity or $p = mv$

$\lambda = h/p$ becomes

$\lambda = h/mv$

This means that if velocity is constant, wavelength is inversely proportional to mass. The wavelength of electrons is approximately 1×10^{-10} m, roughly the spacing of atoms in solids that can therefore diffract them.

The Rutherford–Bohr model of the atom

The quantum theory, and Niels Bohr, had another trick to reveal! The nucleus contains positive charge, electrons are negative – why does electrical attraction between them not cause the atom to self-destruct?

Rutherford's original model of the atom could not explain this mystery. It could not show why electrons do not lose energy continuously and spiral into the nucleus. Bohr's use of quantum theory, with fixed electron orbits, solved the problem.

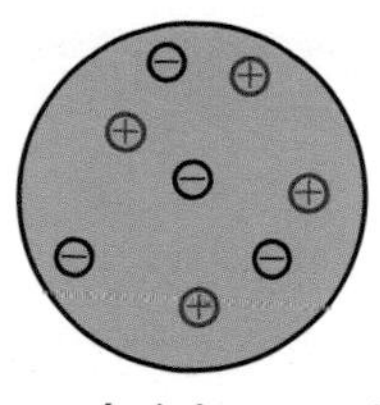

Thomson's 'plum-pudding' model of the atom (1898)

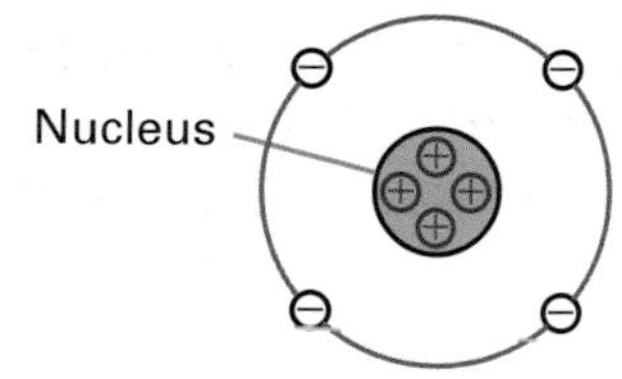

Rutherford's model of the atom (1911) included orbiting electrons

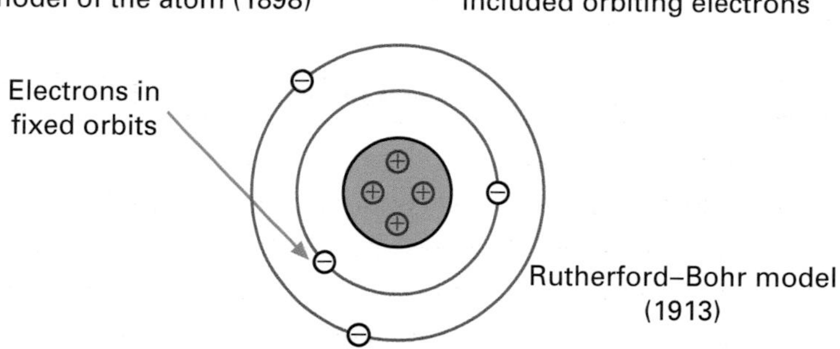

Rutherford–Bohr model (1913)

Models of the atom developed over a number of years. The main points about each model shown are as follows:

- *J. J. Thomson* viewed the atom as a ball of positive charge containing electrons dotted inside, like plums in a pudding!
- *Rutherford* knew that the atom was mostly empty space with positive charge at its centre. He thought that the much lighter electrons would orbit the nucleus like planets around the Sun.
- *Bohr* refined Rutherford's model using quantum theory. As energy could only be absorbed or emitted in well defined packets, he proposed that electrons in higher orbits would have more energy than those below. Energy coming in defined quanta meant that only certain energy jumps could be made, allowing a limited number of electron orbits. This became known as the Rutherford–Bohr model.

Other advances

- Rutherford discovered positive protons in 1919.
- Bohr's theory worked only for hydrogen. In 1925, Schrödinger developed a wave-mechanics model to solve this problem.
- In 1932, James Chadwick discovered neutrons.

Why electrons do not spiral into the nucleus

Electrons can be considered as waves. If an electron is to squeeze towards the nucleus, its length (or wavelength) must decrease causing its energy to increase. If small enough to fit inside the nucleus, an electron would be too energetic for the protons to hold it!

Checkpoint 2

In what way do the Thomson and Rutherford models agree? In what way do they disagree?

Action point

Sketch the Rutherford–Bohr model of the atom as shown here. Update it by including appropriately labelled protons and neutrons.

Links

To find out more on Rutherford (and alpha-particle scattering), see *the atom and its nucleus*, pages 44–5.

Exam practice answers: page 137

(a) Why would it be unlikely for one to observe matter behaving as waves in everyday life? Why do atoms not self-destruct? Draw the Rutherford–Bohr atomic model and a corresponding energy-level diagram. (20 min)

(b) Calculate the speed and de Broglie wavelength of electrons of energy 1.0 eV. (8 min)

Answers
Waves and oscillations

Types of waves and their properties

Checkpoints

1 Both would *travel faster, causing them to bend or refract.*
2 *Waves transfer energy.*
3 *Speed in* m s^{-1}, frequency in Hz, and wavelength in m.

Exam practice

(a) Vibrations at 90° to the direction of the wave motion cause transverse waves. Vibrations moving in the same direction as the wave motion cause longitudinal waves.
(b) It is not possible to polarize longitudinal waves because they only vibrate in the direction of the wave's motion. If vibrations were blocked in this direction, there would be no wave.
(c) Any electromagnetic wave, e.g. light, is transverse. Sound waves are longitudinal.
(d) See page 126 for diagrams showing how transverse standing waves (microwaves) and longitudinal standing waves (sound waves) can be produced.

Examiner's secrets

Try to pick out the keywords used by the examiner to tell you what to do. In this question, the examiner is expecting you to spend longer on parts (a) and (b), as parts (c) and (d) only require examples to be given.

Electromagnetic spectrum

Checkpoint

1 The only real difference is the method of their production.

Exam practice

Physics is an evolving subject and so there are many possible answers to this question. Suggestions include:
Forces and motion You could discuss the ideas of Aristotle, Galileo, Newton and Einstein.
Energy and heat You could compare the caloric and kinetic theories. You could describe the work of Rumford and Joule. Brownian motion could be included. You could bring this timeline up to date by mentioning the quest to reach 0 K.
Waves and particles You could explain that Newton's particle model of light was preferred to Huygens' wave model until after Young's experiment. You could describe the roles of Einstein, Bohr and de Broglie in the development of the wave–particle duality theory.

The *nature of radioactivity*, the *development of particle physics* or a discussion of *advances in astronomy* could all form the basis of an excellent answer to this question.

Examiner's secrets

This type of question, requiring knowledge that is developed over an entire course, is called *synoptic*. Synoptic questions are a feature of all A2 courses.

Reflection and refraction

Checkpoints

1 The Moon does not produce any light. We can only see it if light from stars, e.g. the Sun, bounces off it and into our eyes. By day, our view of the sky is dominated by the Sun so the Moon is not visible.
2 Both would still slow down, but they would travel straight on without bending.
3 It would bend away from the normal.

Exam practice

(a) (i) 3.40°, (ii) 10.14°
(b) The frequency of the light in both media is 7.5×10^{14} Hz. The wavelength of violet in the unknown medium is 333 nm.

Grade booster

Get a list of all the formulae and constants that will be provided in your exams, it will help reassure you during revision. With regular revision, you will need to use it less and less.

Applications of reflection and refraction

Checkpoints

1 41.8°
2 60.5%
3 It changes direction, slows down and its wavelength decreases.
4 The wavelength is reduced by the same proportion as the speed, i.e. by $^2/_3$ in this case.

Exam practice

(a) Try to maintain a balance in essay-style questions that contain multiple parts. In this case you could choose two points relating to technological changes (perhaps in medicine where doctors use endoscopy to see inside the body and in telecommunications where a single optical fibre cable can carry up to half a million telephone calls at the same time).

Keyhole surgery made possible by fibre optics is much simpler than conventional surgery and can often mean that patients spend much less time in hospital which is more cost effective.

The use of fibre optic technology has contributed to cheaper telephone call charges. Fibre optics have opened the way for digital television which will allow many more television channels. This may persuade people to spend even more time watching TV. Cheaper telephone call charges may allow people to keep in closer contact with friends and family living abroad.
(b) Fibre optics can be used to light road signs, they are found in some security fences, the speed of information transfer along fibre optics makes the internet accessible to many people and fibre optics can even be used for decoration!
(c) See the diagrams on page 118.

> **Examiner's secrets**
>
> Read each question carefully. It would be easy to do part (a) of this question and leave nothing to write about in part (b).

Diffraction

Checkpoint

1 This phenomenon would help you!

Exam practice

(a) The entrance to harbours acts in the same way as a gap in a ripple-tank experiment. The incoming water waves are diffracted, spreading out in a semicircular pattern. Each time that the resulting semicircular wavefronts reach the shoreline they erode it. It is as if they are taking semicircular 'bites' from the shoreline so that it becomes semicircular. See the upper diagram on page 120.

(b) The metal grid prevents microwaves from escaping. As the gaps in the grid are much smaller than the wavelength of microwaves, they cannot squeeze through. The wavelength of light is much smaller than the grid gaps so light can pass in and out unaffected.

> **Examiner's secrets**
>
> If more time had been allocated to this question a more detailed answer would be expected. Explaining how Huygens' construction predicts the formation of curved secondary wavefronts, i.e. diffraction, could extend part (a). Part (b) is similar to something you may have come across at GCSE: hot short-wavelength rays from the Sun can squeeze into greenhouses but as the rays cool, their wavelengths increase and they cannot get out again. Just be careful not to stray too far off the point!

Superposition

Checkpoints

1 A supercrest would be formed.

2 The path difference between wavelets from A and C is a whole wavelength, between A and B it is half a wavelength.

Exam practice

(a) The principle of superposition predicts that if a crest is added to a similarly sized trough, zero displacement will result – destructive interference. If we were considering sound waves, this would result in silence. Noise is defined as any unwanted sound. To combat noise pollution, a microphone could be used to sample the unwanted noise. An amplifier could be used to reverse the sound; the result could be called antisound. If the antisound is played over the original, the original can be cancelled.

(b) Establish a line of zero disturbance by moving your hand in a horizontal plane at a certain height. Show that the result of adding a crest (move one hand up) to a trough (move the other hand down) would be zero disturbance (referring back to our original horizontal sweep). Extend the answer by explaining how supercrests and supertroughs could be formed.

> **Grade booster**
>
> Examiners are impressed when candidates can link different parts of their course. Synoptic assessment will make this skill even more important. For example, 'antisound was unsuccessfully tested in the car industry. Instead the engineers chose to avoid materials whose resonant frequencies would be reached during normal use of the car, thus reducing boominess.'

Interference

Checkpoints

1 0.33 m

2 1×10^{-4} m

Action point

4 000

Exam practice

(a) Coherence means constant phase difference. To get constructive interference it is essential that the combining waves are coherent – that they always travel in step. For example, if two crests always reach a point at the same time a supercrest will be formed. But if a crest only occasionally arrives with another crest, no clear result will emerge. Even if the two waves are always out of phase by the same amount, say one wave is always ahead of the other by half a wavelength, a clear interference pattern will form. The essential requirement to get a clear interference pattern is that there must be a constant phase difference. Obtaining coherent wave sources can be difficult and so interference patterns are not very common in everyday life.

(b) Laser light is coherent. Light from an ordinary light bulb is incoherent.

(c) Use of $d \sin \theta = n\lambda$

$0.4 \times 10^{-3} \sin \theta = 2 \times 600 \times 10^{-9}$

$\sin \theta = 3 \times 10^{-3}$

$\theta = 1.72°$

> **Grade booster**
>
> Try to minimize spelling mistakes. Some words, like *synchronous*, are often incorrectly spelt. Why not build up a private word bank of difficult spellings?

Standing waves

Checkpoint

1

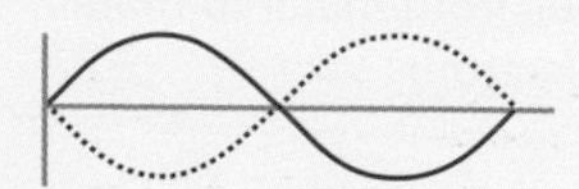

2 Wavelength = 0.64 m, velocity of sound = 327.68 m s^{-1}.
By examining the diagram of sound standing waves on page 127, you can see that the standing-wave pattern extends into the air beyond the top of the glass tube. (The air above the glass tube acts as an extension to the tube itself.) This means that when we measure the length of the glass tube above the water level, we are underestimating the length of the standing-wave pattern.

Exam practice

A good answer would be to describe how to measure the wavelength of sound waves as detailed on page 127. This would allow you to mention end corrections as described in the answer to checkpoint 1, see above. Having calculated wavelength and noted the frequency f of the tuning fork, use $v = f\lambda$ to get velocity v.

You could then go on to describe using Kundt's dust tube. Dust is evenly deposited inside a glass tube. One end of the tube is closed and a loudspeaker is attached to the other end.

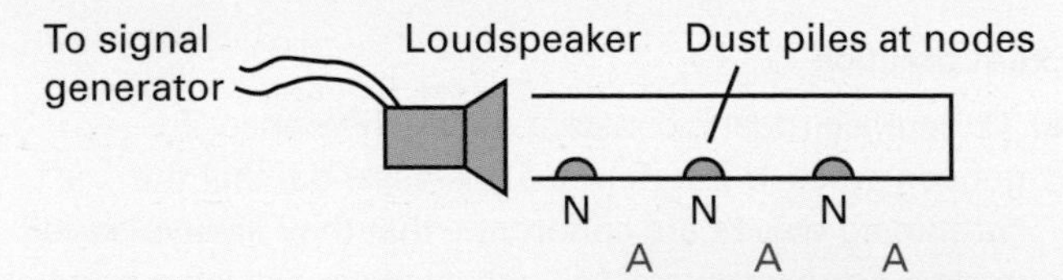

A signal generator is connected to the speaker so that sound waves can be sent down the tube. When a standing-wave pattern is set up, the dust vibrates violently at the antinodes (A). This causes the dust to build up at the nodes (N). The position of the nodes and antinodes can be observed and measured to give an accurate value for the wavelength of the sound waves. The frequency can be read off the signal generator and the velocity can be calculated as before.

Grade booster

Check which experiments are specifically mentioned on your syllabus and learn them!

Photoelectric effect

Checkpoints

1 Metals that hold onto their electrons more strongly have larger work functions. This means that more energy is required to allow them to escape from the metal. $E = hf$ suggests that energy E is directly proportional to frequency f. As there is a minimum amount of energy that is required to let electrons escape, and as this value varies from metal to metal, there is also a minimum frequency of light required. This also varies from metal to metal.

2 Delocalized electrons.

Exam practice

(a) The work function is the minimum amount of energy required to release an electron from a metal. If a photon delivers the work function exactly, it will be able to let the electron reach the surface of the metal. If less than the work function is delivered, not even this will happen, but if the photon delivers more energy than the work function any excess will reappear as kinetic energy as the electron moves away from the surface of the metal. Two electrons could leave the metal with different kinetic energies. This could be because photons of different energies released them or perhaps they were bound more or less tightly to the metal.

(b) Einstein's theory was that each photon delivered its packet of energy to just one electron. If the first photon to strike the metal had sufficient energy, it could release an electron immediately. There was no need to build up enough energy, which would take longer.

(c) Brighter more intense light does not mean that the light photons are more energetic. It simply means that there are more photons. As each electron receives just one photon, more photons mean that more electrons can escape, not that those that do escape will be more energetic. So intensity does not affect the maximum kinetic energy of photoelectrons released from metals in the photoelectric effect.

Grade booster

Try to use the correct scientific terminology wherever possible. When used appropriately, this technique will not only make your answer more professional but it could also save you time. So be specific! Don't say *electron* when you could mean *free electron* or *photoelectron*.

Atomic line spectra

Checkpoints

1 Flame tests are used to identify chemicals on the basis of the colour they produce on burning. Metal ions can be identified in this way. For example, Na^{+} burns with an intense golden yellow colour. A Nichrome or platinum wire is cleaned by repeatedly dipping it in hydrochloric acid, then heating it in a roaring bunsen burner flame. When no colour is given to the flame by the wire we assume that the wire is clean. It is moistened with dilute hydrochloric acid and then used to pick up a sample of the compound, which is then held in a colourless flame for identification.

2 $f = 2.46 \times 10^{15}$ Hz (remember to convert the energy of the photon in eV into J before using $E = hf$). Wavelength = 122 nm.

Exam practice

(a) and **(b)** Emission spectra show the light emitted by hot gases. Photons are emitted when electrons in hot gases fall from higher to lower energy levels. Only certain energy levels are available to the electrons, so only certain energy falls are possible. The energy levels and possible energy falls are characteristic of the element in question. The energy of the photon is equal to the difference in the two energy levels. The energy (and wavelength) of the emitted photons thus identify the element that produced the light. Absorption spectra are formed when white light emitted from the very hot core of a star passes through cooler outer layers. The atoms in the cooler regions can only absorb photons of particular energy and therefore of particular wavelength. It is the electrons within the target atoms that actually absorb photons. These electrons have only certain energy levels available to them. If the incident photon can deliver exactly the right amount of energy to lift an electron to a higher energy level, it will be absorbed. If it cannot, the photon will not be absorbed. As the energy levels available to the electrons are characteristic of the target atom, by examining which photons are absorbed astronomers can identify which elements are present within the star under observation. Refer to the energy-level diagrams on page 130.

Examiner's secrets

Examiners often get ideas for questions from the science that is reported on the TV, or the radio, or in newspapers. They set the questions well in advance of the date that you will sit the examination so it may well be worthwhile to keep yourself informed of what is happening throughout your course. Astronomy is frequently reported by the media, for example.

De Broglie's equation and atomic models

Checkpoints

1 The wave and particle models complement one another. It is not possible to use only the wave model, or only the particle model, to describe matter or radiation. Both models are required.

2 The Thomson and Rutherford models agree that atoms contain both positive and negative charges. However, the Thomson model has positive and negative charges evenly distributed within the atom. The Rutherford model has the positive charge concentrated in a central nucleus with electrons orbiting around the outside.

Exam practice

(a) De Broglie suggested that all particles might have a dual nature. His equation, $\lambda = h/p$, linked a wave property (wavelength λ) with a particle property (momentum p). From this equation we can see that momentum is inversely proportional to wavelength as h is a constant – Planck's constant. So, as momentum increases, wavelength decreases. The momentum of an object that we might see in our everyday life is so big that its associated wavelength is incredibly small. The wavelength of a 100 m sprinter would be about 10^{-36} m for example. To see the sprinter exhibit a wave property, like diffraction, would require him/her to squeeze through a gap of about 10^{-36} m. That's unlikely to happen and so we are unlikely to observe matter behaving as waves in everyday life.

Without an understanding of wave–particle duality, one might assume that negatively charged electrons should spiral towards the protons in the nucleus as opposite charges attract. However, electrons have a wavelength as they can be considered to be waves. Squeezing an electron in towards the nucleus would shorten its wavelength but this would make the electron too energetic for the protons to hold! (Energy is proportional to $1/\lambda^2$.)

See page 133, for a diagram of the Rutherford–Bohr atomic model. The corresponding energy-level diagram looks like this:

(Note the diagram on page 133 shows E_0 and E_1 only.)

Energy levels	Orbits
E_3	$n = 4$
E_2	$n = 3$
E_1	$n = 2$
E_0	$n = 1$ Ground state

(b) $\frac{1}{2}mv^2 = 1.6 \times 10^{-19}$ J

$$\therefore\ v = \frac{3.2 \times 10^{-19}}{9.11 \times 10^{-31}}$$

$$= 5.9 \times 10^6\ \text{ms}^{-1}$$

$$\lambda = \frac{6.6 \times 10^{-34}}{9.11 \times 10^{-31} \times 5.9 \times 10^6}$$

$$= 1.22 \times 10^{-10}\ \text{m}$$

Fields

Non-contact forces (sometimes called at-a-distance forces) can be described using the concept of a field. In this section, gravitational forces and fields are considered but there are also strong links to magnetic fields. The idea of field strength and potential is also important and is used to explain interactions that occur between objects that have mass and/or charge.

Exam themes

- Mathematical description of waves, e.g. the wave–speed equation.
- Force between masses
- Satellites and orbits
- Calculation of g for planets of different masses and radii
- Use of vector algebra to find resultant force and field
- Field strength between capacitor plates
- Motion of charged particles
- Combinations of electrical and gravitational fields
- Similarities and differences between fields
- Calculation of potential and potential difference

Topic checklist

	Edexcel		AQA/A		AQA/B		OCR/A		OCR/B		WJEC		CCEA	
	AS	A2	AS	A2	AS	A2	AS	A2	AS	A2	AS	A2	AS	A2
Newton's law of universal gravitation		●		●		●		●		●		●		●
Gravitational fields		●		●		●		●		●		●		●
Electric forces and fields		●		●		●		●		●		●		●
Electric potential* and charged particle acceleration		●		●		●		●		●		●		●
Comparisons: gravitational and electric fields		●		●		●		●		●		●		●
Synoptic skills		●		●		●		●		●		●		●

* *Note*: the amount of detail that you are expected to know about electric potential depends on the syllabus that you are following. Make sure that you are familiar with the content of yours and revise accordingly.

Newton's law of universal gravitation

"Truth is ever to be found in the simplicity, and not in the multiplicity and confusion of things."

Sir Isaac Newton

Newton's law of universal gravitation allows you to calculate the force of attraction that occurs between *any* objects that have mass. It was proposed almost 350 years ago after Newton had pondered upon why Kepler's laws of planetary motion worked and also more mundane matters such as what made an apple fall to Earth! He realized that they could both be explained using the same idea.

Grade booster

Learn this law – it often comes up and can give you 'cheap' marks in an exam.

Statement of Newton's law of universal gravitation

- Every particle in the Universe attracts every other with a force which is directly proportional to the product of their masses and inversely proportional to the square of their separation.

The law is usually summarized in a diagram like this:

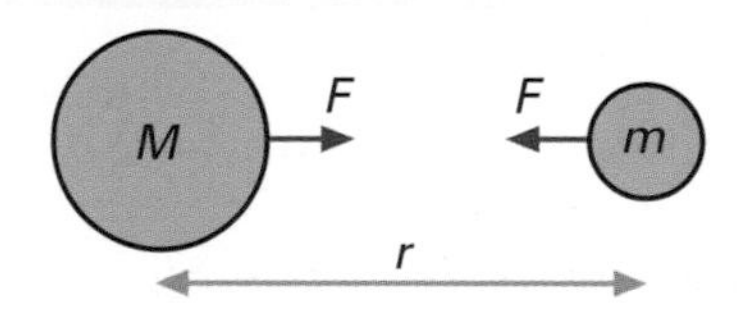

Equation

The law gives this equation:

$$F = G\,\frac{Mm}{r^2}$$

Where M and m are the two masses, r is the separation, F is the force and G is a constant of proportionality, known as the **universal constant of gravitation** ($= 6.67 \times 10^{-11}$ N m^2 kg^{-2}). Note that:

- the force is always attractive and the range is infinite
- strictly, the equation applies only to point masses, but you can use it in all examples you will come across at AS- or A2-level
- r is the distance between the *centres* of the bodies
- G is *very* small – hence gravitational forces are very small, unless one (or both) of the masses is huge, as with a planet

The jargon

Some syllabuses use m_1 and m_2 instead of M and m – get used to yours.

Checkpoint 1

Be able to work out why G has these units – this is another common exam question.

Watch out!

Remember that forces are vector quantities. There is a convention: attractive forces are negative and repulsive forces are positive.

Inverse square laws

Newton's law of universal gravitation is an example of an **inverse-square law**. This simply means that the size of the force is *inversely* proportional to the *square* of the separation of the objects:

- if the separation is doubled (*multiplied* by 2), the force is quartered (*divided* by 2^2)
- if the separation is made 10 times *bigger*, the force is 10^2 (= 100) times *smaller* etc.
- inverse-square laws generate graphs which have a characteristic shape, as shown on the opposite page.

Links

Coulomb's law, see page 144, is another inverse-square law.

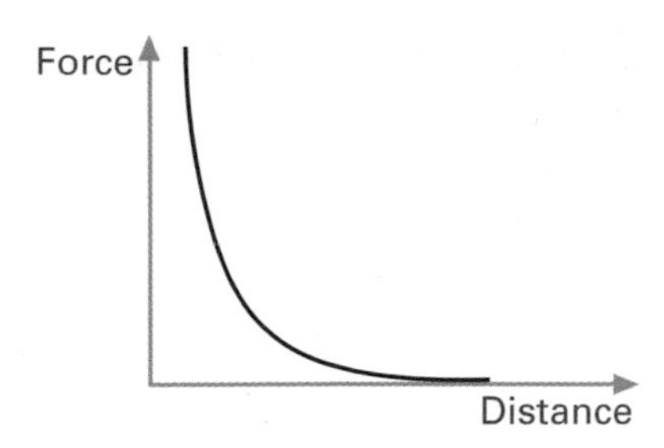

Note With inverse-square laws, the force drops off very rapidly with distance.

Motion of planets and satellites

For an object in orbit, the centripetal force required to keep it moving in a circle is provided by the gravitational force, given by the equation opposite. Thus:

$$G\frac{Mm}{r^2} = m\omega^2 r$$

Where M is the mass of the object *being orbited*, m is the mass, v is the velocity and ω is the angular velocity of the *orbiting* object, r is the radius of the orbit. Note:

- m can easily be eliminated from the equation, showing that the mass of the orbiting object is not relevant
- the equation can be rearranged to apply to particular problems
- in particular, it can generate the equation which relates the *radius* r of the orbit to its *period* T:

$$T^2 = \frac{4\pi^2}{GM}r^3$$

Which takes you back to Kepler's third law and the challenge originally taken up by Newton.

Links

See *circular motion*, pages 34–5.

Action point

It was the motion of the planets that Newton was trying to explain when he formulated his law of universal gravitation – read about Tycho Brahe's observations (and how he lost his nose in a duel!) and the laws that Johannes Kepler worked out to describe them.

Checkpoint 2

Try to derive this equation from the one above it.

Exam practice answers: page 152

1 What is the gravitational pull between a 3 kg mass and a 5 kg mass placed 0.15 m apart? (5 min)

2 Express G in SI *base* units. (5 min)

3 State Newton's law of universal gravitation and explain how it applies to the Earth/Moon system, drawing a labelled diagram to show the forces which act on the two bodies. (5 min)

4 Communication satellites are usually parked in *geostationary* orbits. Explain what is meant by this and show that the height of such an orbit is approximately 36 000 km above the Earth's surface. (15 min)

Gravitational fields

Don't forget

Gravitational fields are *created* by *all* masses and *felt* by *all* masses.

The jargon

The arrows always point towards the object – since the force is always attractive. *Equipotential surfaces* are explained on the opposite page.

Grade booster

All field lines (gravitational, electric and magnetic) must never touch or cross and they always enter and leave surfaces at right angles to them.

Checkpoint 1

Is M the mass of the object *causing* the field or *feeling* it?

Checkpoint 2

Can you show that $N\,kg^{-1}$ is identical to $m\,s^{-2}$?

Watch out!

You need to take *direction* into account when you combine field strengths.

A gravitational field exists around *any* mass, no matter how large or small. Other masses in this region will feel a force, which is *always* attractive: a repulsive gravitational force has never been identified.

Gravitational field lines (lines of force)

Gravitational fields are always drawn using gravitational **field lines**, which show the direction of the force on a mass placed at any point in the field. Around a spherical mass the lines look like this:

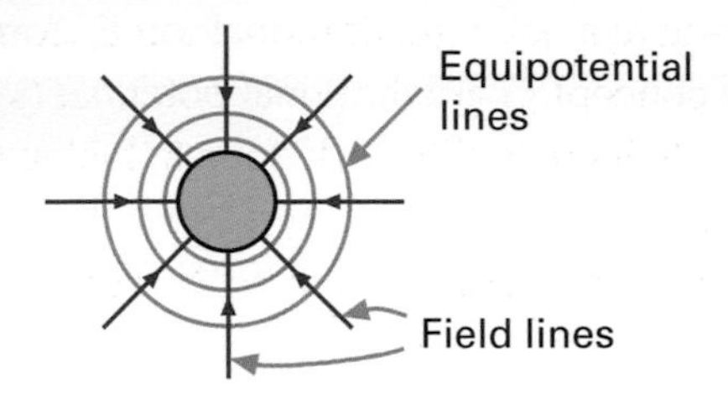

Near the Earth's surface, the field is nearly uniform and the lines are evenly spaced:

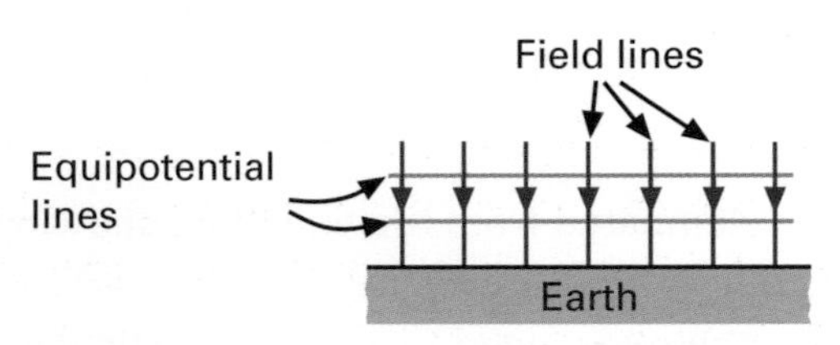

Gravitational field strength *g*

The gravitational field strength g at a point in a field is the force per unit mass on an object placed at that point:

$$g = \frac{F}{m}$$

It follows from Newton's universal law of gravitation that, around a spherical mass M:

$$g = G\frac{M}{r^2}$$

Where r is the distance from the *centre* of the mass. Note that:

- ➜ g has units of $N\,kg^{-1}$
- ➜ g may also be regarded as the acceleration due to gravity (in $m\,s^{-2}$) at the point
- ➜ g is a *vector* quantity
- ➜ the variation of g with distance from a mass is another *inverse-square* relationship

Earth's gravitational field

- ➜ Its strength is 9.81 $N\,kg^{-1}$ at the surface of the Earth.
- ➜ It falls off with distance above and below the surface as shown in the graph on the opposite page.

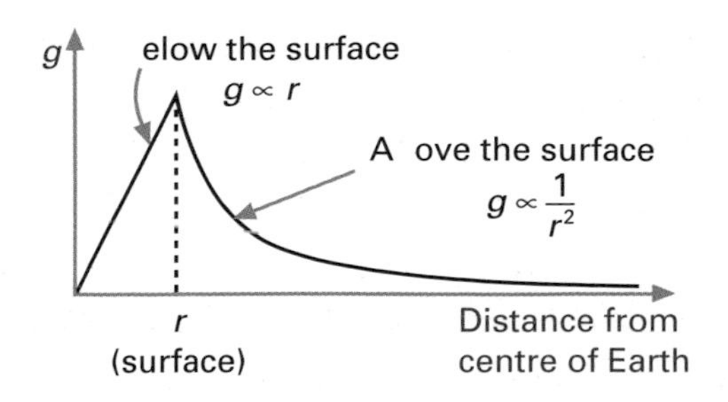

Gravitational potential U

When a mass is moved against a gravitational force, *work* is done. This is described using the concept of gravitational potential (symbol *U*, but *V* in some text books), defined as the work done in taking unit mass from infinity (∞) to a point in a field.

Thus, at a distance *r* from a mass *M*, it can be shown that:

$$U = G\frac{M}{r}$$

Note:
- → *U* is a *scalar* quantity
- → *U* is zero at infinity
- → at all other places, *U* is negative (since a 'negative' amount of work has to be done to move a mass against an attractive force)
- → the gravitational field strength is equal to the *potential gradient* (the *slope* of a graph of *U* against *r*)

Escape velocity

If a mass at the Earth's surface can be given a kinetic energy equal to its gravitational potential energy (= *mU*), it will escape completely from the Earth's gravitational field (and end up at infinity!). The velocity required to do this is called the **escape velocity**, is not dependent upon the mass and is given by:

$$v = \sqrt{\left(\frac{2GM}{r}\right)} = \sqrt{(2gr)}$$

Equipotential surfaces

Points that are at the same potential lie on **equipotential surfaces**. Around a spherical mass, these are concentric spheres – or circles in two dimensions – see opposite. Equipotential surfaces are *always* perpendicular to field lines.

Checkpoint 3

The linear variation in *g* below the surface of the Earth is an approximation, based on the (incorrect) assumption of uniform density. Why is the graph this shape *above* the Earth's surface?

Grade booster

Be familiar with your own syllabus: *learn* any definitions necessary (like this one) and try to make sure that you know when to use all the equations.

Checkpoint 4

What would a graph of *U* against *r* look like? Would it be different from *g* against *r*?

Checkpoint 5

What does each of the quantities in this equation stand for? They have been used frequently in the last four pages.

Grade booster

Very often gravitational potentials are described in terms of gravitational *wells*. To escape from the surface of the Earth, a space craft must climb out of the Earth's potential well.

Exam practice answers: page 152-3

1 Explain what you understand by escape velocity and use the approximate values of $r_E = 6.4 \times 10^6$ m and $g = 10$ N kg^{-1} to show that the escape velocity from the Earth is close to 11 km s^{-1}. (5 min)

2 It is proposed that a black hole with a mass equal to that of the Earth would have a radius of 1 cm. What is the gravitational field strength at the surface of the black hole?
How far away would you have to go for the gravitational field strength to have the same value as that on the surface of the Earth? What object has approximately this size? (10 min)

Electric forces and fields

Coulomb's law allows you to calculate the force between charged objects; it is very similar in form to Newton's law of universal gravitation and leads logically to the concept of electric field and field strength.

Coulomb's law

→ The force between two charges is directly proportional to the product of the charges and inversely proportional to the square of their separation.

The law is usually summarized in a diagram like this:

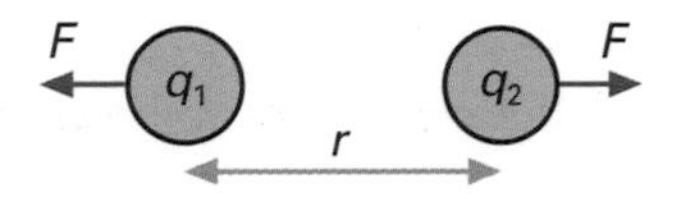

The law gives this equation:

$$F = k\frac{q_1 q_2}{r^2}$$

Where q_1 and q_2 are the charges and r is their separation. k is a constant of proportionality (= 9.0×10^9 N m^2 C^{-2} in a vacuum and effectively the same in air).

Note:

→ if the charges are alike (both positive or both negative), the force is repulsive (as shown above)

→ if the charges are unlike (one positive and one negative), the force is attractive

→ the charges should be small in comparison with their separation

→ r should be measured from the *centre* of the charges

Action point

The form of this equation should be compared with that of *Newton's law of universal gravitation*, pages 140–1. What is the major difference between the two laws?

Permittivity of free space

If the charges above are in a vacuum, the constant k in the equation is usually expressed as:

$$k = \frac{1}{4\pi\varepsilon_0}$$

Where the 4π is included in order to simplify equations derived from Coulomb's law and ε_0 is known as the permittivity of free space (= 8.85×10^{-12} F m^{-1}). If the medium between the charges is different, then ε_0 is replaced by ε, the permittivity of the medium.

Links

ε_0, ε (and ε_R) are described in *capacitors*, pages 88–9.

Electric fields

An electric field is a region around a charge in which another charge feels a force. Unlike gravitational fields, the forces can be either attractive or repulsive.

Electric field lines

As with gravitational fields, lines are used to represent electric fields. The

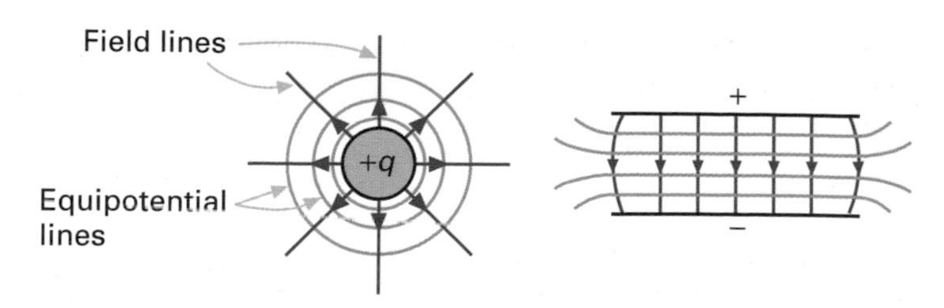

shape of common fields is shown on the opposite page.
Note:

- at any point, the lines show the direction of the force on a positive charge placed at that point
- as a consequence, field lines around a point positive charge and a point negative charge are in different directions
- the field between parallel plates is approximately uniform

Checkpoint 1

Sketch the field lines and equipotential lines around a *negative* charge.

Checkpoint 2

How would you observe electric field lines in the laboratory? (*Hint* One common method uses materials normally found in the kitchen and the garden shed!)

The jargon

Uniform means constant in both magnitude *and* direction because the electric field is a vector quantity.

Electric field strength

The electric field strength E at a point in a field is defined as the force on unit charge placed at that point, and so if a charge q feels a force F, then:

$$E = \frac{F}{q}$$

It follows from Coulomb's law that around a point charge Q

$$E = k\frac{Q}{r^2}$$

Checkpoint 3

What is k in this equation?

Note:

- E has units of N C^{-1} (or V m^{-1}) and is a *vector* quantity
- the direction of E is the direction of the force on a positive charge
- inside a hollow conductor, E is zero
- E varies with distance from a charge according to an inverse-square law and therefore the range of E is infinite.

Grade booster

Be able to show that N C^{-1} and V m^{-1} are equivalent units.

Between parallel plates, the field is uniform and E is given by:

$$E = \frac{V}{d}$$

Where V is the potential difference between the plates and d is their separation – hence the alternative unit for E, above.

Exam practice

answers: page 151

1 Express ε_0 in SI *base* units. (10 min)

2 Draw a free-body force diagram for a tiny particle carrying a charge equal to that of two electrons which is held stationary in an electric field between parallel plates 5 mm apart and with a p.d. of 1 MV between them. Show the polarity of the plates on your diagram.
If $g = 9.81$ N kg^{-1}, calculate the mass of the particle. What assumptions have you made? What practical problems might be associated with this arrangement? (15 min)

Electric potential and charged particle acceleration

Grade booster

Potential difference (voltage) features strongly in current electricity – if you understand potential you will be well on the way to understanding all of electricity.

Grade booster

Make sure that you convert distances to metres before you use this equation.

Checkpoint 1

Sketch a graph showing the variation of V with r.

Checkpoint 2

What is the definition of a volt?

Electric potential is an important quantity as it brings in the fundamental concept of energy (work). It also relates electrostatics to current electricity and can be used to predict how charged particles will be accelerated.

Electric potential

If a charge is moved in an electric field, then work is done/energy is converted.

The potential V at a point in a field is the work done W per unit charge q in taking *positive* charge from infinity (where the force, field and potential are all zero) to that point, i.e. $V = W/q$.

Around a spherical conductor, carrying charge Q:

$$V = k\frac{Q}{r}$$

Where r is the distance of the point from the (centre of) the charge Q.

Note:

- V is a *scalar* quantity
- the electric field strength is equal in magnitude to the potential gradient
- positive charges move *down* a potential gradient (from *high* to *low* potential)
- negative charges move *up* a potential gradient (from *low* to *high* potential)
- the relationship between V and distance is inversely proportional

Equipotential surfaces

Points that are at the same potential lie on equipotential surfaces. Around a spherical charge, these are concentric spheres – or circles in two dimensions (see the diagrams on pages 142 and 145). Equipotential surfaces are *always* perpendicular to field lines.

Motion of charged particles

Acceleration

The idea of potential can be used to describe the acceleration of charged particles. It follows from the definition of the volt (see page 76) that the kinetic energy gained by a particle carrying a charge q when accelerated through a potential difference of V volts is given by:

$$\tfrac{1}{2}mv^2 = qV$$

Where m is the mass of the particle and v its subsequent speed.

This equation can be used to calculate the speed of an electron as it emerges from an electron gun, as used in televisions, for instance. The *deflection* of these electrons across the screen is also caused by electric fields.

Deflection

Charged particles are deflected by electric fields and it follows from the definition of electric field strength (see page 145) that the force F on a particle carrying a charge q in a field of strength E is given by:

$$F = Eq$$

If the field is provided by a pair of parallel plates the shape of the path is as shown below and the equation becomes:

$$F = \frac{Vq}{d}$$

Where V is the potential difference between the plates and d their separation.

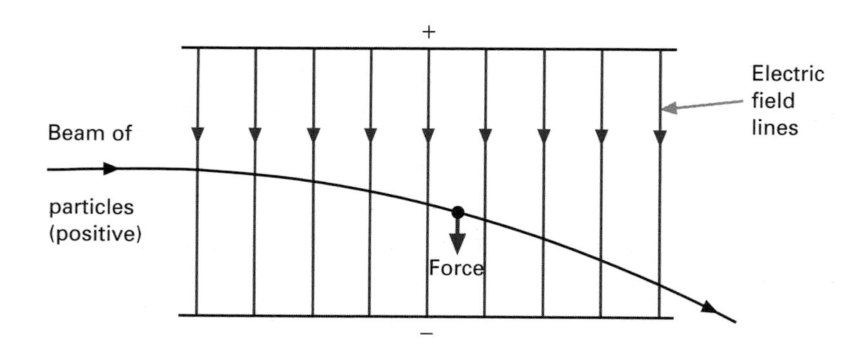

Note:

- ➜ the force is always parallel to the field lines
- ➜ the path of the particles is a parabola while in the electric field
- ➜ the size of the force does not depend on the speed of the particles
- ➜ deflection only occurs when the particles are within the field – elsewhere they travel in straight lines
- ➜ positive and negative charges are deflected in opposite directions
- ➜ we ignore the effects of gravity

Links

Charged particles are also deflected by magnetic fields (see pages 90–1), but *only if they are moving.*

Check the net

As well as televisions, ink-jet printers also work on this principle. Visit www.waythingswork.com/

Checkpoint 3

How is the shape of the path followed by charged particles different in a magnetic field?

Action point

The path of a charged particle in an electric field is a parabola. If you consider the vertical and horizontal motions separately, you should be able to prove that the particle gains KE.

Examiner's secrets

There is an opportunity for examiners to get you to compare the path of a moving charge in an electric field with that in a magnetic field. Make sure you are ready for it!

Exam practice answers: page 151-2

1 A point charge of +4 nC is brought from infinity to a point 30 mm away from a charge of +8 nC. How much work is done? How much more work must be done to bring the charges 10 mm closer together? (10 min)

2 An electron (charge 1.6×10^{-19} C) is accelerated through a potential difference of 2.5 kV and enters an electric field that is perpendicular to the direction of motion of the electron. How fast are the electrons travelling when they enter the field?
Explain why this horizontal velocity does not change as it passes through the field.
If the plates are 5 cm long, for what length of time does the electron stay in the field?
If the field strength is 10 kV m^{-1}, how big is the deflecting force on the electron? (15 min)

Comparisons: gravitational and electric fields

There are numerous similarities between gravitational and electric fields – and some important differences. The comparison is easiest to make when presented in the form of a table.

Action point

For each of these similarities, say whether there is an equivalent effect for a magnetic field.

Similarities

	Gravitational property	*Electrical property*
For the force to act . . .	contact not needed – force acts at a distance	contact not needed – force acts at a distance
Range of field	infinite – size of force decreases as distance increases, but in theory never falls to zero	infinite – size of force decreases as distance increases, but in theory never falls to zero
Field lines	can be used to describe the field – direction is the direction of the force on a *mass*	can be used to describe the field – direction is the direction of the force on a *positive charge*
Field strength	defined as the force on unit mass: $g = \frac{F}{m}$	defined as the force on unit charge: $E = \frac{F}{q}$
Force between two objects	given by Newton's law of universal gravitation: $F = G\frac{Mm}{r^2}$ (inverse-square law)	given by Coulomb's law: $F = k\frac{Qq}{r^2}$ (inverse-square law)
Potential	can be used to describe work done in moving masses: $U = -\frac{GM}{r}$ (inverse law)	can be used to describe work done in moving charges: $V = k\frac{Q}{r}$ (inverse law)
Potential energy	gravitational potential energy = mU	electrical potential energy = qV
Relationship between potential and field strength	field strength is (negative of) potential gradient: $g = -\frac{\mathrm{d}U}{\mathrm{d}r}$	field strength is (negative of) potential gradient: $E = -\frac{\mathrm{d}V}{\mathrm{d}r}$
Kinetic energy	calculated from: $\frac{1}{2}mv^2 = mU$	calculated from: $\frac{1}{2}mv^2 = qV$

Test yourself

What are the units of field strength? Is it a scalar or vector quantity?

Test yourself

What are the units of potential? Is it a scalar or vector quantity?

Checkpoint 1

How is this expression for gravitational potential energy related to the more usual version – namely *mgh*?

Grade booster

Questions comparing gravitational and electric forces and fields come up frequently. Learn as much as you can of the information in these tables.

Differences

	Gravitational property	*Electrical property*
Origin	produced by and act upon *masses*	produced by and act upon *charges*
Effect	cause attraction *only*	can cause attraction *or* repulsion
Shielding	not possible – no material has been found that is able to shield gravitational forces	possible – shielding possible using devices such as Faraday cages
Comparative size (This difference is brought out in Question 2, below – try it!)	insignificant unless one (or both) of masses is huge	much bigger

Exam practice answers: page 152

1 List two similarities and two differences between gravitational and electric fields. (5 min)

2 An electron has a mass of 9.11×10^{-31} kg and a charge of 1.60×10^{-19} C. Calculate the gravitational and the electrical force on two electrons 1.00×10^{-10} m apart in a vacuum.
Calculate the ratio of the two forces and comment on your answer.
Other than the magnitude, state one other difference between these forces. (15 min)

3 What is meant by a field line or line of force? Explain in terms of both gravitational and electric fields, stating how the field lines differ in these cases. (10 min)

Synoptic skills

All A-level specifications (a modern name for the syllabus) must include a minimum of 15% synoptic assessment. It applies only if you go on to study subjects to A2, but it covers AS work too. Synoptic assessment involves the drawing together of knowledge and understanding from different areas of the course. Some synoptic skills are taught through case studies, others via making comparisons (analogies).

Synoptic skills

In theory, synoptic skills should be:

- based on parts of the specification common to all students (including AS)
- the same for all students regardless of specification
- based on a range of different types of question

In practice, synoptic assessment in physics may ask you to:

- make connections between different areas of physics
- apply knowledge and understanding in new situations
- analyse and interpret data
- read a passage of text and answer questions based on it
- produce an extended piece of writing
- answer structured questions with connected calculations

Watch out!

Synoptic papers usually contain questions about other areas of the course. So make sure you check the assessment details of ***your*** specification.

Progressing from GCSE

When you first begin studying A-levels, especially physics, it can sometimes be a large jump up from GCSE level. In fact, you may find that what you learned for your GCSEs may not exactly fit with what you are now expected to know for A-level. The science you needed for GCSE was not wrong; it's just that it was simplified to make it easier to understand. At A-level, you study subjects to a much greater depth. You will probably find it is the same if you go on to study physics at university.

"Only connect! . . . Only connect the prose and the passion and both will be exalted . . ."

E M Forster

Making connections

Studying A-level physics can seem rather daunting at first and students often complain that they don't know how things fit together. Making connections between different areas of physics will take time, but it does happen and usually just in time for the synoptic assessment at the end of A2!

There are lots of ways you can make connections between different areas of physics and these connections will help you revise for your examinations too.

Mind maps

There are different types of so-called *mind maps* and all can be effective in studying and revising. A simple mind map, or *spider diagram*, can be just a series of ideas with a few links between them. *Concept maps* are similar, but the connection between ideas has much more meaning. Begin by *brainstorming* to bring up ideas within a particular topic, then link the ideas together. If you can label the connections with words or phrases it will help you remember how the ideas go together.

Card notes

Make up notes on cards using bullet points. If you get into the habit of doing this as soon as you start your AS course, you will have an invaluable set of memory cards or *aide mémoire*. By the time you get to take your A2 examinations you will have a small library of cards from which to revise.

Don't forget

Keep your notes well organized and indexed. You can't afford to waste time looking for notes near to the examination.

Data analysis

Working scientists communicate many of their ideas through tables and graphs. Physics concepts and laws come alive through measured data. Usually they are produced from your own experiments in class. Being able to analyse unfamiliar data is an important skill. The synoptic examination may ask you to read off data from graphs, process raw data or plot graphs in order to draw conclusions.

Background reading

Reading scientific journals and articles from the Internet is an extremely good way of keeping up to date with current developments in physics. You will certainly become more scientifically literate by reading around the subject.

Action point

Scientific American, *New Scientist* and *Physics Review* are all good sources of background reading.

Practical work

Making measurements is an important aspect of physics and you should keep a clear and up-to-date record of the practical work you do in class. Synoptic questions may well link directly to expected experimental work. You should develop a clear and concise way of recalling experimental arrangements, especially details of how to reduce errors.

Synoptic questions often deal with the differences between data collected on a small scale in the laboratory and that obtained on an industrial scale.

Examiner's secrets

Many synoptic questions are set using articles from popular scientific journals. So, you never know how useful your background reading will be!

Answers

Fields

Newton's law of universal gravitation

Checkpoints

1 The equation can be rearranged to give:

$$G = \frac{Fr^2}{Mm}$$

from which it is apparent that the units of G must be N m^2 kg^{-2}.

2 The derivation uses the relationship $\omega = 2\pi/T$:

$$\frac{GMm}{r^2} = m\omega^2 r$$

$$r^3\omega^3 = GM$$

$$r^3\left(\frac{2\pi}{T}\right)^2 = GM$$

$$r^3 = \frac{GM}{4\pi^2}T^2 \text{ or } T^2 = \frac{4\pi^2}{GM}r^3$$

Exam practice

1 $F = \frac{6.67 \times 10^{-11} \times 3 \times 5}{0.15^2} = 4.45 \times 10^{-8}$ N

2 By rearranging Newton's equation:

$$F = \frac{GMm}{r^2}$$

G can be shown to be given by:

$$G = \frac{Fr^2}{Mm}$$

and, for this equation to be homogeneous, the units of G must be the units of the right-hand side.

These are:

$$\frac{\text{N m}^2}{\text{kg}^2}$$

or N m^2 kg^{-2}.

But these are not yet *base* units and N must be replaced by its base equivalent (kg m s^{-2}), giving units for G of kg^{-1} m^3 s^{-2}.

3 Statement of law as in spread.

Grade booster

With this sort of question, always adopt the approach above – namely, rearrange the equation to make the quantity that you are asked to find the *subject* of the equation, then sort out the units.

For the Earth/Moon system:

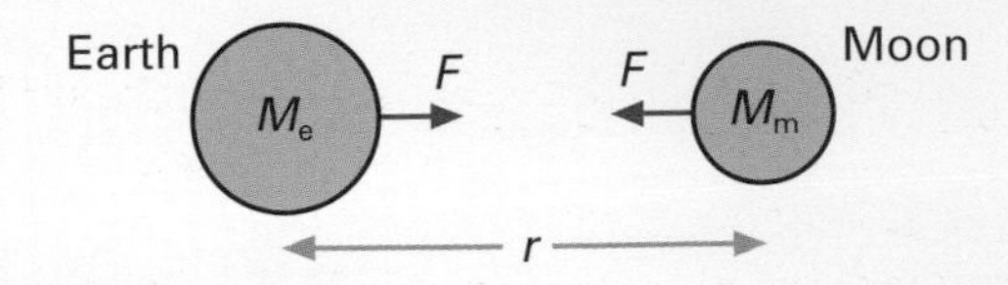

Here the forces labelled F are gravitational forces and are equal and opposite, given by:

$$F = \frac{GM_eM_m}{r^2}$$

4 For a geostationary orbit, the period of the rotation must be exactly the same as that of the Earth – namely, 24 hours. This allows satellites to be *parked* in such an orbit, where they will stay directly above the same point on the Earth's surface, though this point must be directly above the equator. The height of such an orbit can be calculated by equating the centripetal force to the gravitational force at that height. If the height from the centre of the Earth is r, this gives:

$$m\omega^2 r = \frac{GMm}{r^2}$$

Where ω is the angular velocity associated with the rotation. Note that the m can be divided from both sides, showing that the mass of the satellite is not relevant.

Since $\omega = 2\pi/T$ (where T is the period):

$$\frac{4\pi^2}{T^2}r = \frac{GM}{r^2} \qquad r = \sqrt[3]{\left(\frac{GMT^2}{4\pi^2}\right)}$$

Substituting values of $G = 6.67 \times 10^{-11}$ N m^2 kg^{-2}, $T = 24$ hours $= 8.64 \times 10^4$ s and $M_e = 6.0 \times 10^{24}$ kg, gives $r \approx 42\,000$ km and, since the radius of the Earth is 6 400 km, the orbit is 36 000 km above the surface of the Earth.

Gravitational fields

Checkpoints

1 M is the mass of the object *causing* the field.

2 N kg^{-1} = (kg m s^{-2}) kg^{-1} = m s^{-2}.

3 The graph is the shape shown because the relationship between g and distance is an inverse square.

4 The graph of U against r looks like this:

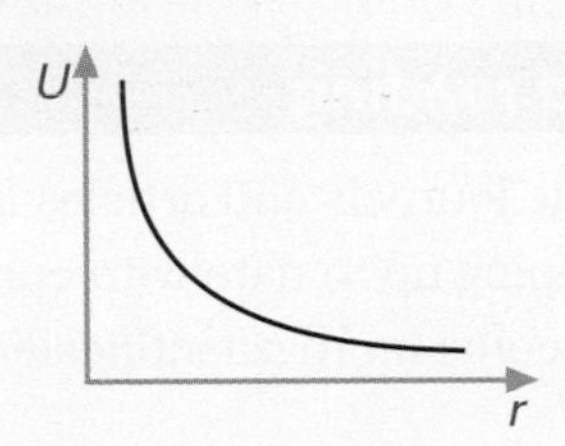

It is different from g against r because it is an inverse relationship, rather than an inverse square.

5 G is the universal constant of gravitation, M is the mass of the Earth (or other planet), g is the gravitational field strength, and r is the radius of the Earth (or other planet).

Exam practice

1 The escape velocity is the minimum velocity that an object must acquire in order to escape completely from a gravitational field.

Using:

$$v = \sqrt{(2gr)}$$

gives:

$$v = \sqrt{(2 \times 10 \text{ N kg}^{-1} \times 6.4 \times 10^6 \text{ m})}$$
$$\simeq 11 \times 10^3 \text{ m s}^{-1} \simeq 11 \text{ km s}^{-1}$$

2 Using:

$$g = \frac{GM}{r^2}$$

gives:

$$g = \frac{6.67 \times 10^{-11} \text{ N m}^2 \text{ kg}^{-2} \times 6.0 \times 10^{24} \text{ kg}}{(1 \times 10^{-2})^2 \text{ m}^2}$$
$$= 4.0 \times 10^{18} \text{ N kg}^{-1}$$

The equation above can be rearranged to give:

$$r^2 = \frac{GM}{g} \quad \text{or} \quad r = \sqrt{\left(\frac{GM}{g}\right)}$$

and for $g = 9.81\ \text{N kg}^{-1}$:

$$r = \sqrt{\left(\frac{6.67 \times 10^{-11}\ \text{N m}^2\ \text{kg}^{-2} \times 6.0 \times 10^{24}\ \text{kg}}{9.81\ \text{N kg}^{-1}}\right)}$$

$$= 6.4 \times 10^6\ \text{m}$$

This is very similar to the radius of the Earth.

Electric forces and fields

Checkpoints

1 Field lines and equipotential surfaces around a negative charge look like this:

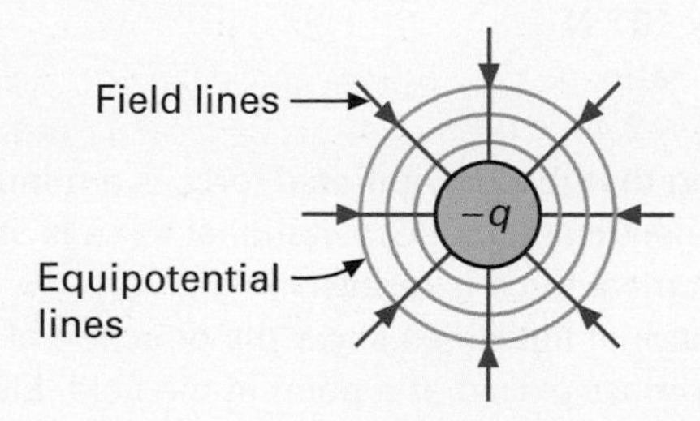

2 To observe electric field lines in the laboratory you could use a high-voltage supply connected to electrodes separated by castor oil; if either grass seed or semolina is sprinkled into the castor oil, the seeds/grains align themselves along field lines. Alternatively, conducting paper can be used to plot equipotential surfaces.

3 k in the equation is $1/(4\pi\varepsilon_0)$.

Exam practice

1 The equation for Coulomb's law:

$$F = \frac{1}{4\pi\varepsilon_0} \times \frac{Qq}{r^2}$$

can be rearranged to give:

$$\varepsilon_0 = \frac{Qq}{4\pi r^2 F}$$

and hence the units of ε_0 must be

$C^2\ (\text{kg m s}^{-2})^{-1}\ \text{m}^{-2}$

or $C^2\ \text{kg}^{-1}\ \text{m}^{-1}\ \text{s}^2\ \text{m}^{-2}$

or $\text{kg}^{-1}\ \text{m}^{-3}\ \text{s}^2\ C^2$

2 Here we have

Equating the electrical and gravitational forces gives:

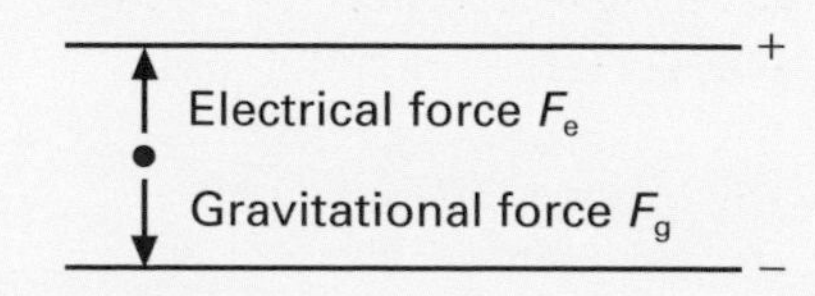

$$\frac{Vq}{d} = mg$$

Hence:

$$m = \frac{Vq}{gd} = \frac{1 \times 10^6\ \text{V} \times 3.2 \times 10^{-19}\ \text{C}}{9.81\ \text{N kg}^{-1} \times 5 \times 10^{-3}\ \text{m}}$$

$$= 6.5 \times 10^{-12}\ \text{kg}$$

It has been assumed that the upthrust on the particle can be ignored.

The problem associated with this arrangement is that the electrical field strength between the plates is huge ($2 \times 10^8\ \text{V m}^{-1}$). This is greater than the breakdown potential for air and therefore the air would begin to conduct.

Electric potential and charged particle acceleration

Checkpoints

1 V varies with r according to an inverse relationship, like this:

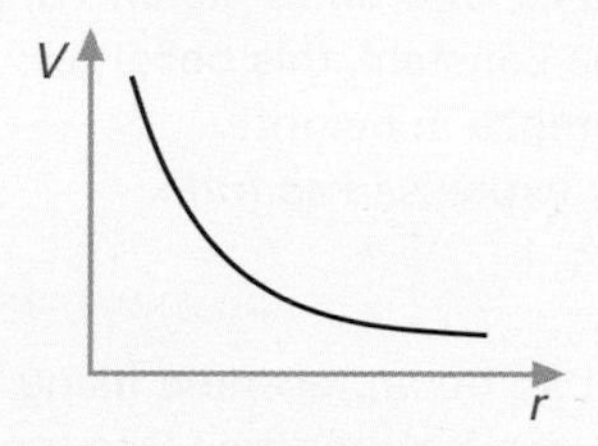

2 If 1 J of energy is converted when 1 C of charge flows between two points, then the potential difference between those points is 1 V.

3 In a magnetic field, moving charged particles follow a *circular* path, since the direction of the force is given by the left-hand rule and is always perpendicular to the direction of motion.

Exam practice

1 The work done in bringing unit charge from infinity to a point a distance r from a point charge is the potential V at that point. To bring a charge q:

work done $= Vq$

$$= \frac{8 \times 10^{-9}\ \text{C}}{4\pi\varepsilon_0 \times 30 \times 10^{-3}\ \text{m}} \times 4 \times 10^{-9}\ \text{C}$$

$$= 9.59 \times 10^{-6}\ \text{J}$$

To take the same charge to a distance of 20 mm:

$$\text{work done} = \frac{8 \times 10^{-9}\ \text{C}}{4\pi\varepsilon_0 \times 20 \times 10^{-3}\ \text{m}} \times 4 \times 10^{-9}\ \text{C}$$

$$= 1.44 \times 10^{-5}\ \text{J}$$

Hence, additional work $= 4.79 \times 10^{-6}$ J

2 The kinetic energy of the electron is equal to the work done by the potential difference:

$\tfrac{1}{2}mv^2 = qV$

Hence,

$$v = \sqrt{\left(\frac{2qV}{m}\right)} = \sqrt{\left(\frac{2 \times 1.6 \times 10^{-19}\ \text{C} \times 2.5 \times 10^3\ \text{V}}{9.11 \times 10^{-31}\ \text{kg}}\right)}$$

$$= 2.96 \times 10^7\ \text{m s}^{-1}$$

Here the force is always vertical (in the direction of the field lines) and so the horizontal velocity is unaffected and the time will be given by

time = distance/velocity

$$= \frac{5 \times 10^{-2}\ \text{m}}{2.96 \times 10^7\ \text{m s}^{-1}}$$

$$= 1.69 \times 10^{-9}\ \text{s}$$

The force on a charged particle q in a field of strength E is given by:

$F = Eq$

$= 10\,000\ \text{N C}^{-1} \times 1.6 \times 10^{-19}\ \text{C}$

$= 1.6 \times 10^{-15}\ \text{N}$

Comparisons: gravitational and electric fields

Checkpoint

1 The change in gravitational potential energy $\Delta(\text{GPE})$ is given by:

$\Delta(\text{GPE}) = m\Delta U$

Since $U = gr$ and, over short distances, g can be assumed to be constant, this becomes:

$mg \times (\text{difference in height})$

and is usually expressed as mgh.

Exam practice

1 *Similarities* Can be represented using field lines; field strength is inversely proportional to the square of the distance etc (see table in spread)

Differences Gravitational fields act on masses, electric fields act on charges; gravitational fields cause attractive forces only, electrical fields cause attractive or repulsive forces.

2 Gravitational force is given by:

$$F_g = \frac{GMm}{r^2}$$

$$= \frac{6.67 \times 10^{-11}\ \text{N m}^2\ \text{kg}^{-2} \times 9.11 \times 10^{-31}\ \text{kg} \times 9.11 \times 10^{-31}\ \text{kg}}{(1.00 \times 10^{-10})^2\ \text{m}^2}$$

$= 5.54 \times 10^{-51}\ \text{N}$

Electrical force is given by:

$$F_e = \frac{Qq}{4\pi\varepsilon_0 r^2}$$

$$= \frac{1.60 \times 10^{-19}\ \text{C} \times 1.60 \times 10^{-19}\ \text{C}}{4\pi \times 8.85 \times 10^{-12}\ \text{Fm}^{-1} \times (1.00 \times 10^{-10})^2\ \text{m}^2}$$

$= 2.30 \times 10^{-8}\ \text{N}$

Hence, ratio:

$F_g/F_e = 2.41 \times 10^{-43}$

showing that the gravitational force is a minute fraction of the electrical force. Gravitational force is attractive but electrical force is repulsive.

3 Gravitational field lines show the direction of the force on *unit mass* placed at a point in the field. Electric field lines show the direction of the force on *unit positive charge* placed at a point in the field.

Options

The commonest seven options are summarized in this section: astrophysics, medical physics, materials, applied physics, further electricity, energy and turning points in physics. Check your syllabus for details.

Exam themes

- *Astrophysics* Measuring astronomical distances, stellar surface temperatures and spectra, imaging methods, the Big Bang theory (and evidence for it), the fate of the Universe (and what it depends on).
- *Medical physics* Diagnosis, treatment, the eye and seeing, the ear and hearing, the heart.
- *Materials* Macroscopic behaviour of metals, ceramics and polymers and its interpretation on an atomic level.
- *Turning points in physics* The discovery of the electron and wave particle duality, Einstein's theory of special relativity, electromagnetic waves
- *Energy and the environment* The effect of burning fossil fuels on the Earth's atmosphere, alternative energy sources
- *Applied physics* Extending ideas in dynamics and thermodynamics
- *Further electricity* More about electromagnetism and alternating currents

Topic checklist

	Edexcel		AQA/A		AQA/B		OCR/A		OCR/B		WJEC		CCEA	
	AS	A2	AS	A2	AS	A2	AS	A2	AS	A2	AS	A2	AS	A2
Astrophysics 1*		●*		●	O*			●*			O	●*		
Astrophysics 2		●*		●	O*			●*						
Astrophysics 3				●										
Astrophysics 4		●*		●	O*			●*	O*			●*		
Medical and health physics 1				●	O*				O*			●	O*	
Medical and health physics 2	O*			●	O*	●*		●*	O*			●	O*	
Medical and health physics 3					O	●*		●*				●		
Materials 1	O*		O*				O*					●		
Materials 2												●		
Turning points in physics 1	O*		O*	●	O*		O*		O*				O*	
Turning points in physics 2				●		●*						●		
Energy and the environment 1					O*							●		
Energy and the environment 2					O*							●		
Applied physics 1				●		●*								
Applied physics 2				●		●*								
Further electricity												●		

*material found in core module, not an option

Astrophysics 1

The further we look, the smaller we seem. The distances and time scales dealt with here are astronomical!

- *Astronomical Unit* (AU) The average distance from the Earth to the Sun. 1 AU = 1.496×10^{11} m.
- *Parsec* (pc) The distance to a star that subtends an angle of 1 second at an arc of length 1 AU (see diagram below). 1 pc = 3.086×10^{16} m.
- *Light-year* (ly) The distance travelled by a ray of light in 1 year. 1 ly = 9.46×10^{15} m.

Links

Using radio waves, a pulse can be reflected off a planet or asteroid and the time taken for it to return can be used to work out distance using the speed of travel, c.

Checkpoint 1

How many light-years are there in a parsec?

Using parallax to measure distances to nearby stars

(1° = 60 minutes = 3 600 seconds)

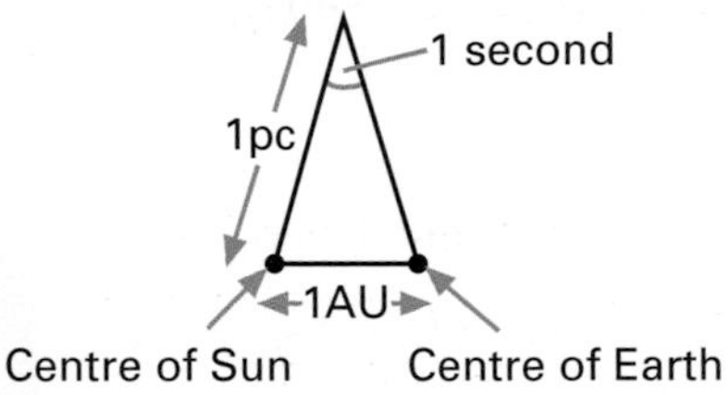

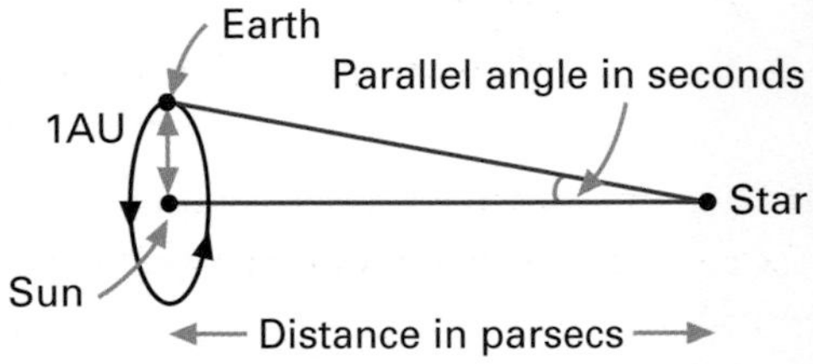

Parallax works only for nearby stars (up to about 100 pc).

The jargon

You observe parallax when travelling in a car for example – the closer an object is to you, the faster it moves relative to a distant object.

Checkpoint 2

(a) How many light-minutes in a light-year?

(b) How many minutes does it take light to reach us from the Sun?

Checkpoint 3

The angle subtended by a person's eyes at a distant gate post is 1.2°. Given that the person's eyes are 10 cm apart, calculate the distance to the post.

Using the inverse-square law to measure greater distances

The intensity of light from a star obeys an inverse-square law:

$$I = L/(4\pi R^2)$$

Where I is intensity in W m^{-2}, L is power output (or *luminosity*) in watts and $4\pi R^2$ is the area the radiation is spread over at a distance of R.

Checkpoint 4

Calculate the Sun's luminosity if the intensity of solar radiation reaching the surface of the atmosphere (the *solar constant*) is 1360 W m^{-2}.

The black-body spectrum

Stars are black bodies.

Black bodies emit a range of wavelengths. The hotter the object, the higher the peak, and the shorter its wavelength.

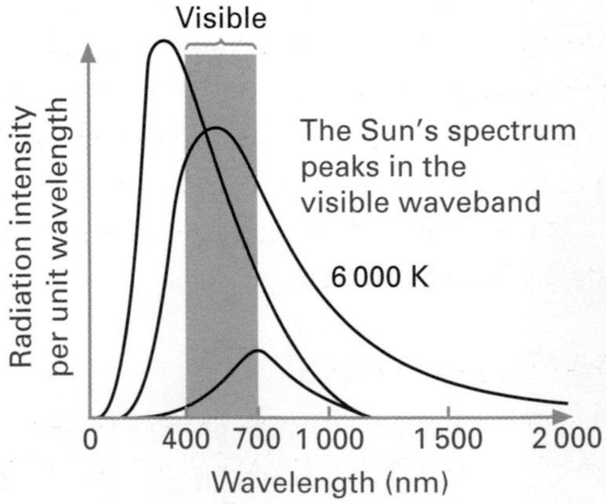

Wien's displacement law

The position of the emission peak in a black-body spectrum is given by:

$$\lambda_{max}T = 2.89 \times 10^{-3}$$

T is surface temperature in kelvins and λ_{max} is given in metres. The higher the temperature, the shorter the peak wavelength.

Links

See *Planck's constant*, pages 60, 129 and 132. The search for a mathematical description of the black-body spectrum first led Max Planck to treat electromagnetic radiation as quanta – discrete chunks (photons) of energy.

The Stefan–Boltzmann law (applied to stars)

A star's luminosity L (in watts) is a measure of the total power it emits:

$$L = 4\pi r^2 T^4$$

Where $4\pi r^2$ is the surface area of the star (of radius r), σ is the Stefan-Boltzmann constant and T is the star's surface temperature (K).

$$\sigma = 5.67 \times 10^{-8}\ \text{Wm}^{-2}\text{K}^{-4}$$

Using magnitudes to work out distances

Apparent magnitude m is a measure of how bright a star appears from the Earth. **Absolute magnitude** M is a measure of how bright a star actually *is*. (M is the magnitude a star would be if viewed from a distance of 10 pc.)

- Both m and M are measured on *logarithmic scales*.
- The brightest stars have the most negative magnitudes, the dimmest stars have the most positive magnitudes!
- The difference between m and M tells us how far away the star is:
  $m - M = 5 \log (d/10)$
  Where d is the distance in parsecs. Rearranging the equation, we get:
  $d = 10 \times 10^{(m-M)/5}$
- This technique works well up to distances of around 10 Mpc.

Examiner's secrets

Check your syllabus to see how much you need to know about magnitudes of stars.

Watch out!

Magnitudes are a (difficult) relic from earlier days of astronomy. Since stars have been painstakingly classified by magnitude, we have to learn how to use the data. $m = 1$ for the brightest stars visible with the naked eye, $m = 6$ for the dimmest stars visible with the naked eye.

Standard candles

Astronomical objects whose absolute magnitude is known are called **standard candles**. *Cepheid variable stars* pulsate periodically, with readily measured changes in brightness. The period of the pulsation is directly related to the star's absolute magnitude M, and so the distance to the star and its galaxy can be calculated. Some types of supernova can also be used as standard candles.

- The greater the distance being measured, the greater its uncertainty.

Checkpoint 5

What are the sources of uncertainty in using absolute and apparent magnitudes to calculate distances to far-off stars? Why do these uncertainties increase with distance?

Stellar absorption spectra

Dark lines in a star's emission spectrum are the result of absorption of radiation by atoms in the cooler outer layers of stellar gas. The Balmer series of absorption lines for electrons in hydrogen originate from electrons in energy level 2 absorbing photons. For a cool star (surface temperature 4000 K) most electrons in hydrogen will be in their ground state so the Balmer lines will be missing. Similarly for a hot star (20 000 K) as the hydrogen electrons will be in higher levels. For temperatures of 10 000 K there will be many electrons in energy level 2 so there will be strong Balmer absorption lines.

Links

See *atomic line spectra*, pages 130–1.

Checkpoint 6

(a) Why are absorption lines so wavelength specific?
(b) Why do absorption lines from spinning stars spread?

The Doppler effect

If a star is moving away from us, the light it emits is stretched out to longer wavelengths (*red shift*). If a star is moving towards us, the waves are bunched closer together (*blue shift*).

Exam practice

answers: page 192

1. Sirius is 2.7 pc away. (a) How long does light from Sirius take to get here? (b) If Sirius' apparent magnitude is –1.46, what is its absolute magnitude? (10 min)
2. Describe how Hydrogen Balmer absorption lines are produced in the spectrum of a star. (10 min)

Astrophysics 2

Links

See *Newton's law of universal gravitation*, pages 140–1.

Links

See *binding energy and mass defect*, pages 54–5 for more information on energy release by nuclear fusion.

Checkpoint 1

Where would you expect to find most of the hydrogen in a giant star – in the core, or around the edges? Explain.

Checkpoint 2

(a) Why do collapsing stars heat up (ignoring fusion)?
(b) Why must white dwarfs fade?

Checkpoint 3

(a) Calculate the Schwarzschild radius of
(i) the Earth,
(ii) the Sun,
(iii) a 5 solar-mass star
(mass of Earth = 6.0×10^{24} kg; mass of Sun = 2.0×10^{30} kg).
(b) Explain why the Earth and Sun will not form black holes.

Studying stars is an ancient art; understanding them is very modern and still far from complete.

Birth of a star

Huge gas clouds are stellar nurseries. Gravity pulls gas particles together, heating them as they accelerate inwards. When they reach **ignition temperature** (between 10^6 and 10^7 K), collisions begin to result in nuclear fusion. Fusion releases huge amounts of energy, heating up the plasma further and causing a **radiation pressure** which halts the **gravitational collapse**.

Death of a star

When a star begins to run out of hydrogen, radiation pressure drops and the star starts to collapse again under gravity. As it collapses, its core heats up; helium ignites, releasing more energy, raising the pressure and causing the star to expand. It becomes a red giant – a huge, unstable star. Gravity dominates at the core, but radiation pressure sometimes dominates around the edges causing continued expansion. The outer layers of a dying red giant may spread out to form an expanding **planetary nebula**; the exhausted core collapses.

White dwarfs, neutron stars and black holes

The fate of a star depends on its mass. The biggest stars (>8 solar masses) explode as **supernovae** which can briefly outshine entire galaxies. Smaller stars live longer and die in a less spectacular fashion.

→ *Small stellar cores* (<1.4 solar masses) will collapse to form **white dwarfs** which slowly fade.
→ *Medium-sized cores* (1.4 to 2.5 solar masses) will collapse further, forming **neutron stars** – entire stars as dense as atomic nuclei.
→ *Large stellar cores* (>2.5 solar masses) undergo complete, perpetual gravitational collapse, forming **black holes**.

Black holes

If a star's escape velocity is greater than the speed of light, then no light can escape it! The radius of such a black hole is given by:

$$R = 2GM/c^2 \quad (G = 6.67 \times 10^{-11} \text{ N m}^2 \text{ kg}^{-2})$$

Karl Schwarzschild was first to solve Einstein's equations of general relativity to give the above solution. R is therefore known as the *Schwarzschild radius*.

Hertzsprung–Russell diagram

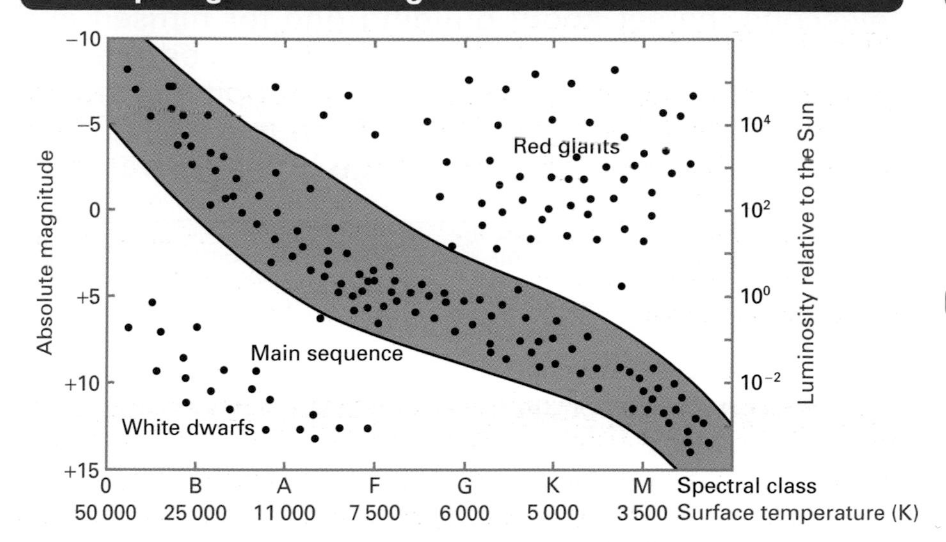

This diagram shows that some hot stars are unusually dim (white dwarfs) and some cool stars are unusually bright (red giants).

Rocket science

- Dynamics deals with objects that have constant mass.
- Newton's laws also apply if the mass changes as in rocket motion.
- In rockets, fuel is oxidised and the exhaust gases are ejected at high speed from the rear of the spacecraft.
- An equal and opposite reaction, called **thrust**, pushes the rocket forwards.
- Gravity changes as the rocket leaves the Earth.
- There are no drag forces outside the Earth's atmosphere.
- The change in speed of a rocket, v_f, after the burning fuel is given by the rocket equation:

  $v_f = v_e \ln (m_0 / m_f)$

  where v_e is the speed of the exhaust gases, m_0 is the initial mass of the rocket and m_f is the final mass of the rocket.

Examiner's secrets

The H–R diagram is a favourite topic for exam questions. Be sure you can trace the life of a typical star (our Sun) from forming a *proto star* through its main sequence life to its *red-giant* phase and subsequent death as a slowly fading *white dwarf.*

Checkpoint 4

Explain why the unusually dim hot stars must be dwarfs and why the unusually bright cool stars must be giants.

Links

See *Newton's laws of motion,* pages 20–1 for revision of dynamics.

Take note

This simple form of the rocket equation assumes there are no gravitational forces acting on the rocket.

The jargon

The Shuttle is a **multistage** rocket where the fuel is stored in smaller separate tanks which are discarded when empty. This means that energy is not used to accelerate empty tanks.

Exam practice

answers: page 192

The speed of the satellite in orbit needs to be changed by 2.3 m s^{-1} as part of the manoeuvre to make the orbit geostationary. Fuel is ejected from the satellite at 95 m s^{-1} to accomplish this. If the final mass of the satellite after the gas ejection is 1 800 kg, show that about 44 kg of gas needs to be ejected.
(5 min)

Astrophysics 3

When Galileo heard about the invention of the telescope, he set about building one for himself and made improvements to the design. He used them to discover the craters on the Moon, Jupiter's moons and the phases of Venus. Since then telescopes have greatly accelerated our knowledge of the Universe.

Refracting telescopes

A simple refracting telescope has two converging lenses – an objective lens and an eyepiece lens. Their separation is the sum of their focal lengths ($f_O + f_E$). The objective lens forms a real image of a distant object, such as a star, at its focal length from the lens. The eyepiece produces a magnified inverted virtual image of the star at infinity.

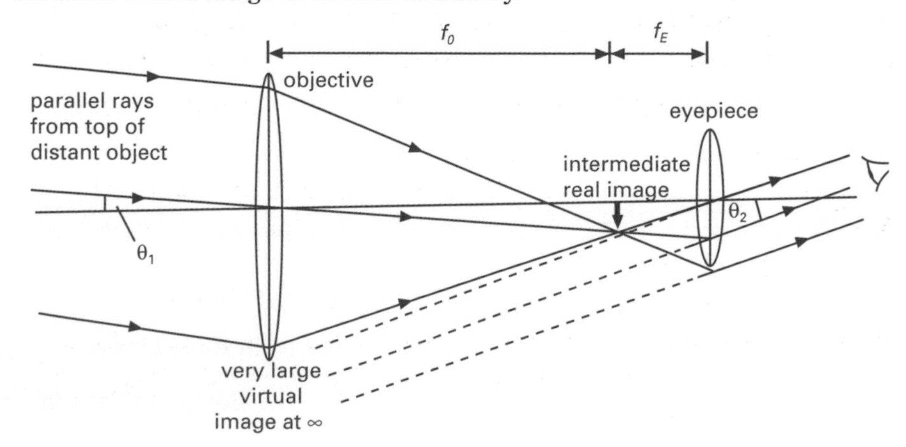

The angular magnification, M, is defined as:

$$M = \frac{\text{angle subtended by the image at the eye } (\theta_2)}{\text{angle subtended by the object at the eye without the telescope } (\theta_1)}$$

From the geometry, this is also $M = f_O/f_E$.

Resolving power

As well as having a large magnification, a telescope needs to have a large diameter eyepiece. This is to reduce the effects of diffraction which limit the ability of the telescope to resolve detail. Rayleigh's criterion is used: two objects can be seen as separate if their angular separation in radians is given by $\theta = \lambda / D$ where λ is the wavelength of the radiation and D is the diameter of the aperture. So the larger the diameter, the finer the detail that can be resolved.

A large diameter objective also makes a brighter image.

Reflecting telescopes

Early refracting telescopes produced coloured fringes around objects because different coloured light was refracted by different amounts. Isaac Newton realised what was happening and produced a reflecting telescope to overcome this.

- ➜ concave mirrors reflect parallel light to a point called a *focus*.
- ➜ parabolic mirrors make a much sharper image.

- → A value of 65 km s^{-1} Mpc^{-1} for H implies the maximum age of the Universe is 15 billion years.

Cosmological principle

On a big scale, the Universe would look pretty much the same from any position. Hubble's law would hold anywhere.

The Universe can be modelled as the surface of a balloon. As it inflates, the stars separate. Hubble's law holds from every viewpoint. The expansion of space also accounts for cosmological red shift and the stretching of the remnant radiation from the Big Bang.

Fate of the Universe and the omega factor

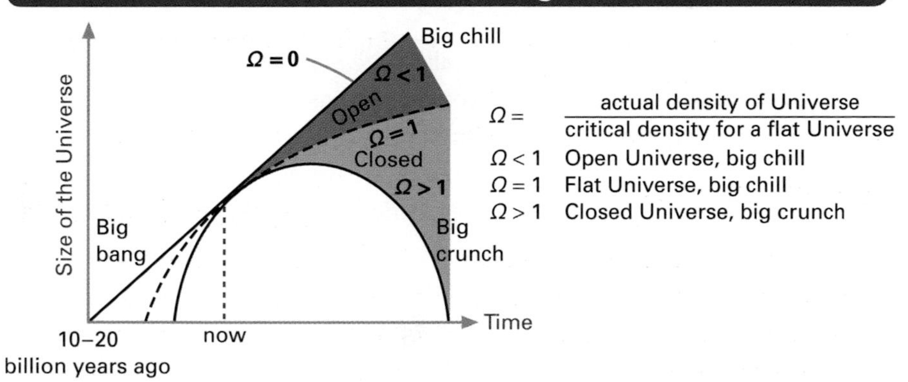

The critical density is about 5 hydrogen atoms per m^3. In a flat Universe, the density is exactly sufficient to eventually (when time reaches infinity!) halt the Universe's expansion. Many cosmologists believe the universe is flat.

The Big Bang theory

The Universe was very hot in the beginning; the radiation pressure was immense. The Universe expanded and cooled rapidly. When its temperature had fallen to around 3000 K (after about 300 000 years), the radiation pressure could no longer prevent the formation of atoms. Matter condensed and the Universe cleared; light could at last travel freely. Typical radiation at 3000 K might have a wavelength of 1 μm. **Remnant microwave background radiation** (MBR) still bathes us from all directions, with a black-body temperature of 2.7 K and a peak wavelength of around 1 mm. Since clearing, the Universe has expanded and cooled by a factor of around 1000.

Evidence for the Big Bang theory

1. The Universe is expanding.
2. Remnant MBR.
3. The proportion of helium in old stars matches the amount the theory predicts would have been created in the Big Bang.
4. Radio astronomy shows the distant (old) Universe is different from the near, present Universe.

Exam practice answers: page 192–3

The hydrogen absorption spectrum from quasar 3C273 is red shifted by 15%. (a) Calculate how fast it is receding from us. (b) Assuming $H = 65$ km s^{-1} Mpc^{-1}, work out how far away it is. Convert your answer to light-years.

(15 min)

Checkpoint 4

Outer stars (only) in galaxies move too fast. Their motion can be explained by the presence of *dark matter* in the outer realms of galaxies. Explain why this dark matter must lie far from the centre of the galaxy, but within the orbit of these fast outer stars. (Difficult.)

The jargon

Ω is called the *density parameter*.
Ω is the ratio of the actual density of the Universe to the critical density. The critical density, $\rho_0 = 3H^2/8\pi G$ where G is the gravitational constant.

Action point

Find out more about why many cosmologists, particularly those with a background in particle physics, favour a flat Universe.

Checkpoint 5

Use Wien's law (page 156) to check the peak wavelengths of black-body radiation at 3 000 K and at 2.7 K. What are the corresponding regions of the electromagnetic spectrum?

Grade booster

Many syllabuses require you to give evidence that the Universe is expanding. Much of the research evidence changes constantly. Make sure you keep up to date – read scientific journals and stay in touch!

The jargon

Quasar stands for QUASI-stellAR object. Quasars are intense sources of electromagnetic radiation with high red shifts implying that they are very distant objects.

Medical and health physics 1

The jargon

The minimum distance from the eye an object can be in focus, is called the *near point*.

Myopia and *hypermetropia* are the technical terms for short sight and long sight respectively.

The retina retains an image for a fraction of a second, called *persistence of vision*, which is why we see smooth motion from a movie or cartoon that is made of lots of stills frames.

The *fovea* is the part of the retina where the image of what we are looking directly at falls. Here is the greatest concentration of cones with hardly any rods. Rods cannot distinguish different colours, but can pick up lower light levels than the colour-sensitive cones. Rods are useful by night, cones by day.

Checkpoint 1

An optician prescribes a lens of -0.25D for a person's right eye. What does this mean? What is the focal length of the lens? The lens for the left eye has -0.50D underneath the letters Cyl. What defect does the left eye have?

Checkpoint 2

What would happen if light from an object fell on the blind spot in the retina?

Action point

Sketch the following:

➜ rays from a distant object entering a short-sighted eye

➜ rays from a distant object entering a long-sighted eye

For each one, add an appropriate lens and show how each defect is corrected.

Watch out!

It is always worthwhile to use the correct scientific terms, but make sure that you spell them correctly!

Physicists can help doctors in diagnosis and treatment. However, many doctors died as a result of using X-rays shortly after their discovery, before their true nature was understood.

The eye and seeing

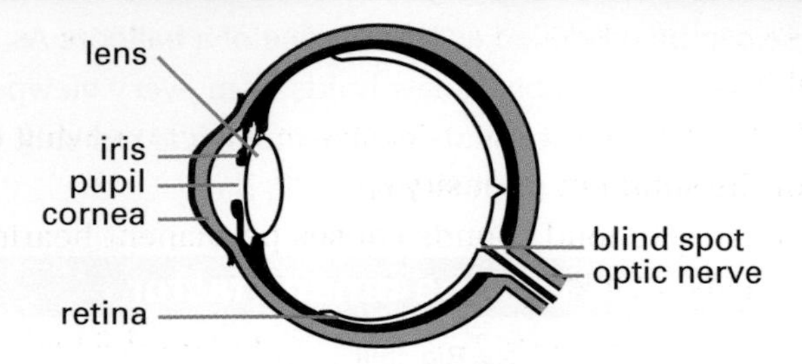

Light enters the eye through a hole called the *pupil* which is covered by the cornea, a transparent protective layer. It then passes through a clear salt solution. The *iris* covers more or less of the pupil to control the amount of light entering the eye. About two thirds of the refraction of the light occurs at the cornea. The *lens* makes fine adjustments to the focusing of light onto the *retina*. The *ciliary muscles* contract to make the lens thinner and relax to let it get fatter, allowing us to see both distant and near objects. This automatic focusing is called *accommodation*. Light next passes a clear jelly before forming an image on the retina – a layer of light-sensitive cells (*rods* and *cones*). Rods cannot distinguish different colours, but can pick up lower light levels than the colour-sensitive cones. Rods are useful by night, cones by day. The *optic nerve* carries electrical impulses from the retina to the brain.

Eye defects and their correction

➜ *Short sight* The eye cannot focus clearly on distant objects because the lens is too strong for the eyeball.

➜ *Long sight* The eye cannot focus clearly on nearby objects because the lens is too weak for the eyeball.

➜ *Astigmatism* Images are seen clearly in only one plane, e.g. the vertical plane. This can be corrected with lenses which have a cylindrical curvature.

The ear and hearing

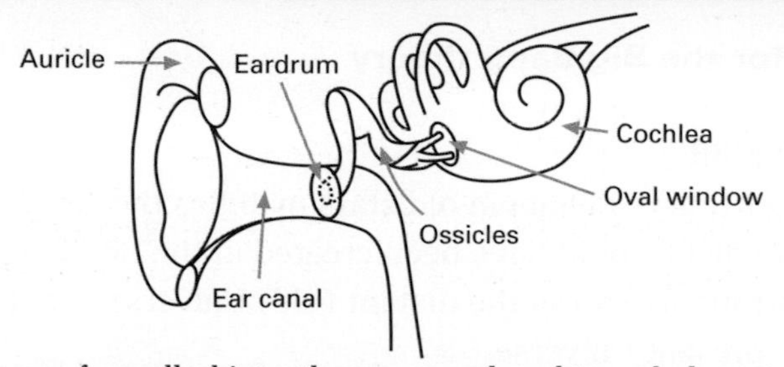

Sound waves are funnelled into the ear canal and travel along to the eardrum, making it vibrate. The vibrations are transferred to the tiny bones or ossicles, which change a small vibration of a large area into a large vibration of a small area - the oval window. This forces fluid in the cochlea to move. Then tiny, sensory hairs in the cochlea vibrate, sending nerve impulses to the brain where they are interpreted as sounds.

The ear's response

- A sound's intensity I is defined as power per unit area (W m^{-2}).
- The minimum detectable intensity (threshold of hearing) I_0 = 10^{-12} W m^{-2} and the highest safe intensity (threshold of pain) is 10 Wm^{-2}.
- Because the range of intensities that the human ear can detect is so large, a *logarithmic* scale is used to measure loudness, L, in decibels (dB).
- $L = 10 \log10(I/I_0)$, where I is the intensity of a sound having the same frequency of the sound of intensity I_0.
- Frequent exposure to loud sounds causes permanent hearing loss (most pronounced at 4 kHz).
- As people get older, they lose the ability to detect the higher frequencies of sound.

The heart

In the circulatory system: deoxygenated blood from the body goes into the right atrium, then to the right ventricle to the lungs. Oxygenated blood from the lungs goes to the left atrium, then to the left ventricle to the body. The heart acts as a double pump:

- Blood enters the atria which then contract sending blood into the ventricles.
- The ventricles contract and send the blood out of the heart again.
- The valves only let the blood flow one way through.

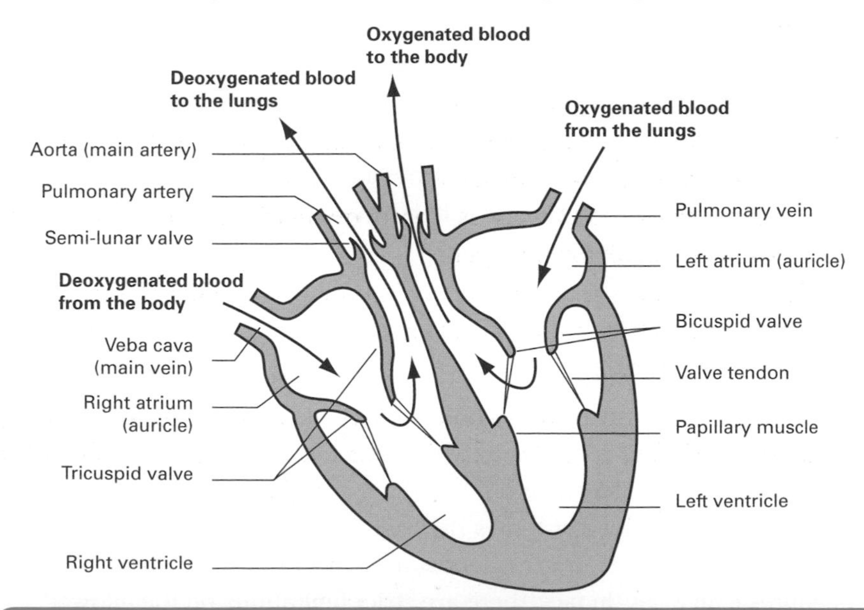

Exam practice

answers: page 193

Use these graphs to describe the ear's response to frequency and intensity. (20 min)

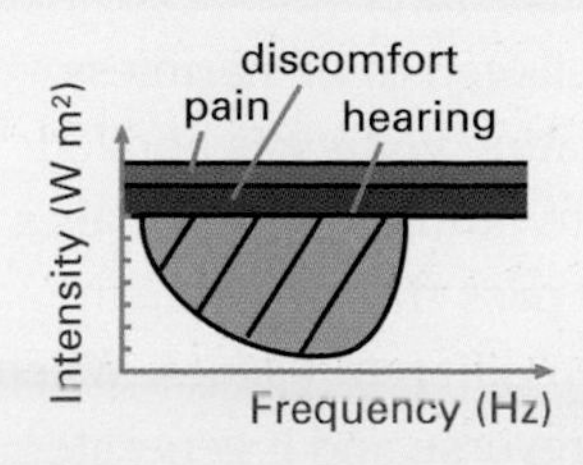

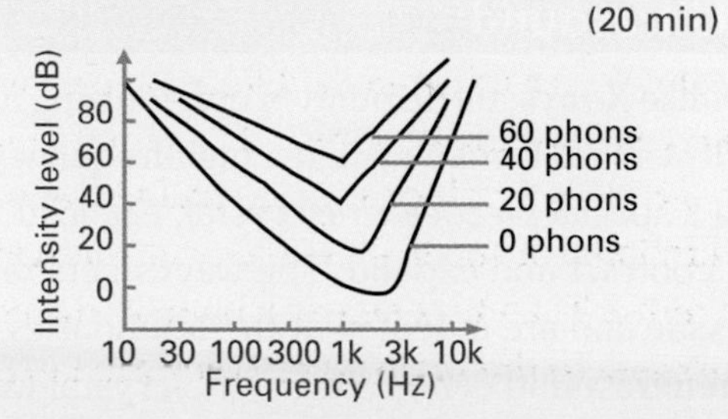

Checkpoint 3

Show that a doubling of intensity corresponds to a change in intensity of 3 dB.

The jargon

Loudness is an individual response to intensity and frequency: the same sound, played to two people, can seem louder to one person than the other. The *dBA scale* (adjusted decibel scale) is more sophisticated than the dB scale as it takes account of both intensity and frequency.

Take note

The normal range of audible sounds for a healthy young person is 20 Hz to 20 kHz.

Checkpoint 4

Why does your own voice sound so strange when you hear it on an audio or video tape recording.

Checkpoint 5

After attending a loud music concert you hear ringing in your ears. What does this mean?

The jargon

The pacemaker controls the heart beat. The *action potential* is a sudden change in p.d. in the pacemaker that causes muscle fibres to contract. The pacemaker is called the *sinoatrial* (*SA*) node. The *atrioventricular* (*AV*) node delays the pulse for about 0.1 s before passing it on to the ventricles.

Medical and health physics 2

> **The jargon**
>
> *Exponential changes* occur when the next value in a sequence depends upon the previous one. E.g. the number of bacteria that will be present in a colony tomorrow depends upon how many are there today. The amount of energy that will be able to travel deeper than 50 cm inside a human body depends upon how much energy got as far as 50 cm.

Medical physics began with the use of X-rays and radiotherapy. It now also uses ultrasound, light, infrared and radio-frequency radiation to assist other health professionals to help patients.

Medical diagnosis

X-rays

X-rays are high frequency electromagnetic waves produced when fast electrons hit a metal target. The beam intensity can be increased by increasing the filament temperature of the electron gun. The X-ray photon energy can be increased by increasing its accelerating voltage. Most of the electron beam energy becomes thermal energy so the target is cooled and rotated rapidly.

X-rays can be detected when they react with photographic film. As they pass through the patient, dense structures, e.g. bones, absorb most energy, and so cast dark shadows on a photographic film. If soft tissue, e.g. the stomach, is to be X-rayed, a contrast medium (such as barium sulphate) can be used.

X-rays are ionising so damage body tissue. The X-ray machine is enclosed in a thick lead shield with a small window to let through a narrow beam. This helps to produce a sharp image as well as a small target and a lead grid to reduce scattered radiation.

> **The jargon**
>
> Computer tomography (CT) scans use narrow X–ray beams to produce a 3D image of a cross-section of the body. An X–ray tube is moved around the patient. Detectors pick up transmitted X–rays and information is sent to a computer to reconstruct and display the image.

A fluorescent intensifying screen of crystals that absorb X-rays and emit light helps to develop the photograph. The intensity after passing through material is given by $I = I_0 e^{-\mu x}$ where x is the thickness of the material and μ is its linear absorption coefficient.

Because the absorption is exponential, there is a constant thickness of material which halves the intensity of the beam, called the half-thickness, $x_{1/2} = \ln 2/\mu$

The mass attenuation coefficient is given by: $\mu_m = \mu/\rho$

The table below shows typical half-thickness for 200 keV X-rays.

Concrete	Lead
25 mm	0.5 mm

When you have an X-ray these days the nurse or dentist will keep well away from the X-ray machine. It's nothing personal, but they may even leave the room! In either case they will be shielded from the harmful X-rays by standing behind leaded glass.

This reduces the dose of X-rays they get, but what about the patient? The difference is that a nurse or dentist may carry out many X-ray procedures every day. In fact, there are strict limitations on the allowed dose to medical staff and patients too.

> **The jargon**
>
> *Ultrasound waves* have a frequency above the limit of normal human hearing i.e. greater than 20 kHz. The frequencies used in medical diagnosis have frequencies from 1 to 20 MHz.

Ultrasound

Unlike X-rays, ultrasound is non-ionizing and is believed to be harmless. Ultrasound is made when a high-frequency alternating voltage is applied to a special *piezoelectric* crystal, e.g. lead zirconate titanate, which makes it contract and expand. The waves generated can pass through human tissue and are reflected at the boundaries between different tissues. When an ultrasound echo returns, the crystal works in reverse and it generates an alternating voltage. Information from the scan can be displayed on a cathode-ray oscilloscope (an amplitude or A-scan) which gives the dimensions of internal organs or on a TV screen (a brightness or B-scan) which gives a 2D image.

The acoustic impedance of a material is given by, $Z = \rho v$ and the ratio of the reflected beam, I_r, to the incident beam, I_i is given by $I_r/I_i = (Z_1 - Z_2)^2/(Z_1 + Z_2)^2$ when ultrasound passes from material 1 to material 2.

A coupling gel is applied to the patient's skin to reduce the amount of reflection as ultrasound enters the body (by matching the acoustic impedance of the skin).

The speed of blood flow in a blood vessel can be measured by the double **Doppler shift** in a pulse of ultrasound that is reflected from red blood cells.

The jargon

Acoustic impedance describes how difficult it is for sound waves to pass through a certain medium.

Magnetic Resonance Imaging

MRI maps the position of protons in the human body. Hydrogen nuclei each contain a proton so MRI maps the position of hydrogen nuclei. As the human body is 70% water (H_2O) and 20% fat (CH_3 and CH_2), MRI is ideal for imaging soft tissue.

Protons spin and as they are charged they act like magnets. If they are placed within another large magnetic field, these tiny magnets line up, either parallel or antiparallel to the external magnetic field. Another oscillating magnetic field makes the protons flip from parallel to antiparallel and vice versa. When this field is switched off the protons return to their original state emitting energy which is picked up and used to generate an image.

MRI scanners use superconducting magnets cooled with liquid helium so they are expensive. They are not thought to be harmful but the process is lengthy, noisy and some people find it claustrophobic.

The jargon

Magnetic resonance imaging (MRI) was once called nuclear magnetic resonance (NMR) imaging. The N in NMR referred to the nuclei being scanned. The name was changed because of fear associated with the word nuclear.

The jargon

The protons don't stay still in the magnetic field but they wobble – called *precession*. The frequency of the wobbling is called the *Larmor frequency* and it is in the radio frequency range. Applying an external field of this frequency makes the protons *resonate* (see pages 32–3 for more information on resonance.)

Endoscopy

Endoscopes use optical fibres to look inside the body. Light bounces off the inside of optical fibres by total internal reflection. TIR means that the light pipe does not have to be straight (so it can be pushed along cavities that are not straight, like blood vessels). Light is sent down a bundle of optical fibres, reflects off the inside of the body and returns along a second bundle.

Checkpoint 1

Light reflected back to the doctor travels along a bundle of fibres. What would happen to the image if the fibres became twisted over one another (non-coherent)?

Electrocardiographs

The changing p.d.s in the heart muscles can be detected at the body surface and used to produce a voltage-time graph which can be used to check the function of the heart.

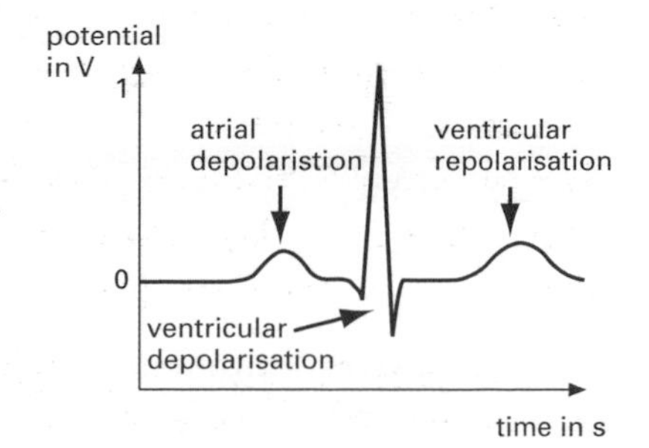

Electrodes are placed on the chest and limbs after removal of hair and use of a conductive gel to make good contact.

Checkpoint 2

Why are the electrodes for an ECG never put on the right leg?

Action point

Construct a table to compare the good and bad points of the main medical imaging techniques. You should think about benefits, safety implications, cost, exposure time, for example.

Exam practice answers: page 195

Describe how ionizing radiation is used in radiotherapy. (25 min)

Medical and health physics 3

"O Roentgen, then the news is true, And not a trick of idle humour, That bids us each beware of you, And of your grim and graveyard humour."

G. E. M. Jauncey

X-rays and gamma rays can be used for medical diagnosis but they are ionising radiations and can also be used to kill cells. Their use has to be monitored carefully.

Radioisotopes

Radioactive isotopes (radioisotopes) can trace the passage of substances through the body, e.g. doctors can use radioactive technetium to monitor blood flow through the brain. Iodine is essential for the production of a hormone in the thyroid gland. It can be made into two radioactive isotopes for medical use: iodine-123 and iodine-131. These can be swallowed by patients with a suspected thyroid problem and some will end up in the thyroid gland. I-123 decays by gamma ray emission which does not cause much ionisation and can be used as a tracer. A gamma camera is used to pick up the radiation emitted from the I-123 and gives an image of the thyroid gland. I-131 emits beta radiation and is used to destroy overactive or cancerous thyroid tissue. The half-life of the radioisotope has to be long enough to allow measurements to be taken but short enough so that it does not linger! These radioisotopes emit gamma rays, then decay into stable daughter products.

Checkpoint 1

By referring to relative ionizing powers and ranges, explain why gamma rays are more useful tracers than alpha or beta particles.

Positron Emission Tomography

PET uses a radioisotope that decays producing a positron (positive electron). The isotope is injected and designed to collect in the part of the body being examined. The positrons ejected on decay annihilate electrons producing a pair of gamma ray photons. These are detected by a gamma camera and used to display the structure and function of the organs. PET scans are also performed to detect cancer.

A gamma camera consists of:

- A collimator which is a thick lead sheet with holes drilled through. The holes only allow photons travelling in certain directions into the camera which enables the place of their origin to be more accurately located.
- A crystal that emits a flash of light or scintillates when a gamma ray photon strikes it
- A photomultiplier which produces an electron from the flash of light and then amplifies the electrons to produce a pulse of charge.

Speed learning

Visualizing a process often allows one to remember it. E.g. if a deep-seated tumour is to be attacked by external radiation, careful planning can avoid unnecessary damage to surrounding healthy tissue. Several beams can be directed at the tumour from different directions, rather like several needles piercing an apple. The needles are directed towards the rotten core of the apple. Each individual needle inflicts minimum damage as it travels towards the core. But maximum damage is inflicted where all the needles meet, at the core!

Links

See *properties of ionizing radiation*, pages 50–1.

Medical treatment

Radiotherapy uses X-rays, gamma rays or electrons to kill cancer cells. The radiation prevents cell division (which can cause cell death) by attacking the DNA within the cancer cell's nucleus. There are three types of treatment. A beam of radiation can be directed at the patient from the outside, a radioactive source can be placed into, or alongside, the tumour, or a radioactive liquid which is then taken up by the tumour can be injected into, or swallowed by, the patient.

Watch out!

Half-thickness is also known as *half-value thickness*.

Don't forget

Half-thickness applies to gamma rays too.

Early radiotherapy involved using ordinary X-ray machines at slightly higher energies than for diagnosis. More energetic (between 1.17 and 1.33 MeV) gamma ray photons, from cobalt-60, then became available to treat deeper tumours. Today, linear accelerators that can produce X-rays of up to 25 MeV are preferred. They can also produce electrons of varying energies to treat surface tumours.

Absorbed dose (Gy)

The amount of radiation absorbed by the human body is measured in units called **grays** (Gy) where 1 Gy is equivalent to 1 J kg^{-1}. This seems simple enough, but the energy absorbed is not all there is to it.

It turns out that the gray is not good enough in predicting the potential consequences of exposure to radiation. The risk must also take into account the *type* of radiation involved.

Quality factor

The table below gives the quality factor of different ionizing radiation.

Radiation	Quality factor
alpha	20
neutrons	10
gamma	1
X-rays	1
beta	1

The quality factor takes into account the variation in sensitivity of body tissue to different types of radiation. The table shows that exposure to alpha radiation carries the greatest potential risk, whereas X-rays, gamma rays and beta radiation are potentially the least harmful.

Dose equivalent (Sv)

The dose equivalent is the absorbed dose multiplied by the quality factor. To distinguish it from absorbed dose, it has the unit **sieverts** (Sv).
So, 2 Gy from an alpha source leads to an absorbed dose of 40 Sv (2 × 20). For X-rays, an equivalent absorbed dose gives rise to only 2 Sv (2 × 1).

Risk = probability × consequences

Risk from radiation can often be difficult to quantify because it is by no means certain that the potential effects of exposure will actually turn out to be life threatening. The concept of risk combines the probability of an event with the consequences of that event.

The current estimate of risk due to radiation dose is 5% per sievert. This means that 5 in 100 can be expected to contract a fatal radiation-induced cancer if they receive an absorbed dose of 1Sv.

Exam practice

answers: page 193

A new X-ray cancer-screening programme is suggested. Every person over the age of 35 is to have a single chest X-ray that gives them a dose of 50 × background. How many deaths in a population of 20 million may be expected from this programme? Is the risk worth it? (5 min)

Checkpoint 2

If the intensity of an X-ray beam is I0, what would be the intensity penetrating:
(a) 50 mm of concrete?
(b) 1.5 mm of lead?

Watch out!

It is a popular myth that gamma rays are more damaging than alpha particles. This is not actually the case. Gamma rays, because they are uncharged, generally pass straight through the body. On the other hand, alpha particles are strongly ionizing and are absorbed easily in the skin, doing lots of damage.

Grade booster

Dose equivalent in Sv is often just known as **dose.** To distinguish it from **absorbed dose** (Gy) you must learn that the units are different.

Checkpoint 3

Calculate the dose equivalent of a combination of doses totalling 2 mGy, from a source that emits 10% alpha particles and 90% gamma rays.

Checkpoint 4

Background dose equivalent is about 2 millisieverts (2 mSv). How many people out of a population of 1 million may be expected to die from background radiation?

Materials 1

Materials behave differently under stress. When dropped, a wine glass shatters into pieces, a rubber ball deforms then bounces back and a metal can dents!

Links

See *stress, strain and Hooke's law*, pages 30–1.

Take note

X-rays can be diffracted by crystals and give information about the arrangement and sizes of the particles in matter.

Checkpoint 1

Write down the definitions of stress, strain and Young's modulus of a material. Give the units of each.

Stress–strain graphs for different materials

Metal wires in tension

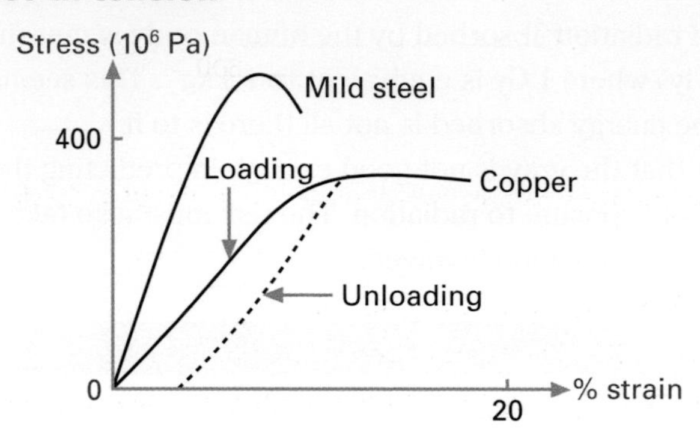

Metals are **polycrystalline** materials – their atoms are arranged as lots of bits of randomly oriented crystals.

- The wire is **elastic** for small strains (about 0.1%) then it becomes **plastic**.
- The **limit of proportionality** is the point after which the line is no longer straight.
- After the **elastic limit** or **yield point**, the wire no longer returns to its original length but is said to have a **permanent set**.

The jargon

The *strength* or *ultimate tensile stress* (UTS) of a material is the greatest stress it undergoes before breaking.
The *yield stress* is the stress when the material begins plastic behaviour.
The *breaking stress* is the stress at the breaking point.

Freshly drawn glass fibre in tension

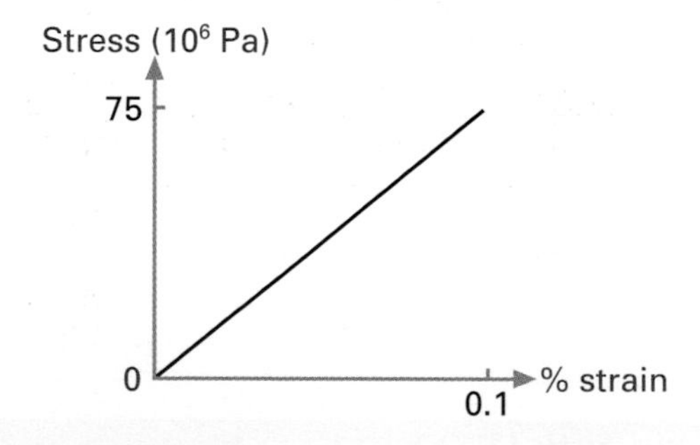

Checkpoint 2

High-carbon steel is more rigid, or stiffer, than mild steel. It has a greater UTS and yield stress and undergoes less plastic strain. Sketch a graph to show how their stress–strain graphs will differ.

Glass belongs to a group of materials called **ceramics,** which includes china and brick. Ceramics have an **amorphous** structure – the molecules have no regular order. Ceramics are strong in compression but weak in tension.

- The fibre is quite strong – it requires a large force to break it.
- It is elastic.
- It breaks cleanly with a brittle fracture without showing plastic behaviour.

Watch out!

The energy stored by an elastic material is the area under the force–extension graph. The area under the stress–strain graph gives the energy stored *per unit volume* of the material. For a material which obey's Hooke's law: energy per unit volume = ½ $\sigma\varepsilon$

Rubber band in tension

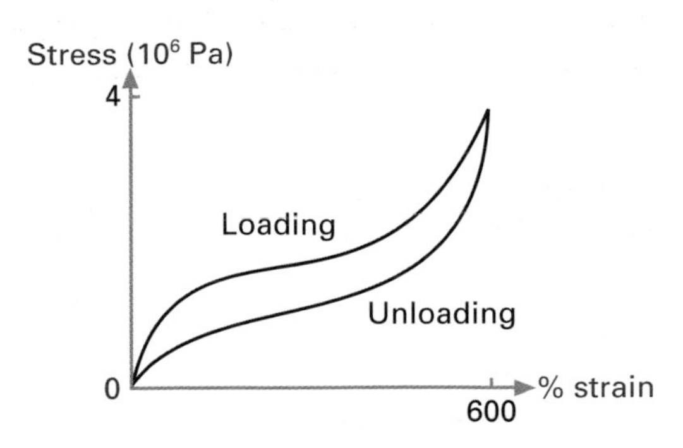

Rubber is a polymer – it consists of long-chain molecules, hydrocarbons in this case.

- Rubber is elastic for very large strains.
- It stretches easily at first but then becomes stiffer.
- The unloading line does not coincide with the loading line.

Composite materials

The term composite materials usually refers to materials that have been engineered from two or more different materials, often fibres in a matrix. Wood is a natural composite of long cellulose fibres held together by a weaker substance called lignin. Composites take advantage of the properties of their component materials.

Values of the Young Modulus and UTS for some materials

Material	*E* (GPa)	UTS (MPa)
high tensile steel	200	350
copper	130	220 to 430
sheet glass	70	4 to 150
rubber		30

Checkpoint 3

The following word pairs have opposite meanings. Give the meaning of each one: stiff/flexible, strong/weak, brittle/tough, elastic/plastic.

Grade booster

You should be able to give examples of materials which are, for example, *weak* and *brittle* or *strong* and *tough*.

Exam practice

answers: page 194

(a) Describe how you would measure Young's modulus of copper using a copper wire.

(b) Explain why it is necessary to use a long thin wire.

(c) Two wires, X and Y, are made from the same material. Wire X is three times as long as Y and has twice the diameter of Y. When a load is suspended from X the wire extends by 8 mm. How much will wire Y extend with the same load?

(15 min)

Materials 2

Scientists now understand how the properties of a material are related to its atomic structure, and so they can improve existing materials or design new ones.

The jargon

In a regular arrangement, each metal atom has six nearest neighbours. This is known as *close packing*.

Point defects are where an atom is missing at a site, or where an impurity atom replaces one of the metal atoms, or squeezes in the gaps of a regular array.

Checkpoint 1

Explain why it is more important to know the yield stress of a metal than its breaking stress if it is to be used for a load-bearing structure, such as the cables in a suspension bridge.

Behaviour of metals

Elastic behaviour occurs when the stretched metal is released and the interatomic forces pull the atoms back again.

Crystalline materials are plastic because of **slip** – planes of atoms (**slip planes**) slide over one another. This happens in a single crystal.

Metals usually contain many defects called **dislocations**. This is an extra layer of atoms in the crystal structure. The dislocation moves easily under tension or compression, giving the material a permanent change in shape.

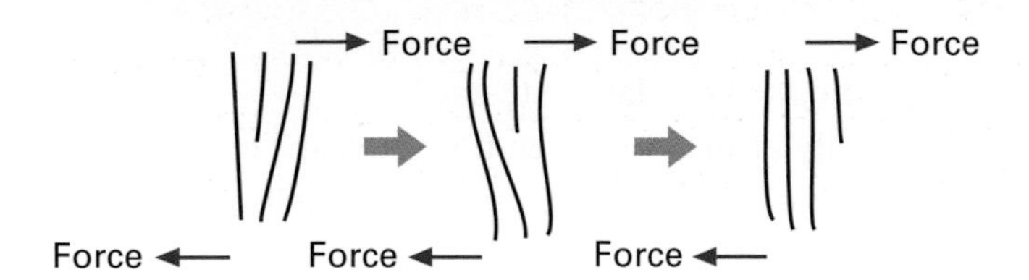

Metals break in a ductile cup and cone fracture after forming a neck.

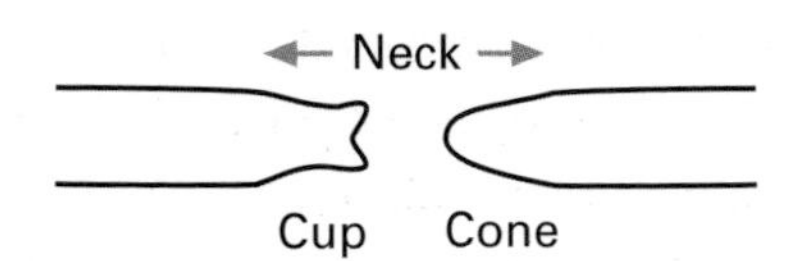

The jargon

A metal which undergoes considerable plastic deformation before breaking, like copper, is said to be *ductile*.

Plastic deformation also allows a metal to be *malleable*, that is it can be hammered into shape.

A *hard* metal is resistant to plastic flow.

Superalloys are metallic materials that have been designed for service at high temperatures eg for jet engine turbine blades.

Reducing plastic flow

- Remove all defects
- Single crystals can be grown with few dislocations. These are strong but impractically small in size.
- Add defects
- Work hardening (or cold working) introduces more dislocations which jam up and prevent plastic flow.
- Dislocations cannot move past the edge or **grain boundary** of a crystal. Heat treatment can add grain boundaries reducing plastic flow. **Quenched** steel has been heated then cooled suddenly so only tiny crystals have time to form. It is hard and brittle.
- Added impurity atoms, such as the carbon atoms in steel, occupy the spaces between the iron atoms and make it harder for dislocations to move.

The jargon

Tempered metal has been quenched then reheated to a lower temperature and cooled slowly. It is tough and elastic.

Annealed metal has been heated then cooled slowly. It is flexible and ductile.

The jargon

Fatigue failure occurs when a metal has been subjected to repeated loading and unloading.

Checkpoint 2

Explain why repeatedly flexing a piece of metal paper clip makes it become stiff and finally break.

Behaviour of glass

There can be no slip with glass, because there is no crystal structure. Glass fails because of tiny scratches in its surface. In tension, the stress at the tip of a scratch or crack is very large and this causes the crack to spread through the material, making a clean break.

Preventing brittle fracture

- Cracks do not propagate in ductile materials since slip occurs which blunts the tip of the crack and reduces stress.
- Pre-stressed or toughened glass is cooled by blowing jets of cold air on the surface. This makes the outside cool and contract while the inside is still molten. When the inside eventually cools and contracts, it holds the outside in compression so it is difficult for cracks to propagate.
- Fibre glass is an example of a composite material. It consists of thin glass fibres glued in a resin called the matrix. A crack in one fibre is stopped at the resin boundary.

The jargon

A *tough* material is resistant to the propagation of cracks and requires a relatively large amount of energy to break it. The work done is the area under the force–extension graph, and so this area will be large for a tough material.

Checkpoint 3

Concrete is amorphous. Explain why it is weak in tension. Pre-stressed concrete has concrete mix poured around metal rods held in tension. When the concrete is set, the rods are released. How does this make the concrete stronger?

Behaviour of rubber

The long-chain molecules of rubber are tangled at room temperature. When a force is applied, the chains untangle and line up. On release, thermal energy causes the chains to become tangled again.

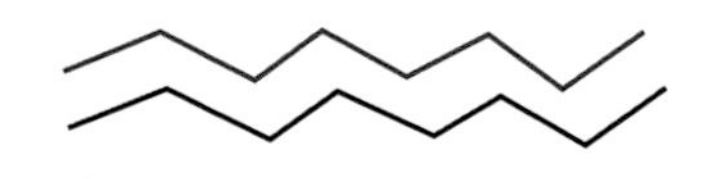

Rubber undergoes elastic hysteresis and this is why the unloading line does not coincide with the loading line. When rubber is stretched and released, not all of the work done in stretching it is recovered. Some of the work done increases its internal energy, making it hotter.

The jargon

Creep is when a loaded material gradually extends with time. In rubber this is due to the plastic flow of the molecules. The process of vulcanization prevents plastic flow by causing the chains to be crosslinked by sulphur atoms. Vehicle tyres are made from vulcanised rubber because it is harder and more durable than natural rubber.

Other polymers

Thermoplastics can be moulded when warm.

- Below their **glass transition temperature**, they are flexible and rubbery.
- Above this temperature, they are rigid and glassy.
- Near the melting point they are **viscoelastic** – the chains slide past one another – behaving like a viscous liquid, although a rapidly applied stress causes elastic behaviour.
- Perspex is amorphous. It is vacuum moulded to produce crash helmets, for example.
- Polythene and nylon are semicrystalline.

Checkpoint 4

Bouncing putty is a silicone-based polymer. Explain why, when rolled into a ball and dropped, it bounces, but it flows into a thin disc if left alone.

Thermosets do not soften on heating as the long chains are crosslinked.

- They are rigid and brittle.
- Bakelite and melamine are used to make electrical fittings.

Exam practice answers: page 194

Explain why (a) copper is ductile, (b) glass is brittle and (c) rubber is very elastic. (15 min)

Turning points in physics 1

Can you imagine life without electricity? You might take electricity for granted but your grandparents, or greatgrandparents, were probably very excited when 'the electric' came to their home. The discovery of the electron shed new light on the structure of the atom and much more!

The discovery of the electron

The word atom is derived from a Greek word atomos meaning indivisible. Radioactivity was discovered in 1896. This discovery meant that earlier ideas about atoms were wrong. Atoms could be broken up!

Radioactivity was discovered in 1896. This discovery meant that earlier ideas about atoms were wrong. Atoms could be broken up!

The jargon

Electricity comes from the Greek word for amber, *elektron*.

J. J. Thomson's discovery of the electron

In 1897 J. J. Thomson was studying particles produced in cathode-ray tubes by a process called thermionic emission. Thomson could not measure the charge and mass of cathode-ray particles (or corpuscles as he called them) separately, but he could change their direction using an electric field and deduced that they had a negative charge. He was also able to bend them using a magnetic field.

Thomson used an electric field to move the corpuscles in one direction and a magnetic field to move them in the opposite direction. He was able to calculate their charge to mass ratio (e/m) and using a value for their charge e taken from electrolysis experiments, Thomson concluded that the corpuscles were much smaller than atoms.

The jargon

Cathode-ray tubes are similar to modern TV tubes. They contain gas at low pressures. Cathode-ray particles (electrons) are emitted from a cathode (negative terminal) and accelerated towards an anode (positive terminal). The kinetic energy of the electrons is given by KE = $\frac{1}{2}mv^2 = eV$

The jargon

Thermionic emission is a process in which metals emit electrons. The metal is heated so that some of its free electrons gain enough kinetic energy to leave its surface.

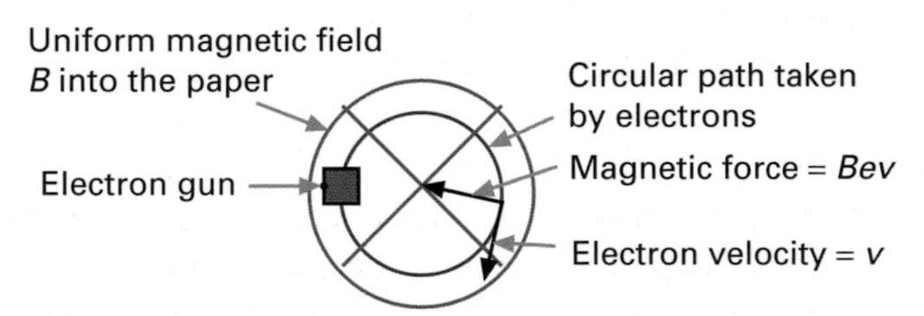

This diagram shows a way to find e/m. A pair of Helmholtz coils produce a uniform magnetic field that makes electrons follow a circular path. A centripetal force (mv^2/r) is provided by, and equal to, the magnetic force (Bev). It can be shown that as $Bev = mv^2/r$, $e/m = v/(Br)$.

Thomson was the first to discover a subatomic particle. Others called it the electron because it had an electric charge. Thomson found that the value of e/m was not affected by the gas contained in the cathode-ray tube. This implied that all atoms contain identical corpuscles, or electrons. J. J. Thomson received the Nobel prize in 1906.

Checkpoint 1

Thomson calculated e/m to be -1.76×10^{11} C kg^{-1}. The specific charge Q/M for hydrogen ions was known to be 9.65×10^{7} C kg^{-1}. Thomson assumed that the charge on an electron e was equal and opposite to the charge on a hydrogen ion Q. Explain what this meant about the mass of an electron compared with the mass of a hydrogen ion.

Checkpoint 2

As $Bev = mv^2/r$, prove that $e/m = v/Br$.

Millikan's oil-drop experiment to find a value for *e*

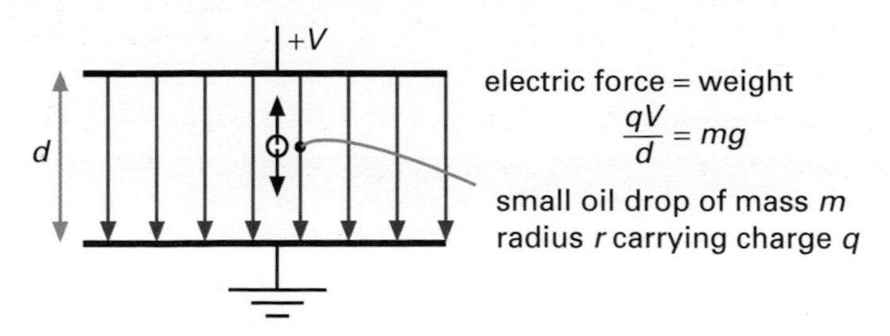

If the equipment is adjusted so that the oil drop remains still,

Force due to electric field = weight

$$qV/d = mg = 4\pi r^3 \rho g/3$$

so the charge q on the oil drop of density ρ equals $4\pi r^3 \rho g d/3V$. To find q, the radius of the drop r had to be calculated. To do this the electric field was switched off. The drop accelerated at first but then reached its terminal velocity. At this point the drop's weight equalled its viscous force $6\pi\eta r v$. η is called the coefficient of viscosity of air.

q was always found to be a whole number multiple of -1.6×10^{-19} C, suggesting charge is carried in packets of -1.6×10^{-19} C (by electrons).

Checkpoint 3

As the volume of a sphere = $^4/_3\pi r^3$, obtain an expression for the weight of a spherical oil drop of density ρ, radius r when acceleration due to gravity = g.

Links

Revise *moving through liquids*, pages 24–5 if you need reminding about terminal velocity and Stokes' law.

Links

See *photoelectric effect* on pages 128–9.

Wave–particle duality and microscopes

By the 1930s biologists wanted powerful microscopes to view structures (nuclei, mitochondria, etc.) within cells, which are too small to be resolved by light. Electron microscopes are based on the discovery that electrons have wave properties and gave the required magnification (×10 000). Electrons can be accelerated by electric fields: the larger their momentum, the smaller the wavelength and the greater the resolution.

Waves or particles?

Isaac Newton favoured the idea that light consisted of particles as diffraction and interference effects were not easily observed (1704). Young's slits experiment showed interference fringes and confirmed Huygens' wave theory (early 19th century). The photoelectric effect was explained by Einstein saying that light energy existed in packets (early 20th century) so light appears to have properties of both a wave, eg wavelength and a particle, eg discrete packets of energy. De Broglie's hypothesis that particles should have wave properties was confirmed for electrons in 1926.

The **transmission electron microscope** (TEM) was the first electron microscope. A focused beam of electrons passes through a sample. Electrons that pass through the sample strike a phosphor screen to produce light. Thick or dense areas of the sample allow fewer electrons through so a darker image results.

Scanning tunnelling microscopes (STMs) were invented in 1981. A very fine probe (with a single atom projecting from its end) scans a small area of the sample's surface. The probe is held at a constant height ($\leq$1 nm) above the surface. Electrons tunnel across the gap. If the probe comes across a raised atom, more electrons can tunnel across so the tunnelling current increases. Variation in the tunnelling current as the probe moves allows an image to be formed.

Checkpoint 4

In what ways is a TEM rather like a slide projector?

Exam practice answers: page 194

Use $\lambda = h/(2meV)^{1/2}$ to estimate the anode voltage needed to produce wavelengths of the order of the size of atoms. Describe two advantages of an STM over a normal electron microscope. (10 min)

Turning points in physics 2

"It's not that I'm smart, it's just that I stay with problems longer."

Albert Einstein

Our understanding of the Universe has developed over many years. Breakthroughs have opened up previously unforeseen technologies. For example, who could have imagined television before the discovery of the electron?

Einstein's theory of special relativity

Albert Einstein provided the first new model of the Universe since Newton, two hundred years before. Einstein's contemporary, Charlie Chaplin, said that people cheered him (Chaplin) because everyone understood him, but that people cheered Einstein because no one understood him.

Measuring the speed of light *c*

Galileo was possibly the first scientist to measure c. Since then c has been measured more often, and more accurately, than any other physical constant. Albert Michelson is the person most closely associated with this work. He spent fifty years measuring c. In 1879, Michelson and Edward Morley measured c to be $299\,910\,000\ \mathrm{m\,s^{-1}} \pm 50\,000\ \mathrm{m\,s^{-1}}$.

Relativity

If you stand by the side of a road as a car drives past at $50\ \mathrm{km\,h^{-1}}$, you see it travelling away from you at $50\ \mathrm{km\,h^{-1}}$. If you jump on a bicycle and pedal after it at $10\ \mathrm{km\,h^{-1}}$, it is now only travelling away from you at $40\ \mathrm{km\,h^{-1}}$. It is travelling at $40\ \mathrm{km\,h^{-1}}$ relative to you. Michelson and Morley found that light always travels at the same speed, no matter whether you are travelling towards it or away from it.

Checkpoint 1

Use the idea of a car overtaking a person cycling along a road, then turning around and driving towards the cyclist, to explain why Michelson and Morley expected the value that they would obtain for c to vary depending on whether light (the car) was travelling towards or away from Earth (the cyclist).

Michelson–Morley experiment

Michelson and Morley intended to measure the speed of Earth travelling through the *ether*. They assumed that light would travel at a constant speed through the ether. They thought that if the Earth was moving in the same direction as the light they studied, their measurement of c would be less than if the Earth was moving towards the light source.

The jargon

In the 19th century, scientists thought that if light is a wave, it must require a medium to travel through. As light can travel from the Sun through the vacuum of space to us, this medium had to be rather special. They called it *ether*. It is now known that ether does not exist.

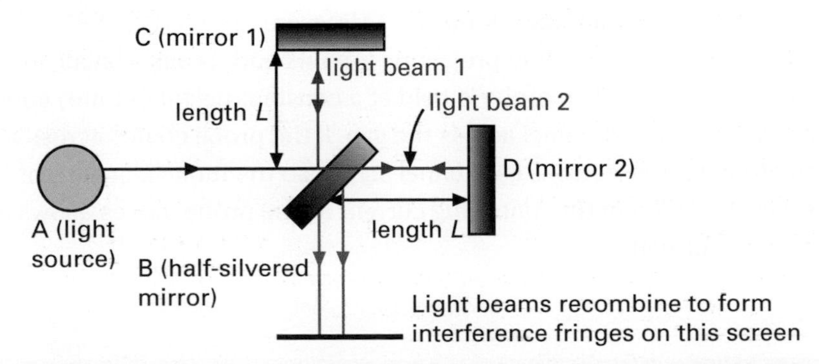

Mirrors 1 and 2 were placed at equal distances L from a half-silvered mirror B. B splits light coming from the light source so that some light, beam 1, travelled towards mirror 1 and some, beam 2, went to mirror 2. After reflection at mirrors 1 and 2, the beams recombined to form interference fringes on the screen. Michelson and Morley expected the two beams to be slightly out of phase when they recombined.

If the equipment (situated on the Earth which they believed to be travelling through the ether) was moving in the same direction as light from A, they predicted that the journey time of ray 1 would be slightly less than that of ray 2. No time difference was ever found; this suggested that the Earth was stationary! Was nature conspiring against them?

"A complete conspiracy is itself a law of nature."

Poincaré

Einstein's postulates

No one accepted that the Earth was stationary. Einstein solved the problem by saying that c was always constant. It did not matter whether the Earth was moving towards or away from the light source, c would always be the same! Michelson and Morley were vindicated. Einstein used two basic assumptions to rewrite Newton's laws:

- the speed of light in free space is always constant
- physical laws have the same form in all inertial frames

"Newton was the greatest genius who ever lived, and the most fortunate, for there cannot be more than once a system of the world to establish."

J. L. Lagrange (1736–1813)

Einstein's predictions

- *Time dilation* If the time between two events was measured, an observer moving at speed v would measure a longer time interval t than the time interval t_0 that a stationary observer would:

  $$t = \frac{t_0}{\sqrt{(1 - v^2/c^2)}}$$

- *Length contraction* If a rod moves in the same direction as its length it appears shorter, length l, than when it is stationary, length l_0.

  $$l = l_0\sqrt{(1 - v^2/c^2)}$$

- *Relativistic mass* The mass of a stationary object m_0 is less than the mass of the same object m moving at speed v.

  $$m = \frac{m_0}{\sqrt{(1 - v^2/c^2)}}$$

- *Energy and mass* Einstein showed that energy and mass were inter-changeable in his most famous equation: $E = mc^2$.

The theory of special relativity has since been confirmed by experiment.

Action point

Confirm that at speeds other than speeds approaching c, Newton's laws of motion are still valid. Do this by substituting $v = 0.1c$ in each of the first three equations shown opposite. You should find that $t \approx t_0$, $l \approx l_0$ and $m \approx m_0$.

Checkpoint 2

How much energy is released when a uranium nucleus of mass $4.046\,865\,3 \times 10^{-25}$ kg decays to produce a thorium nucleus of mass $3.978\,741\,2 \times 10^{-25}$ kg and an alpha particle of mass $6.804\,42 \times 10^{-27}$ kg?

Electromagnetic waves

Maxwell predicted the existence of electromagnetic waves in 1864. He showed that a changing current in a wire creates electromagnetic waves that spread out from the wire with speed $c = 1/\sqrt{(\mu_0\varepsilon_0)}$, where μ_0 and ε_0 are constants.

Hertz is generally given the credit for discovering radio waves in 1888. However, he was only able to interpret his observation, that a spark between two spheres sent out an electromagnetic wave that could be picked up some way away, because of Maxwell's earlier work. Marconi began experimenting with radio waves in 1894. In 1901 he transmitted across the Atlantic, revolutionizing communications.

Exam practice answers: page 194–195

Muons are particles that disintegrate after an average lifetime of 2.2×10^{-6} s. They are created at the top of the atmosphere, some 10 km up. How far could they travel at the speed of light? Bearing this answer in mind, explain why they can be found at ground level. Illustrate your answer with mathematics. (15 min)

Energy and the environment 1

One of the major problems that humans have to face in the 21st century is the increase of greenhouse gases in the atmosphere from the burning of fossil fuels which has led to a rise in the air temperature near the Earth and the oceans. This global warming is likely to have seriously damaging effects on the Earth's ecosystems.

The jargon

Fossil fuels are *non-renewable* since they took millions of years to form from photosynthesis of plants that existed millions of years ago, and they will eventually be used up. *Renewable* sources such as solar cells, tidal, wind and wave power will not be used up.

Links

See *Astrophysics 1*, pages 156–7 to remind yourself that the energy radiated by a black body depends on its surface area and the fourth power of its absolute temperature.

The energy received at a distance R from an energy source is inversely proportional to R^2.

Links

See *electromagnetic spectrum*, pages 114–5 to revise the types of radiation in the electromagnetic spectrum.

Checkpoint 1

Explain how deforestation contributes to the greenhouse effect.

Take note

Venus has an atmosphere of 67% carbon dioxide which results in a surface temperature of 467°C!

Checkpoint 2

How do scientists know how the composition of the Earth's atmosphere has changed?

Energy sources

In 2004 the energy consumption for the UK is shown in the diagram below.

UK Energy Consumption 2004

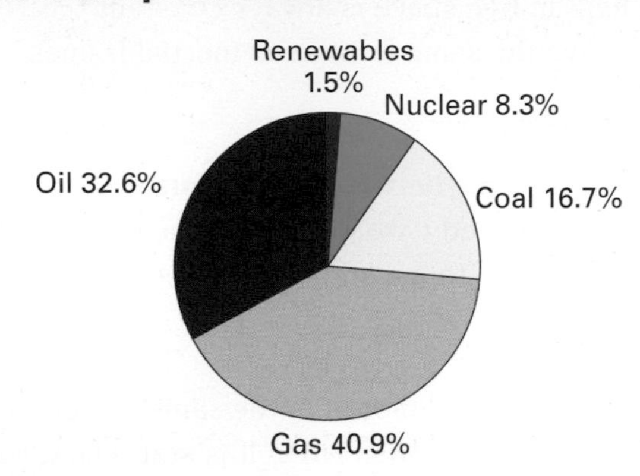

Burning of fossil fuels contributed 90.2% of the total energy supply. The worldwide energy consumption for the same period is not known accurately but is believed to be about 15 TW (1 TW = 1 terawatt = 10^{12} W) with 86.6% coming from burning fossil fuels.

Most of the world's energy resources are either directly or indirectly from the Sun's radiation.

The **solar constant** measures the power per unit area of all the Sun's radiation on a surface at right angles to the radiation. It is measured by satellites above the Earth's atmosphere.

Its value is about 1375 W m^{-2}, but although called a 'constant' it varies by up to 30 W m^{-2} because the activity of the Sun varies and the distance between the Sun and the Earth varies.

The Greenhouse Effect

The Earth and the Moon are similar distances from the Sun yet the Moon is more than 30°C cooler than the Earth.

The Earth's atmosphere is responsible for keeping the Earth at a higher temperature than the Moon, which has no atmosphere.

The Sun's radiation reaches the Earth's atmosphere where some is reflected, some is absorbed and some (mainly infra-red and visible) is transmitted to the Earth's surface.

The transmitted radiation warms the Earth's surface which in turn re-radiates this back into the atmosphere as infra-red radiation.

Some gases in the atmosphere absorb part of this radiation so it does not escape back into space. This absorption of infra-red is called the **greenhouse effect** and the gases responsible, eg, water vapour, carbon dioxide, methane and nitrous oxide are called the **greenhouse gases**.

Without greenhouse gases, the Earth would be hotter during the day and colder at night: the average temperature of the Earth would be –18°C.

Greenhouse gases cause the average temperature of the Earth to be 15°C.

In both situations, the energy absorbed by the Sun is re-radiated back into space so there is equilibrium.

Since the industrial revolution, humans have been burning fossil fuels which have increased the levels of greenhouse gases in the atmosphere.

This means that more infra-red radiation is being absorbed by the greenhouse gases. The Earth will reach a higher temperature before equilibrium is regained.

Checkpoint 3

Explain how (a) melting ice on land and (b) melting sea ice contribute to rising sea levels.

Links

See *nuclear fusion and fission,* pages 56–7.

Nuclear energy

One way of reducing the emission of greenhouse gases is to generate electricity by using nuclear power, but this has disadvantages too. In 2004 the UK got 8.3% of its energy from **nuclear fission**. Here the nuclei of heavy atoms, such as uranium-235, absorb slow moving neutrons and become unstable. They then split into two smaller nuclei releasing more neutrons and a lot of energy. A **moderator**, such as graphite, surrounds the fuel rods to slow down the neutrons so they can be absorbed by other uranium atoms – keeping up a **chain reaction. Control rods** made from boron for example, absorb neutrons and can be raised or lowered into the core of the reactor so the reaction can be controlled. The energy is used in the same way as in a fossil fuel power station by heating water to produce steam which drives turbines to operate generators that generate electricity.

The disadvantage of nuclear power stations is that the waste products are radioactive so the workers, public and environment have to be protected. The spent fuel forms high-level waste which is highly radioactive and very dangerous. It must be kept in pools for several years and then transferred to strong casks and buried for thousands of years.

Nuclear fission is when nuclei of two lighter elements, eg deuterium, fuse releasing energy. Nuclear fusion is how stars generate their energy but as it requires enormous temperatures and pressures, it is difficult to achieve on Earth. However there is an abundant supply of deuterium in sea water and the reaction is not thought to produce a lot of radioactive waste. The experimental fusion reactor (JET) in the UK uses more energy than it releases and a larger and more powerful one (ITER) is being built in France.

Take note

Uranium is called a nuclear *fuel* but in a nuclear reactor the energy is released from a nuclear reaction not from a chemical reaction as in a fossil fuel power station where the fuel is burnt.

The jargon

Natural uranium contains only about 0.7% of uranium-235 so it has to be enriched to about 4% by using centrifuges to remove some of the heavier non-fissile uranium-238.

Reactors that use plutonium-239 are called *fast breeder reactors* because the core is surrounded by uranium-238 which generates more plutonium and does not need moderators to slow down the neutrons.

Nuclear reactors that use slow neutrons are called *thermal reactors.* Making a nuclear reactor safe when it has finished its useful life is called *decommissioning* and is a lengthy and expensive process.

Links

See *radioactive decay,* page 53 to find how radioactivity changes with time.

Checkpoint 4

Why are high temperatures and pressures needed for fusion reactions to occur and how are these achieved on Earth?

Exam practice answers: page 195

(a) Explain the need for (i) a moderator and (ii) control rods in a nuclear reactor.

(b) Energy is released in the core of a nuclear reactor at a rate of 1000 MW. Each fission of uranium-235 produces 200 MeV. Calculate the number of uranium nuclei disintegrating per second.

(c) Show that the mass of uranium-235 that is used each second is 1.22×10^{-5} kg. (Avogadro's constant = 6.02×10^{23} mol^{-1}). (10 min)

Energy and the environment 2

The use of renewable energy resources is increasing as fossil fuel supplies are becoming scarcer and climate change becoming more of a concern.

Renewable resources

Solar cells

Solar cells or photovoltaic cells use solar energy directly to produce electricity. Although they are robust, they are expensive and only about 20% efficient.

Checkpoint 1

Show that $P = ½\pi r^2 \rho v^3$ for the power from a wind turbine.

Wind energy

Wind energy comes indirectly from the Sun as its energy causes convection currents in the atmosphere. Wind turbines convert the kinetic energy of the moving air into electrical energy. The (maximum) power from a wind turbine can be calculated from $P = ½\pi r^2 \rho v^3$ where r is the radius of the blades, ρ the density of air and v the wind speed.

Checkpoint 2

What are the energy changes that occur in a HEP scheme?

Hydroelectric power

Hydroelectric power also comes directly from the Sun's radiation which causes water on land to evaporate. It then returns to the Earth as rainfall. A dam is used to trap water high up in mountains. The water falls down through pipes to drive turbines that drive generators which generate electricity.

The jargon

Large quantities of electricity cannot be stored so demand has to be anticipated. Fossil fuel power stations take hours to reach their peak efficiency but HEP plants respond within seconds. To cope with fluctuations in demand, *pumped storage HEP plants*, such as the one in Dinorwic, are used. These are used to generate electricity when there is a high demand. Then when the demand is low, power is used to pump water back into the reservoir.

Tidal power

Tides get their energy from the gravitational pull of the Moon and to a lesser extent the Sun. To extract energy from the tides, a barrage or dam is built across an estuary. The ebb and flow of the tide causes water to pass through tunnels in the barrage driving a turbine as it moves through. So tidal power stations don't generate electricity at a constant rate. Although tidal power is predictable and it does not produce carbon dioxide, changes in water levels affects the plants and animals that live in the estuary.

Wave power

Winds across water causes waves so (most) of the energy of a wave at sea comes indirectly from the Sun. Waves have both gravitational potential energy and kinetic energy and various different methods of extracting this energy are being tried experimentally.

Fuel cells

A fuel cell converts chemical energy to electrical energy. A fuel such as hydrogen passes into the cell and is ionised at the anode. Electrons form the electric current. Oxygen atoms pick up the returning electrons and the oxygen ions recombine with the hydrogen ions making water which leaves the cell as a waste product. Catalysts are used to assist the reactions.

Action point

Draw up a table listing the advantages and disadvantages of different energy sources.

Thermal energy transfer

Another approach to reducing greenhouse gas emissions is to reduce heat loss from buildings so an understanding of thermal energy transfer is needed.

Conduction

- is the passage of thermal energy through a solid by particles passing on increased vibrational energy to neighbouring particles
- metals are the best conductors because free electrons contribute to the flow of thermal energy
- for a material of cross-sectional area A, in steady state conditions and no heat lost through the sides, the heat flow rate in *W* is given by $\Delta Q/\Delta t = -AK\,\Delta\theta/\Delta x$, where *K* is the **thermal conductivity** of the material and $\Delta\theta/\Delta x$ is the **temperature gradient**
- the **U-value** of a window for example is the heat flow rate per square metre produced by a 1 K temperature difference between its two surfaces
- so the power loss though the window is given by $P = UA\Delta\theta$ where *A* is the area of the window and $\Delta\theta$ the temperature difference across it.

Checkpoint 3

What are the units of *K* and *U*?

Convection

- is how thermal energy is transferred in liquids or gases (fluids)
- hot fluid is less dense than cold, so it rises and cold fluid sinks, giving rise to circulating **convection currents**
- convection can be reduced by restricting the convection currents: the fluid is trapped in smaller spaces eg cavity wall insulation fills the space between two walls with a foam which has small pockets of air
- convection that takes place in this manner is **natural convection**: when fans assist the removal of the hot fluid it is called **forced convection** and the rate of loss of heat is increased
- for forced convection, the rate of heat loss is proportional to the temperature difference (Newton's law of cooling) which means that the temperature falls exponentially with time
- the **half-cooling time** is the time taken for the temperature difference between the object and its surroundings to become half of the initial temperature difference (analogous with half-life in radioactive decay).

Checkpoint 4

A window is made from a single pane of glass 1 m wide and 2 m high and has a temperature difference of 11 K between the inside and outside surfaces. What is the energy transmitted through the window each second if the U-value for the glass is 5W m^{-2} K^{-1}? What would be the effect on the U-value if the single pane was replaced by two panes of half the original thickness with a layer of air in between?

Radiation

- all objects emit infra-red radiation which depends on the nature and temperature of the object
- polished silver surfaces are the worst emitters and best reflectors of i-r
- matt black surfaces are the best emitters and worst reflectors of i-r.

Exam practice answers: page 195

The table gives the power output of a wind turbine for increasing wind speeds.

wind speed/m s^{-2}	0	2	4	6
power output/MW	0	0.0023	0.0190	0.0630

(a) Show that the data suggest that the power is proportional to the cube of the wind speed.

(b) The power output of the turbine depends on the mass of air striking the turbine blades per second. State and explain the effect on the power output if the diameter of the circle swept out by the turbine blades is doubled. (10 min)

Applied physics 1

The equations in circular motion, pages 34–35, can be applied whenever the size of the object is small compared to the radius of the circle, so even for planets in orbit around the Sun. But the following theory applies if the size of the object is not small compared to the radius of the circle, eg spinning wheels and rotating planets.

Rotational dynamics

The quantities and equations in rotational dynamics are analogous with those already met in linear dynamics.

Linear motion			Rotational motion		
displacement	s	m	angular displacement	θ	rad
velocity	$v = \Delta s / \Delta t$	m s^{-1}	angular velocity	$\omega = \Delta\theta/\Delta t$	rad s^{-1}
acceleration	$a = \Delta v / \Delta t$	m s^{-2}	angular acceleration	$\alpha = \Delta\omega/\Delta t$	rad s^{-2}
mass	m	kg	moment of inertia	I	kg m^2
force	$F = ma$	N	torque	$T = I\alpha$	N m
momentum	$p = mv$	kg m s^{-1}	angular momentum	$L = I\omega$	kg m^2 s^{-1}
kinetic energy	$E_K = ½\, mv^2$	J	rotational KE	$E_K = ½\, I\omega^2$	J
work done	$W = Fs$	J	work done	$W = T\theta$	J
power	$P = W/t = Fv$	W	power	$P = W/t = T\omega$	W

Angular displacement, velocity and acceleration

Just as linear velocity = linear displacement/time, $v = s/t$

for rotational motion, angular velocity = angular displacement/time $\omega = \theta/t$

A wheel accelerates from ω_1 to ω_2 in t seconds

angular acceleration = change in angular velocity/time

so $\alpha = \dfrac{\omega_2 - \omega_1}{t}$ which gives $\omega_2 = \omega_1 + \alpha t$ which is analogous to $v = u + at$

The other equations of motion are also analogous. Using the table above, these become:

Linear motion	Rotational motion
$v = u + at$	$\omega_2 = \omega_1 + \alpha t$
$s = ut + ½\, at^2$	$\theta = \omega_1 t + ½\, \alpha t^2$
$v^2 = u^2 + 2as$	$\omega_2{}^2 = \omega_1^2 + 2\alpha\theta$

Moment of inertia

Moment of inertia is analogous to mass in rotational dynamics. It depends on the mass of the object and how the mass is distributed about the axis of rotation: a flywheel where the mass is concentrated at the rim needs a larger torque to accelerate at the same rate than a uniform disc of the same mass. (The moment of inertia may be different about different axes of rotation.)

For a point mass, $I = mr^2$.

Links

The work on linear dynamics is in *ways of describing motion*, pages 14–15, *equations of motion*, pages 16–17, *Newton's laws of motion*, pages 20–1, *work, energy and power*, pages 26–7 and *momentum and impulse*, pages 28–9. You should already know the theory here, you just have to learn another set of quantities!

Checkpoint 1

Convert the other two equations of motion from checkpoint 1 on page 10 to their rotational dynamics equivalents: $s = vt - ½\, at^2$ and $s = (u + v)t/2$

Links

See *forces and moments in equilibrium*, pages 12–13 for a reminder of the meaning of torque.

The jargon

Inertia means reluctance to a change in motion, so mass measures the reluctance for linear motion to change and moment of inertia measures opposition to angular acceleration.

A solid object has to be considered as a collection of point masses so it is the sum of the moments of inertia of all the points, $I = \Sigma mr^2$.

You won't have to learn expressions for moment of inertia – they will be given where necessary.

Some examples:

Moment of inertia of a disc about an axis through the centre and perpendicular to its plane

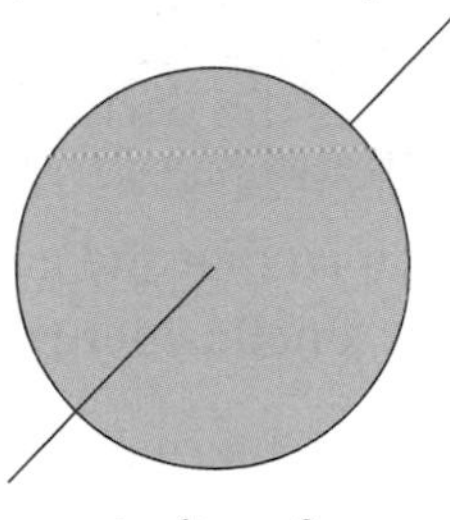

$I = ^1/_2\ mr^2$

Moment of inertia of a thin ring about an axis through the centre and perpendicular to its plane

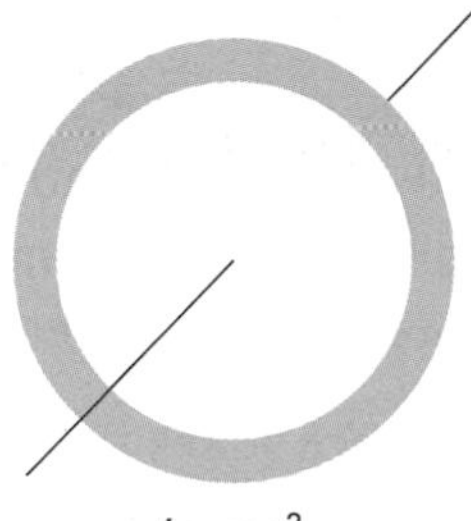

$I = mr^2$

Moment of inertia of a thin ring about an axis through a diameter

$I = ^1/_2\ mr^2$

Checkpoint 2

Use the equation $T = I\alpha$ to show that the units of moment of inertia are kg m^2.

Torque and linear acceleration

Just as a constant resultant force gives a constant linear acceleration, a constant resultant torque gives a constant rotational acceleration. The analogous equation to $F = ma$ is $T = I\alpha$

Work, energy and power

Since work done = force × distance, $W = T\theta$ and power = force × velocity so $P = T\omega$

A resultant torque on a solid object will give it **rotational kinetic energy**, $E_K = ½ I\omega^2$.

A wheel on a moving bike will have both rotational kinetic energy and translational kinetic energy.

Angular momentum

Linear momentum, $p = mv$ and the analogous quantity in rotational motion is angular momentum, $L = I\omega$

Just as linear momentum is conserved in interactions, so is rotational momentum. This can clearly be seen when ice skaters spin: with their arms stretched out they spin more slowly and as they draw their arms closer to their bodies, the rate of spinning increases.

Checkpoint 3

Look at the law of conservation of (linear) momentum in *momentum* and *impulse* on page 22 and rewrite it to apply to rotational motion.

Checkpoint 4

Explain why a spinning wheel eventually slows down and stops.

Checkpoint 5

Use conservation of angular momentum to explain why a spinning skater speeds up when they pull their arms in closer to their bodies.

Exam practice

answers: page 195

Flywheels store energy very efficiently and are being considered as an alternative to battery power.

(a) A flywheel for an energy storage system has a moment of inertia of 0.60 kg m^2 and a maximum safe angular speed of 22 000 rev min^{-1}.
Show that the energy stored in the flywheel when rotating at its maximum safe speed is 1.6 MJ.

(b) In a test the flywheel was taken up to maximum safe speed then allowed to run freely until it came to rest. The average power dissipated in overcoming friction was 8.7 W. Calculate (i) the time taken for the flywheel to come to rest from its maximum speed, (ii) the average frictional torque acting on the flywheel.

Applied physics 2

Links

See pages 100–1 for information on gas laws.

Take note

The first law of thermodynamics is a consequence of the law of conservation of energy.

Thermodynamics leads on from the kinetic theory of gases and developed from the need to improve the efficiency of early steam engines.

First Law of Thermodynamics

The **internal energy** of a gas, U, is the sum of the kinetic and potential energies of all of its particles.

You can increase the internal energy of a gas by

- heating it – the walls of the container get hot, making the particles inside move more quickly so their kinetic energy increases
- compressing it – the walls of the container move in so that the gas particles bounce off more quickly, increasing their kinetic energy.

The **first law of thermodynamics** states that the change in internal energy of a system, ΔU, is equal to the heat energy transferred to the system, Q, plus the work done on the system, W.

$\Delta U = Q + W$

Signs are important:

- $+\Delta U$ means that the internal energy of the system increases, $-\Delta U$ means that it *decreases*
- $+Q$ means that system *gains* heat energy, $-Q$ means the system *loses* heat energy (heat energy always flows from hot to cold)
- $+W$ means that work is done *on* the system, $-W$ means that work is done *by* the system.

P-V diagrams

The graph shows how the pressure of a gas changes at constant temperature for two temperatures where $T_2 > T_1$.

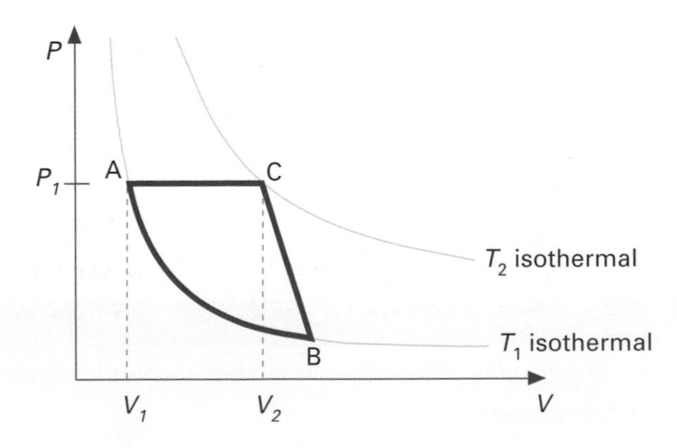

You need to know about three special cases of energy changes.

Checkpoint 1

In an isothermal change, 400 J is added to a gas. How much work is done on or by the system?

1 An isothermal change

- T stays the same, eg from A to B on the graph
- PV = constant
- for an ideal gas constant temperature means constant internal energy
- so $Q = -W$

The gas expands and does work and heat energy is gained from the surroundings to keep the temperature constant.

2 A constant pressure (isobaric) change

➜ P is constant, eg from A to C on the graph

➜ V/T = constant

➜ ΔW is the area under the line AC, so $\Delta W = P_1 (V_2 - V_1) = P\Delta V$

Here heat energy is supplied to the gas, its internal energy increases and the gas does work in expanding.

3 An adiabatic change

➜ no heat energy enters or leaves the gas, eg from B to C on the graph

➜ PV^γ = constant, where γ is a constant for the gas

➜ $Q = 0$, so $\Delta U = W$

The gas is compressed, and its internal energy increases.

The jargon

An *adiabatic change* takes place rapidly in a thick walled poorly conducting container: the expansion of air from a burst tyre is close to an adiabatic change.

An *isothermal change* takes place slowly in a container with conducting walls.

Heat engines

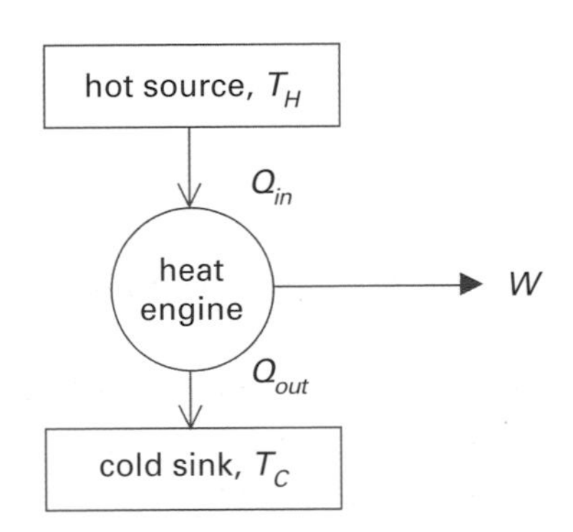

Heat engines are devices for converting heat energy into mechanical work. They use the fact that heat energy flows from a hot 'source' to a cold 'sink'.

From the diagram, efficiency = $W / Q_{in} = (Q_{in} - Q_{out}) / Q_{in}$

The maximum theoretical efficiency = $(T_H - T_C) / T_H$

To get a high efficiency, T_C should be as low as possible (t is often the temperature of the surroundings) and T_H should be as high as possible (it is limited by the properties of the materials involved) so heat pumps are inherently inefficient.

Petrol and diesel (internal combustion) engines and steam engines are heat pumps. They operate on a cycle where the PV diagram is a closed loop and the area enclosed represents the work done by the gas.

Checkpoint 2

Calculate the maximum thermodynamic efficiency of a steam turbine for which steam enters at 900 K and leaves at 300 K. Explain why its overall efficiency will be less than this value.

Four-stroke petrol cycle

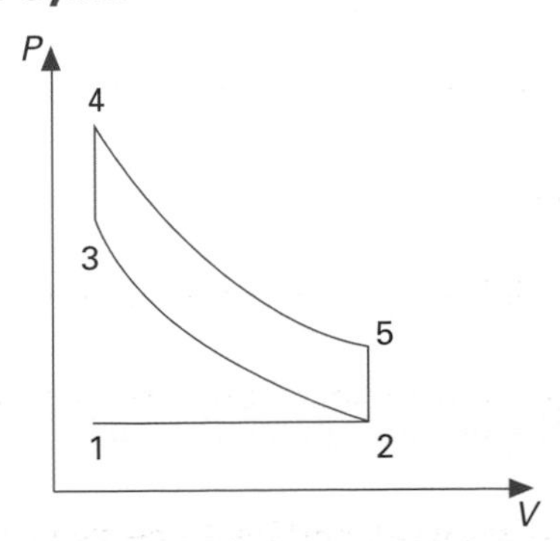

The jargon

A petrol engine is said to be spark ignition while a diesel engine is *compression ignition.*

Check the net

If you type 'animation of a four stroke engine' (or 'animation of a diesel engine') into an Internet search engine you should be able to see an animated diagram of its operation.

The jargon

Heat engines cannot convert all of their input energy into mechanical energy. In a conventional power station, just over half of the input energy is dissipated as waste heat energy in the surroundings. A *combined heat and power* station supplies the heat energy to nearby buildings. In order to produce heat energy that is at a useful temperature, the CHP station does not produce electricity as efficiently as a conventional power station but because there is not so much waste energy, the overall efficiency is higher.

The diagram shows the ideal **Otto cycle.**

1 – 2 **Intake stroke**. Intake valve is open. The piston moves out of the cylinder so the volume increases. Fuel and air is drawn into the cylinder at constant pressure.

2 – 3 **Compression stroke**. Intake valve is closed. The piston moves up and there is adiabatic compression of the fuel/air mixture.

3 – 4 A spark causes the fuel and air to ignite. Combustion occurs quickly at constant volume releasing heat energy.

4 – 5 **Power stroke**. The gas does work on the piston pushing it down. There is an adiabatic expansion.

5 – 2 The exhaust valve opens. The pressure drops and the gas cools at constant volume.

2 – 1 **Exhaust stroke**. The piston goes back and the volume decreases at constant pressure.

Work is done *on* the gas during the compression stroke and work is done *by* the gas during the power stroke. The area enclosed by the graph is the net work done by the gas.

The **indicated power** is the output power of the engine is given by the area of the P-V loop × no. of cycles per second × no. of cylinders.

The **input power** is the calorific value of the fuel (J kg^{-1}) × fuel flow rate (kg s^{-1})

So the **thermal efficiency** of the engine = indicated power/input power.

The output or **brake power** is the power delivered at the crankshaft. Brake power = torque (N m) × angular velocity (rad s^{-1}) = $T\omega$

The **friction power** = indicated power – brake power (measures the power dissipated in overcoming friction in the engine.)

The **mechanical efficiency** of the engine = brake power/indicated power

The **overall efficiency** measures the fraction of the input power that is delivered as useful power: overall efficiency = brake power / input power.

Diesel engine cycle

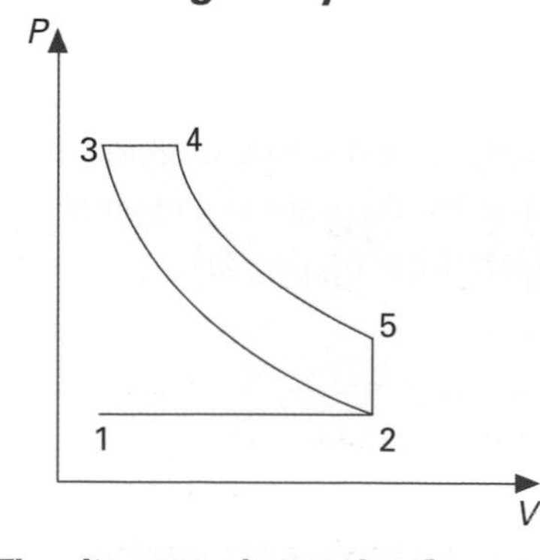

1-2 piston moves down and draws air in
2-3 piston moves up, adiabatic compression
3-4 fuel injection and burns at constant pressu
4-5 adiabatic expansion, work is done
5-1 cooling and exhaust

The diagram shows the **theoretical Diesel cycle.**

The main differences are:

➜ the intake stroke takes in only air: fuel is injected after the compression stroke

➜ there is no spark: the hot air ignites the fuel.

Ideal cycles assume that no heat energy enters or leaves the cylinder during compression and power strokes and there are no friction losses, which does not happen in practice.

Reversed heat engines

Refrigerators and heat pumps are similar to heat engines but they work in reverse. Work is done to transfer heat energy from a cold space to a hot space, against the temperature gradient.

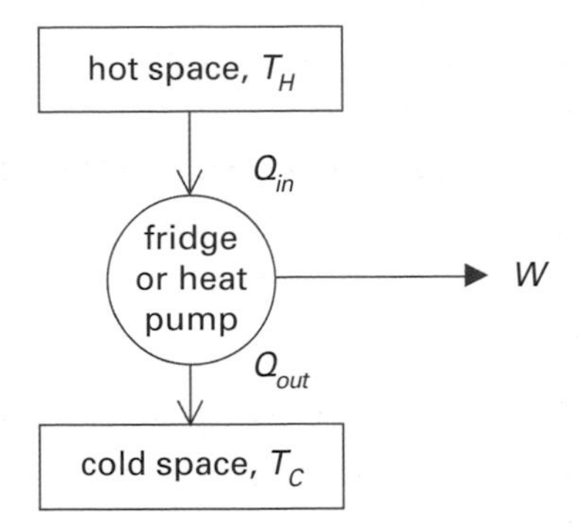

For a refrigerator, the heat energy that is removed from the cold space (the fridge) is important. So efficiency or coefficient of performance $= Q_{out} / W = Q_{out} / (Q_{in} - Q_{out})$

Heat pumps are used to pump heat energy from a cooler ground for example into a warmer building so the heat energy that enters the hot space is important.

So efficiency or coefficient of performance $= Q_{in} / W = Q_{in} / (Q_{in} - Q_{out})$

Exam practice answers: page 195

A pump is used to inflate a dinghy.

A pump is operated quickly so the compression of the air in the cylinder before the valve opens can be considered adiabatic. At the start of a pump stroke, the pump cylinder contains 4.25×10^{-4} m^3 of air at a pressure of 1.01×10^5 Pa and a temperature of 23°C. The pressure of air in the dinghy is 1.70×10^5 Pa.

(a) Show that, when the valve is about to open, the volume of air in the pump is 2.93×10^{-4} m^3. (γ for air = 1.4)

(b) Calculate the temperature of the air in the pump when the valve is about to open. (10 min)

Further electricity

This option builds on the work done on electromagnetic induction and alternating currents.

Links

Look at electromagnetic induction on pages 92–3 to remind yourself about the action of a transformer.

Transformers

A transformer consists of two coils wound on a soft iron core. An alternating current in the primary coil causes an induced alternating EMF in the secondary coil.

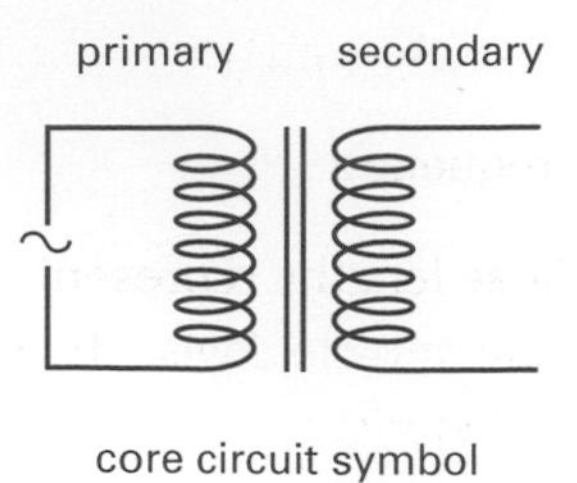

core circuit symbol

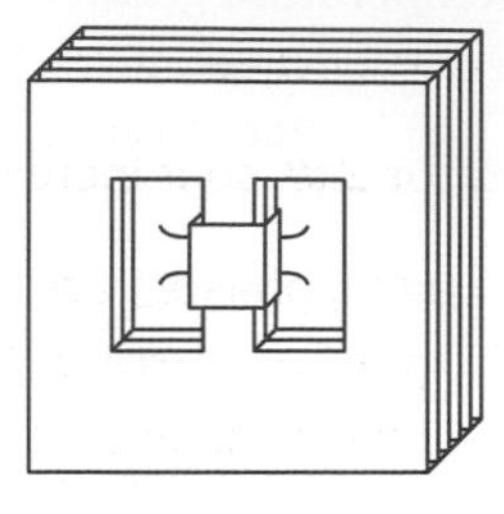
a practical transformer

Checkpoint 1

Explain how a transformer works and why the primary current has to be AC.

In an ideal transformer

- all the flux through the primary links the secondary
- there are no voltage drops across the primary and secondary coils

then $V_1 / V_2 = N_1 / N_2$

And if there are

- no eddy currents in the iron core
- no hysteresis losses in the iron core

then $V_1 I_1 = V_2 I_2$

Take note

The core of a transformer is a closed loop of iron. Practical transformers are of the shape shown in the diagram. The primary and secondary coils are wound on top of each other in the centre of the shape to improve the magnetic flux linkage.

Checkpoint 2

Give reasons for the following:

(a) the coils of a transformer are made from thick copper wire

(b) the iron core is not solid but is made from thin sheets of iron coated in varnish.

Inductors

A changing current in a wire causes changing magnetic flux. The changing magnetic flux will link the wire that produces it causing an induced EMF in itself (often called a **back EMF** as it opposes the change producing it by Lenz's Law).

The induced EMF, $E = -L\Delta I / \Delta t$

Where L is the inductance of the wire in henries, H.

Winding the wire into a coil and putting an iron core inside will greatly increase the inductance.

A coil is said to have an inductance of 1 henry if an EMF of 1 V is induced in the coil when the current changes at the rate of 1 A s^{-1}.

Reactance of an inductor

A pure inductor has no resistance.

The induced EMF is proportional to the *rate of change of current*, so it is greatest when the current is zero and changing at its maximum rate.

In other words, the p.d. across an inductor is out of phase with the current through it (unlike a resistor where the p.d. and current are in phase.)

The **p.d. leads** the current by 90° ($\pi/2$ radians).

The jargon

Self-inductance is when a changing current in one coil an EMF in itself.
Mutual inductance is when a changing current in one coil induces an EMF in a second coil – as in a transformer.
Both are measured in henries.

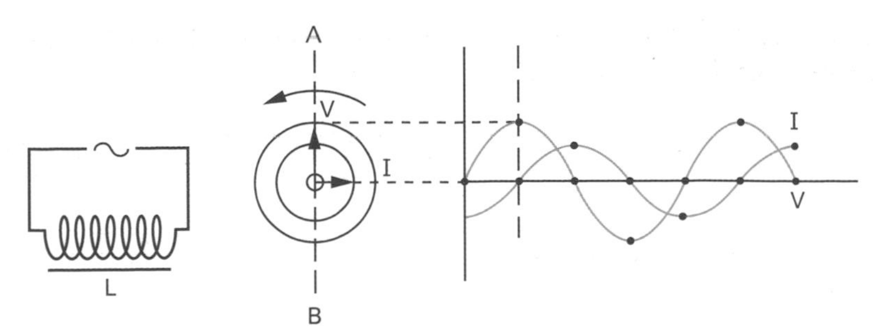

The opposition to ac is called **inductive reactance**, X_L, (in Ω).

$X_L = V_{RMS}/ I_{RMS} = \omega L$ or $2\pi fL$ so it increases with frequency.

In the **phasor diagram**, phasors are vectors whose lengths represent the amplitudes of the p.d. and current. As they rotate anti-clockwise, their projections on AB represent the magnitudes at any instant.

Capacitors and AC

Although a current cannot flow through a capacitor, in an AC circuit, it will be constantly charging and discharging, so a current will flow in the connecting wires.

The charge on a capacitor is proportional to the p.d. across it, but as current is the rate of flow of charge, the current is greatest when the p.d. is zero and changing at its maximum rate.

In other words, the current through a capacitor is out of phase with the p.d. across it.

The **current leads** the p.d. by 90° (π/2 radians).

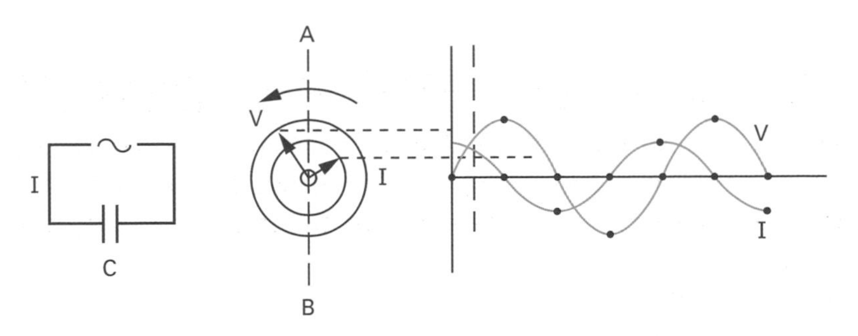

The opposition to AC is called **capacitative reactance**, X_C, (in Ω).
$X_C = V_{RMS} / I_{RMS} = 1/\omega C$ or $1/2\pi fC$ and this decreases with frequency.

Power in L and C circuits

Because there is a 90° phase difference between the current and p.d. in an AC circuit containing just capacitance or just inductance, it turns out that the power dissipated is zero.

LR in series

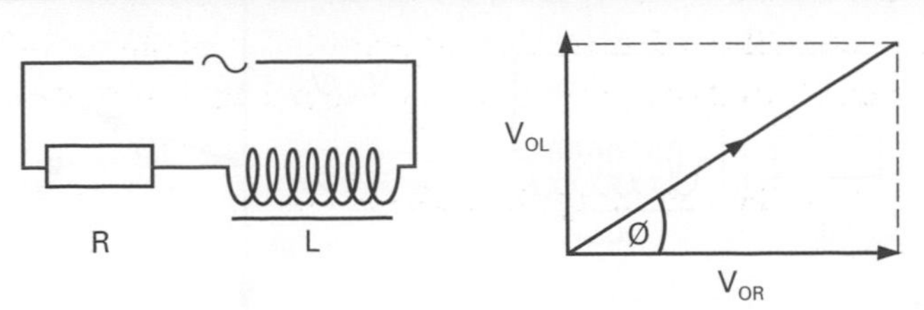

Checkpoint 3

Show how the laws of electromagnetic induction (Faraday's Law and Lenz's Law - see page 92) lead to the expression for self-inductance, $E = -L\Delta I / \Delta t$.

The jargon

An inductance is also called a choke because it blocks or chokes high frequency currents.

Examiner's secrets

The letters of the word **CIVIL** will help you to remember the phase differences for capacitors and inductors: for a capacitor, **C**, the current, **I**, leads the pd, **V**, but this leads the current, **I**, in an inductor, **L**.

For a resistor and inductor in series (in real circuits there will always be some resistance):

- ➜ the p.d. across R is in phase with the current
- ➜ the p.d. across L leads the current by 90°

The resultant voltage can be seen from the phasor diagram to be

$V_0^2 = V_{OR}^{\;2} + V_{OL}^{\;2} = I_0^2 R^2 + I_0^2 X_L^{\;2}$

$V_0 = I_0 \sqrt{(R^2 + X_L^2)}$ where the opposition to AC, $\sqrt{(R^2 + X_L^2)}$ = the **impedance**, Z, in ohms.

The phase angle, ϕ, is given by $\tan \phi = V_{OL} / V_R = I_0 X_L / I_0 R = \omega L / R$

CR in series

For a resistor and capacitor in series:

- ➜ the p.d. across R is in phase with the current
- ➜ the p.d. across C lags the current by 90°

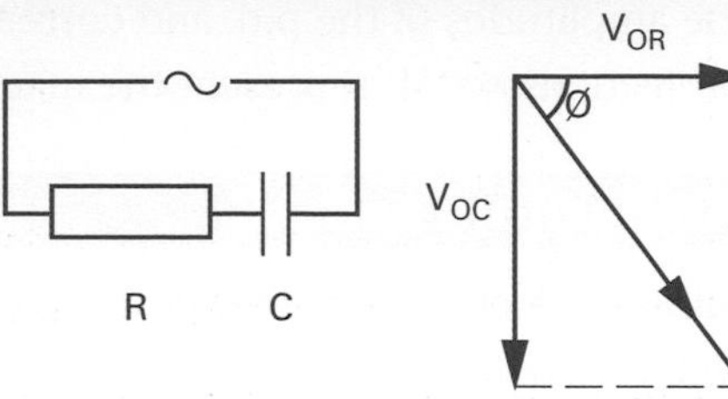

Giving $V_0 = I_0 \sqrt{(R^2 + X_C^2)}$ so $Z = \sqrt{(R^2 + X_C^2)}$

And $\tan \phi = V_{0C} / V_R = -1 / \omega CR$

Filters

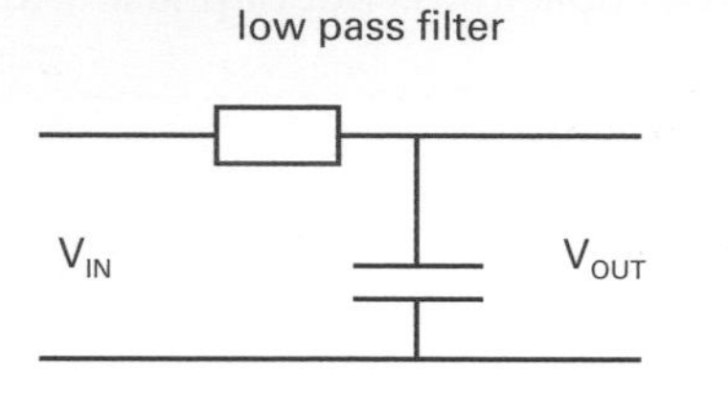

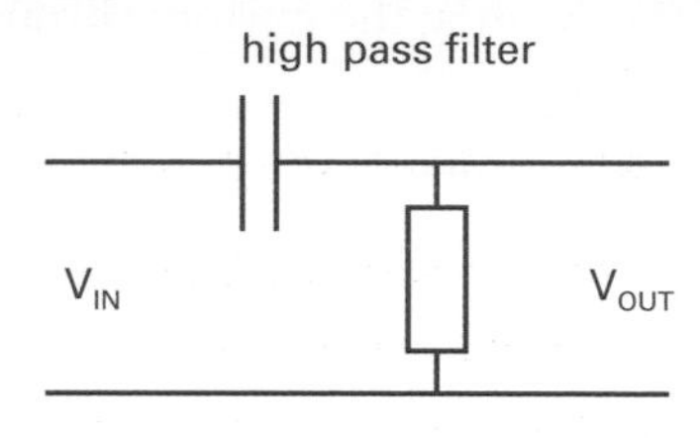

Take note

Low pass filters are used to send low sounds to bass speakers (woofers) and high pass filters are used to send high sounds to high frequency speakers (tweeters).

- ➜ X_C is low for high frequencies and high for low frequencies
- ➜ there will be a small p.d. across C and a large p.d. across R for high frequencies
- ➜ there will be a large p.d. across C and a small p.d. across R for low frequencies
- ➜ in the low pass filter, V_{OUT} will be small for high frequencies which are essentially blocked
- ➜ in the high pass filter, V_{OUT} will be small for low frequencies which are essentially blocked
- ➜ filters are used to send low sounds to bass speakers (woofers) and high sounds to high frequency speakers (tweeters).

LCR in series

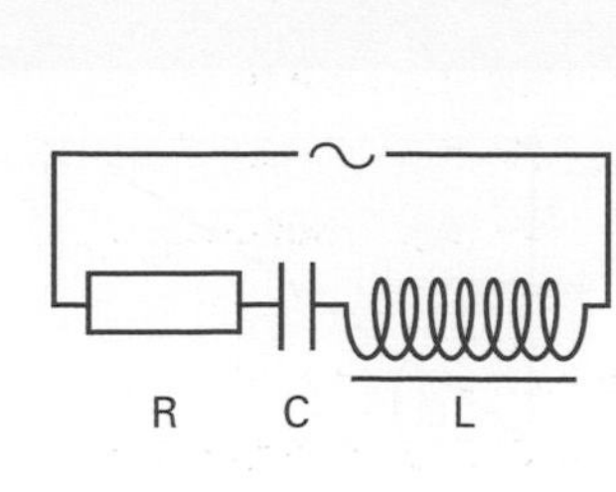

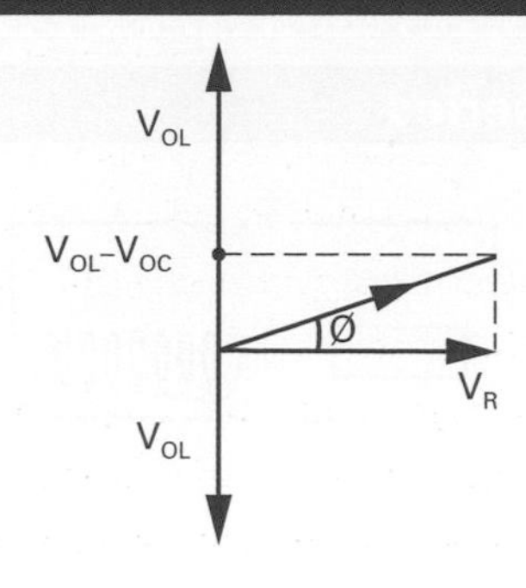

The phasor diagram gives $V_0^2 = V_{0R}^2 + (V_{0L} - V_{0C})^2 = I_0^2 R^2 + I_0^2 (X_L - X_C)^2$

Giving $V_0 = I_0 \sqrt{(R^2 + (X_L - X_C)^2)}$ so $Z = \sqrt{(R^2 + (X_L - X_C)^2)}$

And $\tan \phi = (V_{0L} - V_{0C}) / V_R = (X_L - X_C)/R$

Resonance in LCR circuits

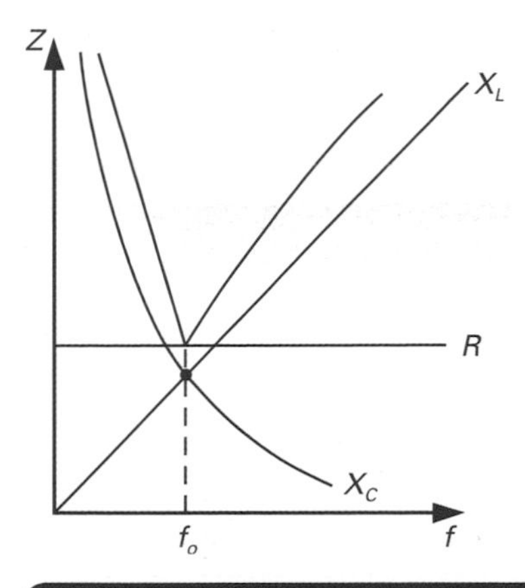

The diagram shows how the R, X_L and X_C change with frequency: X_L increases, X_C decreases and R does not change. At f_0, X_L equals X_C and the impedance has its minimum value of R so the current in the circuit would be largest at this frequency: called the resonant frequency.

Since $X_L = X_C$ then $2\pi f_0 L = 1 / 2\pi f_0 C$ giving $f_0 = 1 / 2\pi\sqrt{(LC)}$

Q factor

In resonance, the sharpness of the peak depends on the degree of damping: a lightly damped oscillation has a sharp resonance peak and a high Q factor.

$$Q = \frac{2\pi}{\text{fraction of energy of system lost per oscillation}} \quad \text{(no units)}$$

Capacitors are almost free from energy losses – the resistance of the coil determines the Q factor.

The fraction of energy lost per cycle is R/fL

So $Q = 2\pi fL/R = \omega L/R$

Exam practice

answers: page 196

In the circuit shown, the input p.d. is always 10.0 $V_{r.m.s.}$, but the frequency can be varied.

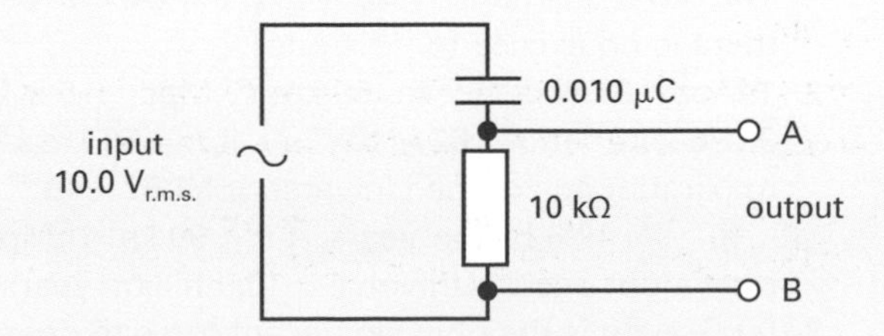

The r.m.s. p.d. between terminals A and B is the output p.d.

(a) Show clearly that, for a frequency of 1000 Hz, the output p.d. is 5.3 $V_{r.m.s}$

(b) The output p.d. approaches 10.0 $V_{r.m.s.}$ at very high frequencies. Explain why this is so.

(c) Sketch a graph of r.m.s. output p.d. against frequency.

(d) The circuit can be used in electronics as a simple 'high frequency pass filter'. Discuss why this name is appropriate. (15 mins)

Take note

Resonance is used in radio tuning circuits. The aerial picks up oscillations of many different frequencies. A variable capacitor is used to change the resonant frequency of the LCR circuit so it matches the frequency of the desired radio transmission. At this frequency a large current flows in the circuit.

Answers
Options

Astrophysics 1

Checkpoints

1 3.26 ly = 1 pc.

2 (a) 525 600; (b) 8.3 minutes.

3 θ (in radians) = arc length/radius. 1.2° is a small angle. $\theta = 1.2 \times 2\pi/360° = 2.09 \times 10^{-2}$ rad.
Distance to post = eye separation/θ in radians = 4.8 m.

4 $L = 3.825 \times 10^{26}$ W.

5 The dimmer the star, the fewer photons captured in a given time, increasing relative errors in the measurement of apparent magnitude. To estimate absolute magnitude, you need a good estimate of surface temperature. A good clear spectrum requires as much light as possible. Observation of dimmer stars is also beset by problems of light pollution and atmospheric turbulence. The furthest stars tend to be dimmest. Use of standard candles depends on these stars behaving as we expect them to. We can only see the brightest very distant stars (e.g. quasars).
Our understanding of them is incomplete etc.!

6 (a) Each absorption line corresponds to a specific quantum leap in electron energy level. (b) Light from one side of a spinning star is red shifted (because that side is moving away from us); light from the other side is blue shifted. The relative velocity varies over the surface of the star, with maximum Doppler shifts at the edges. The faster the spin, the bigger the spread (of each and every absorption line).

Exam practice

1 (a) $2.7 \times 3.26 = 8.80$ years.
(b) $-1.46 - M = 5 \log(2.7/10)$ so
$M = -1.46 - 5 \log 0.27 = 1.38$.

2 Starlight passes through hydrogen gas around the star, if the hydrogen has electrons in the n = 2 level, photons of certain frequencies are absorbed, and re-emitted in all directions so these wavelengths are missing from the spectrum.

Astrophysics 2

Checkpoints

1 Around the edges. Hydrogen is the least dense gas.

2 (a) Because gravitational potential energy is converted to kinetic energy and then heat as the star's imploding atoms accelerate inwards and collide.
(b) Because there is no further energy source; the gravitational collapse has been stopped and the temperature has not been raised sufficiently to cause fusion of the remaining material (or possibly there is
no fuel left, since the fusion chain has reached iron).

3 (a) (i) For the Earth $R_S = 8.9$ mm. (ii) For the Sun, $R_S = 2.95$ km. (iii) For a 5 solar mass star, $R_S = 14.7$ km.
(b) The Earth and the Sun are not massive for gravity ever to crush them sufficiently to form black holes. The minimum mass for a black hole is about 2.5 solar masses.

4 Luminosity depends on surface temperature and surface area. A dim white star must therefore be small and a bright red star must be big.

Exam practice

Using $v_f = v_e \ln (m_o/m_f)$, $2.3 = 95 \ln (m_o/1800)$, $e^{0.0242} = m_o/1800$, m_o = *1844 so the mass of fuel is 44 kg.*

Astrophysics 3

Checkpoints

1 Equal resolution requires that $l/D_{light} = l/D_{radio}$ (5 m). $D_{radio} = 5 \times 0.075/5 \times 10^{-7} = 7.5 \times 10^5$ m (750 km!).
Radio telescopes generally have poor resolution. Arrays of widely spaced telescopes can be linked to give improved resolution.

2 **(a)** 110 cm
(b) 10

Exam practice

The atmosphere is clear in the radio waveband up to 10 m wavelength, so satellite-based telescopes are not essential (the atmosphere is opaque to X-rays and to important IR wavebands, so satellites are essential if these wavebands are to yield any useful information). Also, a radio telescope would need to be big – difficult to launch. (*Note* it may be possible to cheat the Rayleigh criterion by using the motion of the satellite to broaden the base, or by use of several satellites.)

Astrophysics 4

Checkpoints

1 The Universe looks roughly the same in all directions. If the Universe has an edge and we are close enough to see it, then we must be in its centre (which seems unlikely). This breaks the cosmological principle; the Universe would look different from different places. The other solutions already explain Olbers' paradox, so there is no excuse for this one!

2 1 Mpc = 3.086×10^{22} m. 50 km s^{-1} Mpc^{-1} = 5×10^4 m s^{-1}/3.086×10^{22} m = 1.62×10^{-18} s^{-1}. If H = 50 km s^{-1} Mpc^{-1}, maximum age of the Universe = $1/1.62 \times 10^{-18} = 6.17 \times 10^{17}$ s = 19.6 billion years. If H = 90 km s^{-1} Mpc^{-1}, maximum age of Universe = 10.9 billion years.

3 (a) Gravity is the only significant force to operate at great distances and it is a force of attraction, and so the rate of expansion of the Universe should be slowing down (H should be decreasing). (b) $1/H$ is only a good estimate of the age of the Universe if we can ignore gravity and assume constant H. If H was greater in the past than it is now, the Universe must be younger than the Hubble time ($1/H$). For a flat Universe, the age is around ⅔ of the Hubble time.

4 The speed of a star depends on the mass inside its orbit. If the dark matter was in the centre of the galaxy, *all* stars would orbit anomalously fast, not just the outer ones. Any matter beyond a star does not contribute to the gravitating mass pulling it towards the galaxy's centre.
5 When $T = 3\,000$ K, $\lambda_{max} = 0.96$ μm (infrared). When $T = 2.7$ K, $\lambda_{max} = 1.07$ mm (microwave).

Exam practice

(a) At 15% light speed (4.5×10^7 m s^{-1})
(b) $v = Hd$ so $d = 4.5 \times 10^4$ (km)/65 (km s^{-1} Mpc^{-1}), $d = 692$ Mpc $= 2.26 \times 10^9$ ly

Medical and health physics 1

Checkpoints

1 The right eye is short sighted. f = – 4 m. The left lens has astigmatism.
2 The object would be invisible. At the point where the retina disappears down the optic nerve, there is no light-sensitive layer for light to fall on.
3 $L = 10 \log_{10} (2I_0/I_0) = 10 \log_{10}{}^2 = 10 \times 0.3010 = 3$
4 Normally you hear your voice after the sound has passed through the bone and other tissues in your head. Therefore your voice sounds very different to everyone else, or on a recording, as this is not then the case.
5 You have already damaged your hearing.

Exam practice

Different parts of the ear are affected by sounds of different frequencies. Each part has a different natural frequency and so each part resonates when a certain frequency is heard. For example, the outer ear resonates when notes of around 3 300 Hz are heard and the middle ear responds to sounds between 700 Hz and 1 500 Hz. The individual response of all the regions of the ear allows us to hear sounds between20 and 20 000 Hz, though this upper limit declines with age.

The graph on the left shows that the minimum threshold for hearing is about 10^{-12} W m^{-2} between about 1 kHz and 3 kHz. It also shows that the intensity thresholds for discomfort and pain are not frequency dependent. 120 dB cause discomfort irrespective of frequency, for example.

The loudness of a sound (in phons) is the intensity level at 1 kHz that has the same loudness as the sound to a normal ear. The graph on the right shows equal loudness curves. Notice that a 40 dB sound at 100 Hz has the same loudness (0 phons) as a 20 dB sound at 300 Hz.

Medical and health physics 2

Checkpoints

1 Each fibre of the fibre optic carries a small part of the overall image to the eyepiece. If the fibres arve scrambled the image will be too.
2 The right leg is too far from the heart so the signal will be very weak.

Exam practice

See page 166.

Medical and health physics 3

Checkpoints

1 Gamma rays produce less ionisation than alpha or beta so they are less harmful to healthy tissue. Also they are more penetrating so can reach the detector from deep within the body.
2 (a) 50 mm of concrete is 2 × half-thickness, so the intensity is $I_0/4$.
(b) 1.5 mm of lead is 3 × half-thickness, so the intensity is $I_0/8$.
3 Dose equivalent from alpha:
0.10×2 mGy $\times 20 = 4 \times 10^{-3}$ Sv.
Dose equivalent from gamma:
0.90×2 mGy $\times 1 = 1.8 \times 10^{-3}$ Sv.
Total dose = (4 + 1.8) mSv = 5.8 mSv.
4 2 mSv is equivalent to a risk of $5\% \times 2 \times 10^{-3} = 10 \times 10^{-3}\%$. This is a risk of 0.01 in 100 or 100 in 1 million. So 100 people in a population of 1 million may be expected to get cancer.

Exam practice

50 × background dose is 10 mSv.
This is a risk of $10 \times 10^{-3} \times 5\% = 0.05$ %.
So 0.05 in 100, or 5 in 10 000 may get a fatal cancer from this programme. For a population of 20 million, that is 10 000 people. You should consider the consequences of such a programme.

Materials 1

Checkpoints

1 Stress = force/area in Pa (1Pa = 1 N m^{-2})
strain = extension/length (a ratio, so no units)
Young's modulus = stress/strain in Pa.
2

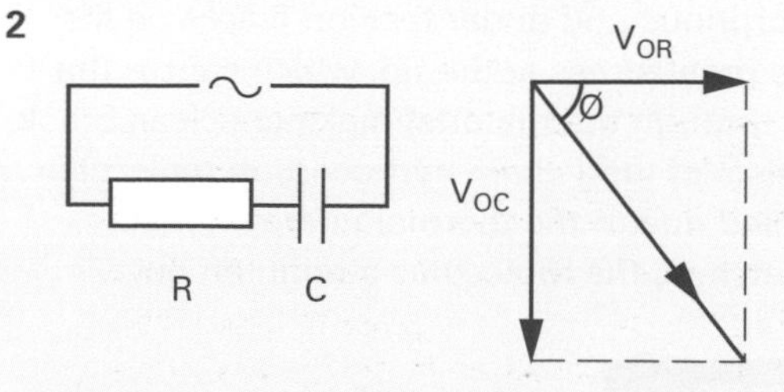

3 *Stiff* needs a large force to stretch it. *Flexible* stretches easily.
Strong needs a large force to break it. *Weak* breaks easily
Brittle shatters easily with a clean break. *Tough* does not shatter easily.
Elastic returns to its original shape when the stress is removed. *Plastic* acquires a permanent set.

Exam practice

(a) Apparatus:

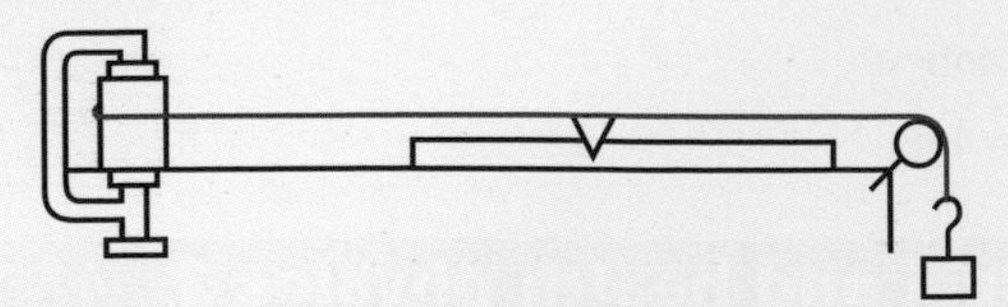

Measure the original length of the wire with a small mass on it to make it taut. Measure the diameter in several places with a micrometer. Add weights to the wire and measure the extension. Plot a graph of force in newtons against extension in metres. Young's modulus of the copper is the gradient of the linear part of the graph multiplied by l/A.

(b) A long thin wire is used so that the extension is long enough to be measured accurately.

(c) Using $e = Fl/EA$, extension is proportional to length, and so Y will have $\frac{1}{3}$ of the extension of X. Extension is proportional to 1/area or 1/(diameter)^2 so Y will have four times the extension of X. Together, extension of Y = $\frac{4}{3} \times 8 = 10.7$ mm.

Materials 2

Checkpoints

1 It is important that a load-bearing structure is not permanently deformed after being stressed.

2 When the paper clip is bent backwards and forwards, it is initially flexible as the dislocations move. After a while, they jam up so the material begins to harden and eventually undergoes brittle fracture.

3 Concrete has an amorphous structure so when pulled apart, cracks will propagate. Stressed rods hold the concrete in compression.

4 A sudden stress causes elastic behaviour as the long-chain molecules do not have time to flow, but left under its own weight it behaves like a viscous liquid.

Exam practice

(a) Copper is polycrystalline and contains many dislocations that move easily causing plastic flow.

(b) Glass is amorphous and under tension cracks on the surface have great stress at the tip, which causes the crack to run through the material making a clean break.

(c) Rubber consists of long-chain hydrocarbon molecules that are tangled due to the thermal energy. When rubber is stretched, the molecules straighten out.

Examiner's secrets

Examiners will be impressed if you use diagrams to help with your explanations.

Turning points in physics 1

Checkpoints

1 As $e/m = -1.76 \times 10^{11}$ C kg^{-1}, $e = -1.76 \times 10^{11}$ m. As $Q/M = 9.65 \times 10^7$ C kg^{-1}, $Q = 9.65 \times 10^7$ M. Thomson assumed that the charge on an electron e would be equal in size and opposite in charge to the charge on the hydrogen ion, Q, i.e. $e = -1.76 \times 10^{11}$ $m = 9.65 \times 10^7$ M. So the mass of an electron m would be very much less than the mass of a hydrogen ion M.

2 $Bev = mv^2/r$. Dividing both sides by Bv gives $e = mv/Br$. Dividing both sides by m gives $e/m = v/Br$.

3 Volume $V = \frac{4}{3}\pi r^3$. Weight $W = mg$. As density, $\rho = m/V$, mass $m = \rho V = \frac{4}{3}\pi r^3 \rho$. $W = mg = \frac{4}{3}\pi r^3 \rho g$.

4 When a projector shines a beam of light through a slide, the light is affected by the slide. The slide 'filters' the light so that only certain parts of the beam pass through certain parts of the slide. In a TEM, an electron beam replaces the light beam and the slide is replaced by a specimen. Thicker or denser parts of the specimen allow fewer electrons through so that part of the image is darker.

Exam practice

Using $\lambda = 10^{-11}$ m gives $V = 1.5 \times 10^4$ V. An STM does not damage the surface of the specimen and it does not need a vacuum to work.

Turning points in physics 2

Checkpoints

1 If a cyclist, travelling at 15 km h^{-1}, is overtaken by a car travelling at 30 km h^{-1}, she sees the car travelling past at 15 km h^{-1}. If the car turns around and drives towards her at 30 km h^{-1} it now appears to be moving at 45 km h^{-1}. Michelson and Morley thought that the speed of light would vary in a similar way.

2 A mass difference of 7.99×10^{-30} kg gives 7.19×10^{-13} J

Exam practice

At the speed of light c muons travel a distance ct_0 in their lifetime t_0 (where $t_0 = 2.2 \times 10^{-6}$ s). This distance is approximately 660 m. Because muons travel at speeds very close to c, from our point of view they seem to live for much longer than 2.2×10^{-6} s. The time they appear to survive for $= t_0/\sqrt{(1 - v^2/c^2)} = 5.047 \times 10^{-6}$ s, long enough for them to reach ground level.

Energy and the Environment 1

Checkpoints

1 Deforestation is the removal of forested areas of land. Plants produce remove carbon dioxide and produce oxygen by photosynthesis. Also the incineration of forest plants increases the amount of carbon dioxide in the atmosphere.

2 Scientists have drilled deep into Antarctic ice and removing cores of frozen snow which contain tiny pockets of air locked in from long ago when the snow was deposited. The analysis of the composition of the air has shown that current greenhouse gas concentrations are higher now than in the last 800 000 years.

3 (a) Water from melted continental ice flows into the sea raising sea levels.
(b) Water expands when it freezes so ice is less dense than water. Ice shelves and ice bergs float in the sea so the upthrust equals the weight of water displaced (Archimedes' principle) which equals the weight of the ice. When the ice melts, its volume is equal to the volume of water displaced so the sea level does not change.

4 Energy has to be supplied to overcome the electrostatic repulsion between nuclei so they can fuse. The material is usually heated so the atoms lose electrons forming a hot plasma of ions. Magnetic fields are used to confine the atoms since they would vaporise solid materials.

Exam practice

(a) (i) To slow down the neutrons in order that they will be absorbed by uranium nuclei.
(ii) To absorb some neutrons and prevent the reaction from increasing out of control.

(b) 200 MeV = $200 \times 10^6 \times 1.6 \times 10^{-19} = 3.2 \times 10^{-11}$ J
No. of atoms = $1000 \times 10^6 / 3.2 \times 10^{-11} = 3.12 \times 10^{19}$

(c) mass = $3.12 \times 10^{19} \times 235 / 6.02 \times 10^{23}$ g

Energy and the Environment 2

Checkpoints

1 E_{power} = KE of air per second = $\frac{1}{2} mv^2/t$ (assuming all the kinetic energy of the wind is converted to electrical power) Mass of air = volume × density = $\pi r^2 l\rho$ where l is the length of air that hits the blades each second
$P = \frac{1}{2}\pi r^2 l\rho v^2/t = \frac{1}{2}\pi r^2 \rho v^3$ since $v = l/t$

2 Gravitational potential energy to kinetic energy to electrical energy.

3 K has units of W m^{-1} K^{-1}, and U, W m^{-2} K^{-1}

4 Energy per second = $2 \times 11 \times 5 = 110$ J, the U-value would increase

Exam practice

(a) Calculate P/d^3 to show the values are similar.

(b) Power output would increase for the same wind speed, the area swept out by the turbine blades would be four times greater, so the mass of air would be four times greater.

Applied physics 1

Checkpoints

1 $\theta = \omega_2 t + \frac{1}{2}\alpha t^2$ and $\theta = (\omega_1 + \omega_2)t/2$

2 $I = T/\alpha$ so the units are N m / rad s^{-2} (rad has no base units as it is a ratio)
So this gives kg m s^{-2} m s^2 or kg m^2

3 The principle of conservation of angular momentum states that the total angular momentum of a system is constant, provided no external torques act.

4 Just as frictional forces oppose linear motion, rotating objects will experience frictional couples. To keep an object rotating at constant angular velocity, an equal and opposite torque must be applied.

5 With arms outstretched, the skater's moment of inertia is increased slightly. When the arms are pulled in, the moment of inertia is reduced so the angular velocity increases so the angular momentum remains the same.

Exam practice

(a) 22 000 rev min^{-1} × $2\pi/60$ = 2300 rad s^{-1}, $E_K = \frac{1}{2} I\omega^2$
= 1.6 MJ

(b) (i) $t = E_K/P = 1.6 \times 10^6/8.7 = 1.84 \times 10^5$ s (51 hours)
(ii) $T = P/(average)\ \omega = 8.7/(2300/2) = 7.5(6) \times 10^{-3}$ N m

Applied physics 2

Checkpoints

1 In an isothermal change, $\Delta U = 0$, so 0 = +400 – W, W = +400 J. So 400 J is done by the system.

2 Efficiency = $(T_H - T_C) / T_H$ = 0.666 (67%) but there will be frictional losses at the turbines bearings for example.

Exam practice

(a) $1.01 \times 10^5 \times (4.25 \times 10^{-4})^{1.4} = 1.70 \times 10^5 \times V^{1.4}$
$V = 2.93 \times 10^{-4}$ m^3

(b) T_1 = 296 K, using $P_1V_1/T_1 = P_2V_2/T_2$
T_2 = 343 K (70 °C)

Further electricity

Checkpoints

1 An alternating voltage is connected to the primary coil causing changing magnetic flux in the core. The changing flux links the secondary coil which causes an induced alternating current to flow.

2 (a) so the resistance and p.d. are low (b) to reduce the size of eddy currents, both of which help to improve the efficiency of the transformer

3 the induced emf is proportional to the rate of change of flux (Faraday's law) which is proportional to the rate of change of current, so $E \propto \Delta I / \Delta t$. The minus sign is because the induced EMF opposes the change producing it (Lenz's law).

Exam practice

(a) X_C = 15.9 kΩ, Z = 18.8 kΩ, $I = 5.3 \times 10^{-4}$A, $V = IR$ = 5.3 V

(b) As f increases, X_C decreases, so V_C decreases and V_{out} increases

(c) Curve from 0, 0 and 1000, 5.3 (see the diagram in *resonance in LCR circuits*)

(d) At high frequencies $V_{out} = V_{in}$, at low frequencies, $V_{out} << V_{in}$ so high frequencies give a high output.

Resources section

This section is intended to help you develop your study skills for examination success. You will benefit if you try to develop skills from the beginning of your course. Modern AS- and A2-level exams are not just tests of your recall of text books and your notes. Examiners who set and mark the papers are guided by assessment objectives that include skills as well as knowledge. You will be given advice on revising and answering questions. Remember to practise the skills.

Exam board specifications

In order to organize your notes and revision you will need a copy of your exam board's syllabus specification. You can obtain a copy by writing to the board or by downloading the syllabus from the board's website.

AQA (Assessment and Qualifications Alliance)
Publications Department, Stag Hill House, Guildford, Surrey GU2 5XJ (www.aqa.org.uk)

CCEA (Northern Ireland Council for Curriculum, Examinations and Assessment)
Clarendon Dock, 29 Clarendon Road, Belfast, BT1 3BG (www.ccea.org.uk)

EDEXCEL
Stewart House, 32 Russell Square, London WC1B 5DN (www.edexcel.org.uk)

OCR (Oxford, Cambridge and Royal Society of Arts)
1 Hills Road, Cambridge CB2 1GG (www.ocr.org.uk)

WJEC (Welsh Joint Education Committee)
245 Western Avenue, Cardiff CF5 2YX (www.wjec.co.uk)

Topic checklist

	Edexcel		AQA/A		AQA/B		OCR/A		OCR/B		WJEC		CCEA	
	AS	A2	AS	A2	AS	A2	AS	A2	AS	A2	AS	A2	AS	A2
Numbers and maths	○	●	○	●	○	●	○	●	○	●	○	●	○	●
Errors and uncertainties	○	●	○	●	○	●	○	●	○	●	○	●	○	●
Studying, revising, passing exams	○	●	○	●	○	●	○	●	○	●	○	●	○	●

Numbers and maths

Basic numeracy is essential for a physicist, but it is not enough. In physics, every measurement is real and imperfect. We need to understand the implications of our imperfect knowledge!

Grade booster

You will be penalized if you give your answer to more significant figures than were used in the question.

Numbers

Every number you write down in any assessed work should be a significant number. No meaningless strings of decimal places!

Significant figures

In physics exams you should think in terms of significant figures rather than decimal places. You should always give your answers to the same number of significant figures as the data provided in the question. For example, the answer to $6.1 \div 3.1$ should be given as 1.0, rounded up to 2 significant figures.

- Significant figures are those numbers we are confident in.
- Every figure you write will be assumed to be significant – so don't give figures that cannot be justified.

$R = 3\ \Omega$ really means that R is $3\ \Omega$ to the nearest ohm; i.e. $R = 3 \pm 0.5\ \Omega$. $R = 3.0\ \Omega$ means R is $3.0\ \Omega$ to the nearest 0.1 ohms; i.e. $R = 3.0 \pm 0.05\ \Omega$ (greater precision) etc.

Test yourself

1 (a) If the p.d. across a component is 6.0 V and the current through it is 1.7 A, what is its resistance?
(b) Repeat the calculation given the following (more precise) measurements: $V = 6.00$ V and $I = 1.700\,0$ A.

2 Write the number 389.455 1 to
(a) 4 sig. fig.
(b) 2 sig. fig.

(*Answers* 1. (a) 3.5 Ω (b) 3.53 Ω.) 2(a) 389.5, or (better) 3.895×10^2 (b) 390, or (better) 3.9×10^2.)

Decimals

If a number has a decimal point, you should be able to infer the precision and the number of significant figures. For example, 13.45 clearly has 4 sig. fig.; 0.004 63 has 3 sig. fig. since leading zeros ignored; 3.80 has 3 sig. fig. since trailing zeros are given only if they are significant.

Whole numbers

Whole numbers are more tricky. For example 2 708 clearly has 4 sig. fig., but 4 600 could have 4, 3 or just 2 sig. fig. There is no way of knowing whether the zeros are significant figures or not. The whole mess is solved by the use of standard form.

Examiner's secrets

Calculator errors are still common. Never buy or borrow a calculator just before an exam. Get to know your calculator in good time for the exams!

Standard form

In standard form, a number is written as a decimal greater than or equal to 1 and less than 10, multiplied by a power of 10. For example:

- 157 804 is written $1.578\,04 \times 10^5$
- 0.000 78 is written 7.8×10^{-4}

Standard form is very useful for:

- Dealing with very big and very small numbers.
- Rounding answers to an appropriate number of significant figures *without ambiguity*.

In standard form, 4 600 would be written 4.6×10^3, 4.60×10^3 or 4.600×10^3 according to the number of significant figures (2, 3 or 4).

- When you use standard form, the number of figures given in the decimal bit is always exactly the number of significant figures.

Algebra

Models and theories are often most concisely expressed as mathematical equations. We need to be able to interpret and manipulate them. Remember:

→ whatever you do to one side of an equation, you must also do to the other side
→ units must equate – you can work out the units of an unknown term, provided you know all other units
→ you should also understand the meanings of inequality symbols: $>, \gg, <, \ll, \geq$ and $\leq$.

Action point

Find out what the inequalities mean.

Trigonometry and geometry

You must be competent at calculating trigonometric functions using degrees or radians as required. You must know the definitions of the trigonometric functions sine, cosine and tangent. Remember:

→ 2π radians = 360°
→ for small angles, θ (in radians) $\approx \sin\theta \approx \tan\theta$

Grade booster

Choosing a suitable scale is not always easy. Neither axis necessarily *has* to start at the origin. You will get a better estimate of a graph's gradient if you stretch out the scale!

Graphs

→ Graphs are a useful way of *showing* relationships.
→ Graphs can be used to *test* relationships.

Straight-line graphs are best! You will often have to manipulate an equation to get it to predict a testable, straight-line graph (of the form $y = mx + c$). For example, the inverse-square law for radiation (of any kind) from a point source predicts that the intensity will be proportional to the inverse square of the distance from the source ($I = k/r^2$). So don't plot I against r (you will get a curve – bad idea); plot I against $(1/r^2)$. You should get a straight line through the origin.

Test yourself

Boyle's law states that the product of pressure and volume of a gas is a constant (provided temperature is constant). If you measured a range of values of P and V, what would you plot to get a straight line?
(*Answer* P = a constant/V, so plot P against $1/V$ and the gradient will be the constant of proportionality.)

Plotting graphs

Marks are awarded for:

→ choosing a suitable scale, labelling axes, giving units, plotting points clearly and accurately, drawing a **line of best fit**

A line of best fit follows the underlying trend and ignores the scatter of experimental measurements. It may be a straight line or a curve. Any predictions should be based on the best fit. It smooths rough data.

If you are asked to plot A against B, then A should be plotted on the y-axis (vertical axis). If you have to decide how to plot your graph:

→ the independent variable should be plotted on the x-axis and the dependent variable should be plotted on the y-axis. (y depends on x)

For example, distance travelled may depend on time (time never depends on distance travelled), and so distance is plotted up the y-axis.

→ Gradient = increase in y-variable/increase in x-variable. Think big: draw big tangents or use big sections of a straight-line graph to find its gradient.

Grade booster

When measuring gradients, credit is given for use of big triangles – large rises and runs. You should also be able to generate graphs using IT.

Errors and uncertainties

Every measurement has an associated uncertainty or error which you are expected to estimate and quote in practical work. Uncertainties are kept to a minimum by good experimental methods, but they never disappear!

Accuracy and precision

- **Accuracy** is closeness to the truth.
- **Precision** is the smallest change in value that can be resolved or measured reliably.

You may have to use your judgement to decide what this smallest resolvable change is – for a particular instrument in a particular experiment. The best most instruments can do is measure to the nearest graduation, but there are exceptions to this rule (thermometers and measuring cylinders can sometimes be read to the nearest half division).

Watch out!

It is quite possible to have an instrument or measurement which is highly precise, but not very accurate! Examples:

- A *micrometer*, which has a precision of 0.01 mm, but has a zero error of 1 mm.
- A *stopwatch*, which gives timings to the nearest 0.01 s, controlled by a person with a reaction time of 0.2 s.

Measurements, tolerances and uncertainties

A measurement should consist of two parts: a *number* (how many) and a *unit* (of what). In addition, all experimental measurements should include an estimate of uncertainty, called **tolerance** (e.g. length ±1 mm).

- Experimental results should be tabulated, with units and tolerances given in column headings, wherever possible.

The minimum acceptable value for a measurement's tolerance is the instrument's precision. As a rule, you should quote the tolerance as ± the smallest graduation on the instrument. Better still, repeat measurements and use the spread in your results to estimate experimental error.

- A good estimate of uncertainty is simply half the range.

The jargon

The terms *experimental error, uncertainty* and *tolerance* are used interchangeably here. Tolerance is normally used to describe the accepted give or take in any particular measurement. Uncertainty means exactly the same thing as error in this context, but it is a fairer term, since it does not imply a mistake has been made (uncertainties are present in *all* data).

Absolute and relative errors

- Instrument tolerances are absolute errors – they have units
- Relative error = absolute error/measurement (a fraction)
- Percentage error = relative error × 100

Random and systematic errors

Systematic errors arise from poor experimental technique and poorly calibrated instruments (i.e. from a poor *system* of measurement). They can be hard to spot, since they do not increase the scatter in the experimental data. Zero errors and parallax (alignment) errors are systematic errors.

- Systematic errors cause *inaccuracy* (consistent over- or under-estimation of a measurement).

Random errors are the errors that cause scatter in a set of data. Random errors can be reduced by repeating measurements and taking averages. (There is less uncertainty in an averaged value than in the raw data.) You can never get rid of random errors completely.

- Random errors cause *imprecision*.

Test yourself

Find the period of a pendulum given the following timings for consecutive single swings: 6.2, 5.4, 6.0, 5.5, 5.1 s. Estimate the tolerance in your answer.
(*Answer* 5.6 ± 0.6 s.)

Errors in calculated quantities

When you *multiply* or *divide* factors to find a quantity, you should *add up* the relative (or percentage) error in each factor used. For example:

- the percentage error in electrical resistance is equal to the percentage error in current *plus* the percentage error in voltage
- to measure the volume of a cylinder, you use the formula $V = \pi r^2 l$, you measure r and l. The percentage error in V is just 2 × percentage error in r (since r^2 is $r \times r$) plus percentage error in l

Never give a calculated answer more significant figures than the least precise data used to calculate it!

Test yourself

You measure a glass bead across three perpendicular diameters, using a micrometer. Your results are: 3.24, 3.15 and 3.10 mm. Calculate the volume of the bead and give the error in your answer.
(*Answer* $1.66 \pm 0.11 \times 10^{-8}$ m^3.)

Practical exams and assessed coursework

Practical skills can be assessed either through coursework or through exams. Both options should assess the same skills.

Coursework will be assessed on the following criteria:

Planning	8 marks
Implementing	7 or 8 marks
Analysing	8 marks
Evaluating	6 or 7 marks

Practical exams will aim to assess the same skills. It is worth reading the coursework mark criteria (given in the subject specifications) – even if you are taking the examined option.

Examiner's secrets

Practical exams require preparation. Be sure you know the form well in advance. Make sure you know instrument tolerances; get some practice with multimeters and CROs etc. (just in case!). You have to follow detailed instructions so don't let adrenaline take over. Read the instructions carefully before you do anything.

What is tested in practical exams and coursework?

- Can you plan an experiment with a purpose?
- Can you take accurate readings and measurements with a range of instruments (micrometers, vernier callipers, multimeters, CROs, etc.)?
- Can you tabulate your results, giving appropriate units and tolerances?
- Are you aware of sources of error – and can you see which sources are most important?
- Do you know how to estimate the magnitudes of errors?
- Can you minimize errors by good experimental technique?
- Can you draw logical conclusions from experimental data?

Watch out!

The list opposite is meant to help, but it is not a definitive guide to everything that should appear in the assessed practical component of your course! Check your subject specifications for greater detail of how marks are awarded.

Graphs

- Do you understand what graphs are for?
- Do you use a line of best fit when appropriate (and do you understand the need for best-fit lines)?
- Can you recognize an equation that should give a straight line?
- Can you manipulate equations and data to give the equation of a straight line?
- Can you choose appropriate scales for your graphs?
- Can you measure (accurately) the gradient of a curve?
- Can you infer correct units?

Test yourself

Why is *extrapolation* (extending a best-fit line to make predictions) more uncertain than *interpolation* (making predictions within the data range)?
(*Answer* Because the underlying relationship may have limits, a small error in gradient can result in a big error in values predicted far beyond the range tested.)

Studying, revising and passing exams

It is time to take control. You cannot learn passively – so get organized, get educated and get qualified!

Studying

Learning is an active process so get *involved.*

Action point

Check your filing system right now. Try to make it as easy to use as possible. *Do not* have a section marked *miscellaneous*; you might as well throw anything deemed miscellaneous into the bin straightaway, because you'll never find it again!

Organize your notes

If you have switched from exercise books to file paper, beware: file paper gets dropped, gets lost, and gets muddled! You *must* get into good habits as soon as possible As a minimum:

- date every sheet of paper
- have a file for each examined unit of study
- subdivide your files into topics

If you are superefficient, you could even have your files ready before you start a topic (very few people do this, but it does help).

Action point

Make or revise your personal timetable. Let friends know about it (then they can organize their free time to suit you!). *Remember* free time is important too.

Organize your time

If you plan your use of time, you will get more done. Here are some suggestions.

- Make a timetable of everything you do on a regular basis.
- Allocate any free periods in the day to a subject or activity.
- Allocate time slots of a suitable length to A-level study (e.g. 4.30–6.00 or 6.30–9.00 p.m. etc.). If you have to see *EastEnders* or life is not worth living, treat yourself – block it into your timetable, but once you have a timetable, stick to it!

Take responsibility for your learning

Watch out!

If you don't understand a topic first time around, you will have to learn it from scratch later – which is far more difficult.

- Make use of all resources at your disposal.
- Every lesson has a point. Ask yourself at the end – did you get it? If not, do something about it!
- Ask and answer questions. Risk getting it wrong. Mistakes are good, we learn from them!
- Never accept that a concept is beyond you. You must assume that you are capable of learning everything. The brain's capacity for information may be infinite so don't impose limits!

Action point

Make a list of sources of further help and information. You may decide to incorporate time into your schedule for surfing the net or browsing books and magazines at the library.

Do all work set as soon as possible

Why?

- It's easier; it takes less time if the subject is fresh in your mind.
- If you get stuck, you'll have time to get help.
- A backlog of work is stressful. It's harder to relax when you're not on top of your work.
- If someone has to drag the work out of you, you never get to feel good about your achievement!
- You may find others come to you for help. Explaining your answers clarifies your understanding.

Revising

- Revision works only if you have studied actively and understood the subject matter when it was covered!

Revision takes time, so don't put it off until the last minute. Ignore any of your 'friends' who claim to do no work; they don't have your best interests at heart (they have their own).

- Plan your revision so that no topic gets missed out. (Don't leave out your least favourite subjects.)
- Don't change your work routine too much when revising.
- Your work space should be well lit, comfortable, uncluttered.
- All necessary resources (books, calculators, pens) must be to hand.
- No distractions (phone off, TV off; 'do not disturb' sign up).

Revise *actively*. Learn, test yourself, take notes and summarize them (use spider diagrams, mind maps, bullet lists, etc. – whatever works for you). Practise exam questions, check your answers for clarity, use of units, etc. Use a highlighter pen to make key facts stand out. This book is meant to be an interactive guide; make use of it!

- Short bursts of intensive revision are most effective. Divide a topic into chunks of suitable length (aim for 20–30 minute spells).
- Take (short) breaks between each burst. Leave the room, run about, speak to someone, have a drink, jump up and down, etc.
- Stop on time. Congratulate yourself and relax.

Don't overdo the coffee – especially late at night. The later you stay up, the more sleepy you will be the next day.

Action point

Buy a highlighter pen (or even a set of them). Start using them – on your notes and on this guide (assuming you own it!).

Action point

Time yourself while revising. You will soon find that 30 minutes is not very long. Don't be tempted to extend revision sessions to fit the content. Work hard to fit the content into the time slot!

Action point

Make twice as much use of your efforts by analysing where you lose marks in exam questions. It is often worth repeating a question you did poorly on. Make sure you learn from your mistakes.

Passing exams

- *Preparation* Make sure you know the format of the exam. Get a copy of the data and formulae you will be given and learn any essential formulae not provided. Learn definitions and units. Do as many practice questions as you can. Find out where you are dropping marks.
- *Answering the question* When you enter an exam, the adrenaline is flowing; you want to get as much information out of your head and onto your answer paper as you can, as soon as possible! You must fight the urge to start writing. Breathe deeply. Calm yourself. Now *read the question*. What is it asking for? (If it totally throws you, take another deep breath and do question 2 first.)
  - Write the question in short form (e.g. 'Find F').
  - Write down the key information (e.g. $m = 24$ kg).
  - Work out the answer one step at a time.
- *Layout of work* There are two essential steps to getting credit in exam answers: the first is to find the answer, the second is to communicate the answer to the examiner! Step two causes most problems. At the end of a difficult calculation you may be so relieved that you forget to check that your brilliant answer is actually legible!
  - *Remember* Strings of calculations are meaningless until you write down what they represent.

Grade booster

- Write clearly, don't scrawl! Examiners are pressed for time. It is in your best interests to make it easy for them. Credit can only be given for a wrong answer if the workings are clear.
- Avoid misusing the '=' sign. (Don't string calculations together; check that the numbers on each side of an equation are actually equal!)
- State the final answer clearly. Underline it if necessary. Give appropriate units.
- Do not cross out workings unless you are replacing them with something better.
- Never hedge your bets by writing two answers in the hopes that you will get credit for whichever one of them is right. Ambiguous answers are wrong answers.

Recommended reading

Breithaupt, J, *Up-grade A-level Physics*, Stanley Thornes (1996).

Brodie, D, *Introduction to Advanced Physics*, John Murray (2000).

Close, F, *The Cosmic Onion*, Heinemann Educational Books (1983).

Eastway, R, and Wyndham, J, *Why Do Buses Come in Threes?*, Robson (1999).

Ehrlich, R, *Turning the World Inside Out*, Princetown University Press (1990).

Fisher, I, *How to Dunk a Doughnut*, Weidenfeld and Nicholson (2002).

Holden, A, *The Nature of Solids*, Dover Publications (1992).

Hollins, M, *Medical Physics*, Macmillan (1990).

Homer, D, *Synoptic Skills in Advanced Physics*, Hodder and Stoughton (2002).

Honeywell, C, *Maths for Physics*, Hodder and Stoughton (1999).

Hunt, A, and Millar, R, *AS Science for Public Understanding*, Heinemann (2000).

Hutchings, R, *Physics*, Nelson (2000).

Johnson, K, et al., *Advanced Physics for You*, Stanley Thornes (2000).

Lockett, K, *Physics in the Real World*, Cambridge University Press (1990).

Muncaster, R, *Astrophysics and Cosmology*, Stanley Thornes (1997).

Powell, S, *Statistics for Science Projects*, Hodder and Stoughton (1996).

Index